15 Math Concepts Every Data Scientist Should Know

Understand and learn how to apply the math behind data science algorithms

David Hoyle

‹packt›

15 Math Concepts Every Data Scientist Should Know

Group Product Manager: Niranjan Naikwadi
Publishing Product Manager: Yasir Ali Khan
Content Development Editor: Joseph Sunil
Technical Editor: Seemanjay Ameriya
Copy Editor: Safis Editing
Project Coordinator: Urvi Sharma
Proofreader: Safis Editing
Indexer: Hemangini Bari
Production Designer: Joshua Misquitta
Marketing Coordinator: Vinishka Kalra

First published: July 2024
Production reference: 2231024

Published by Packt Publishing Ltd.
Grosvenor House
11 St Paul's Square
Birmingham
B3 1RB, UK

ISBN 978-1-83763-418-7
www.packtpub.com

To my wife Clare for her unwavering love, support, and inspiration throughout our life together.

– David Hoyle

Contributors

About the author

David Hoyle has over 30 years' experience in machine learning, statistics, and mathematical modeling. He gained a BSc. degree in mathematics and physics and a Ph.D. in theoretical physics, both from the University of Bristol, UK. He then embarked on an academic career that included research at the University of Cambridge and leading his own research groups as an Associate Professor at the University of Exeter and the University of Manchester in the UK. For the last 13 years, he has worked in the commercial sector, including for Lloyds Banking Group – one of the UK's largest retail banks, and as joint Head of Data Science for AutoTrader UK. He now works for the global customer data science company dunnhumby, building statistical and machine learning models for the world's largest retailers, including Tesco UK and Walmart. He lives and works in Manchester, UK.

This has been a long endeavor. I would like to thank my wife and children for their encouragement, and the team at Packt for their patience and support throughout the process.

About the reviewer

Emmanuel Nyatefe is a data analyst with over 5 years of experience in data analytics, AI, and ML. He holds a Masters of Science in Business Analytics from the W. P. Carey School of Business at Arizona State University and a Bachelors of Science in Business Information Technology from Kwame Nkrumah University of Science and Technology. He has led various AI and ML projects, including developing models for detecting crop diseases and applying Generative AI to innovate business solutions and optimize operations. His expertise in data engineering, modeling, and visualization, alongside his proficiency in LLMs and advanced analytics, highlights his significant contributions to data science. His dedication to data-driven innovation is evident in his book review.

Table of Contents

3

Matrices and Linear Algebra   87

4

Loss Functions and Optimization   135

5

Probabilistic Modeling   165

Part 2: Intermediate Concepts

6

Time Series and Forecasting   203

7

Hypothesis Testing   237

8

Model Complexity   271

Part 3: Selected Advanced Concepts

11

Dynamical Systems   353

12

Kernel Methods   383

13

Information Theory   401

14

Non-Parametric Bayesian Methods   429

15

Random Matrices

Preface

This is not a book about a specific technology or programming language. This is a book about mathematics. And mathematics is a language. It is the language of science, and so it is the language of data science as well. We can say beautiful things with that language. Just as a piece of great literature is more than a large collection of individual letters, a mathematical equation is more than just a collection of symbols. An equation conveys a way of thinking about a data science problem. It conveys a concept or an idea. If you want to fully exploit the power of those ideas and adapt them to your own data science work, you need to move beyond just recognizing the symbols in an equation and move towards understanding what that equation is really telling you.

Many people are not confident in reading and interpreting mathematical equations and mathematical ideas. And yet, as with great literature, once someone guides us through the nuances and subtexts, their beauty is revealed and becomes obvious. That is what this book aims to do.

This book will not make you an expert in every area of mathematics. Instead, it will give you enough skills and confidence to read and navigate mathematical equations and ideas on your own. We do that by walking you through the core concepts that underpin many data science algorithms – the 15 math concepts of the book's title. We also do that by walking through those concepts slowly and in detail. I am not a fan of mathematics books that consist solely of theorems, lemmas, and proofs. Instead, this book is unapologetically long-form math. When we introduce an equation, we will explain what the equation tells us, what its implications and ramifications are, and how it connects to other parts of math. We also illustrate those concepts with code examples in Python.

At the end of the book, you will be equipped to look at the math equations of any data science algorithm and confidently unpack what that algorithm is trying to do.

Who this book is for

This book is for data scientists and machine learning engineers who have been using data science and machine learning techniques, software, and Python packages such as scikit-learn, but without necessarily fully understanding the mathematics behind the algorithms. This could include the following types of people:

Data scientists who have a college/undergraduate degree in a numerate subject and so have a basic understanding of mathematics, but they want to learn more, particularly those bits of mathematics that will be helpful in their roles as data scientists.

Data scientists who have a good understanding of some of the mathematics behind bits of data science but want to discover some new math concepts that will be useful to them in their data science work.

Data scientists who have business or data science problems they need to solve, but existing software does not provide appropriate algorithms. They want to construct their own algorithms but lack the mathematical guidance on how to apply mathematics to the new data science problems.

What this book covers

Chapter 1, Recap of Mathematical Notation and Terminology, provides a summary of the main mathematical notation you will encounter in this book and that we expect you to already be familiar with.

Chapter 2, Random Variables and Probability Distributions, introduces the idea that all data contains some degree of randomness, and that random variables and their associated probability distributions are the natural way to describe that randomness. The chapter teaches you how to sample from a probability distribution, understand statistical estimators, and about the Central Limit Theorem.

Chapter 3, Matrices and Linear Algebra, introduces vectors and matrices as the basic mathematical structures we use to represent and transform data. It then shows how matrices can be broken down into simple-to-understand parts using techniques such as eigen-decomposition and singular value decomposition. The chapter finishes with explanations of how these decomposition methods are applied to **principal component analysis (PCA)** and **non-negative matrix factorization (NMF)**.

Chapter 4, Loss Functions and Optimization, starts by introducing loss functions, risk functions, and empirical risk functions. The concept of minimizing an empirical risk function to estimate the parameters of a model is explained, before introducing Ordinary Least Squares estimation of linear models. Finally, gradient descent is illustrated as a general technique for minimizing risk functions.

Chapter 5, Probabilistic Modeling, introduces the concept of building predictive models that explicitly account for the random component within data. The chapter starts by introducing likelihood and maximum likelihood estimation, before introducing Bayes' theorem and Bayesian inference. The chapter finishes with an illustration of Markov Chain Monte Carlo and importance sampling from the posterior distribution of a model's parameters.

Chapter 6, Time Series and Forecasting, introduces time series data and the concept of auto-correlation as the main characteristic that distinguishes time series data from other types of data. It then describes the classical ARIMA approach to modeling time series data. Finally, it ends with a summary of concepts behind modern machine learning approaches to time series analysis.

Chapter 7, Hypothesis Testing, introduces what a hypothesis test is and why they are important in data science. The general form of a hypothesis test is outlined before the concepts of statistical significance and p-values are explained in depth. Next, confidence intervals and their interpretation are introduced. The chapter ends with an explanation of Type-I and Type-II errors, and power calculations.

Chapter 8, Model Complexity, introduces the concept of how we describe and quantify model complexity and discusses its impact on the predictive accuracy of a model. The classical bias-variance trade-off view of model complexity is introduced, along with the phenomenon of **double descent**. The chapter finishes with an explanation of model complexity measures for model selection.

Chapter 9, Function Decomposition, introduces the idea of decomposing or building up a function from a set of simpler basis functions. A general approach is explained first before the chapter moves on to introducing Fourier Series, Fourier Transforms, and the Discrete Fourier Transform.

Chapter 10, Network Analysis, introduces networks, network data, and the concept that a network is a graph. The node-edge description of a graph, along with its adjacency matrix representation is explained. Next, the chapter describes different types of common graphs and their properties. Finally, the decomposition of a graph into sub-graphs or communities is explained, and various community detection algorithms are illustrated.

Chapter 11, Dynamical Systems, introduces what a dynamical system is and explains how its dynamics are controlled by an evolution equation. The chapter then focuses on discrete Markov processes as these are the most common dynamical systems used by data scientists. First-order discrete Markov processes are explained in depth, before higher-order Markov processes are introduced. The chapter finishes with an explanation of Hidden Markov Models and a discussion of how they can be used in commercial data science applications.

Chapter 12, Kernel Methods, starts by introducing inner-product-based learning algorithms, then moves on to explaining kernels and the kernel trick. The chapter ends with an illustration of a kernelized learning algorithm. Throughout the chapter, we emphasize how the kernel trick allows us to implicitly and efficiently construct new features and thereby uncover any non-linear structure present in a dataset.

Chapter 13, Information Theory, introduces the concept of information and how it is measured mathematically. The main information theory concepts of entropy, conditional entropy, mutual information, and relative entropy are then explained, before practical uses of the Kullback-Leibler divergence are illustrated.

Chapter 14, Bayesian Non-Parametric Methods, introduces the idea of using a Bayesian prior over functions when building probabilistic models. The idea is illustrated through Gaussian Processes and Gaussian Process Regression. The chapter then introduces Dirichlet Processes and how they can be used as priors for probability distributions.

Chapter 15, Random Matrices, introduces what a random matrix is and why they are ubiquitous in science and data science. The universal properties of large random matrices are illustrated along with the classical Gaussian random matrix ensembles. The chapter finishes with a discussion of where large random matrices occur in statistical and machine learning models.

To get the most out of this book

To get the most out of this book, we assume you have at least some familiarity with high-school mathematics, such as complex numbers, basic calculus, and elementary uses of vectors and matrices. To get the most out of the code examples in the book, you should have some experience of coding in Python. You will also need access to a computer or server with a full Python installation and/or where you have privileges to run and install Python and any additional packages required.

Software/hardware covered in the book	Operating system requirements
Python, Jupyter Notebook	Windows, macOS, or Linux

The code examples given in each chapter, and the answers to the exercises at the end of each chapter, are available in the book's GitHub repository as Jupyter notebooks. To run the notebooks, you will need a Jupyter installation.

If you are using the digital version of this book, we advise you to type the code yourself or access the code from the book's GitHub repository (a link is available in the next section). Doing so will help you avoid any potential errors related to the copying and pasting of code.

Download the example code files

You can download the example code files for this book from GitHub at `https://github.com/PacktPublishing/15-Math-Concepts-Every-Data-Scientist-Should-Know`. If there's an update to the code, it will be updated in the GitHub repository.

We also have other code bundles from our rich catalog of books and videos available at `https://github.com/PacktPublishing/`. Check them out!

Conventions used

There are a number of text conventions used throughout this book.

`Code in text`: Indicates code words in text, database table names, folder names, filenames, file extensions, pathnames, dummy URLs, user input, and Twitter handles. Here is an example: "The following code example can be found in the `Code_Examples_Chap5.ipynb` notebook in the GitHub repository."

A block of code is set as follows:

```
map_estimate = minimize(neg_log_posterior,
                        x0,
                        method='BFGS',
                        options={'disp': True})
# Convert from logit(p) to p
```

```
p_optimal = np.exp(map_estimate['x'][0])/ (
    1.0 + np.exp(map_estimate['x'][0]))
print("MAP estimate of success probability = ", p_optimal)
```

Bold: Indicates a new term, an important word, or words that you see onscreen. For instance, words in menus or dialog boxes appear in **bold**. Here is an example: "The name **ARIMA** stands for Auto-Regressive Integrated Moving Average models."

Tips or important notes
Appear like this.

Get in touch

Feedback from our readers is always welcome.

General feedback: If you have questions about any aspect of this book, email us at customercare@packtpub.com and mention the book title in the subject of your message.

Errata: Although we have taken every care to ensure the accuracy of our content, mistakes do happen. If you have found a mistake in this book, we would be grateful if you would report this to us. Please visit www.packtpub.com/support/errata and fill in the form.

Piracy: If you come across any illegal copies of our works in any form on the internet, we would be grateful if you would provide us with the location address or website name. Please contact us at copyright@packt.com with a link to the material.

If you are interested in becoming an author: If there is a topic that you have expertise in and you are interested in either writing or contributing to a book, please visit authors.packtpub.com.

Share your thoughts

Once you've read *15 Math Concepts Every Data Scientist Should Know*, we'd love to hear your thoughts! Scan the QR code below to go straight to the Amazon review page for this book and share your feedback.

https://packt.link/r/1-837-63418-1

Your review is important to us and the tech community and will help us make sure we're delivering excellent quality content.

Download a free PDF copy of this book

Thanks for purchasing this book!

Do you like to read on the go but are unable to carry your print books everywhere?

Is your eBook purchase not compatible with the device of your choice?

Don't worry, now with every Packt book you get a DRM-free PDF version of that book at no cost.

Read anywhere, any place, on any device. Search, copy, and paste code from your favorite technical books directly into your application.

The perks don't stop there, you can get exclusive access to discounts, newsletters, and great free content in your inbox daily

Follow these simple steps to get the benefits:

1. Scan the QR code or visit the link below

https://packt.link/free-ebook/9781837634187

2. Submit your proof of purchase
3. That's it! We'll send your free PDF and other benefits to your email directly

Part 1:
Essential Concepts

In this part, we will introduce the math concepts that you will encounter again and again as a data scientist. These concepts are vital to gain a good understanding of. After a recap of basic math notation, we look at the concepts related to how data is produced and then move through to concepts related to how to transform data, finally building up to our end goal of how to model data. These concepts are essential because you will use and combine them simultaneously in your work. By the end of Part 1, you will be comfortable with the math concepts that underpin almost all data science models and algorithms.

This section contains the following chapters:

- *Chapter 1, Recap of Mathematical Notation and Terminology*
- *Chapter 2, Random Variables and Probability Distributions*
- *Chapter 3, Matrices and Linear Algebra*
- *Chapter 4, Loss Functions and Optimization*
- *Chapter 5, Probabilistic Modeling*

1

Recap of Mathematical Notation and Terminology

Our tour of math concepts will start properly in *Chapter 2*. Before we begin that tour, we'll start by recapping some mathematical notation and terminology. Mathematics is a language, and mathematical symbols and notation are its alphabet. Therefore, we must be comfortable with and understand the basics of this alphabet.

In this chapter, we will recap the most common core notation and terminology that we are likely to use repeatedly throughout the book. We have grouped the recap into six main math areas or topics. Those topics are as follows:

- *Number systems*: In this section, we introduce notation for real and complex numbers

- *Linear algebra*: In this section, we introduce notation for describing vectors and matrices

- *Sums, products, and logarithms*: In this section, we introduce notation for succinctly representing sums and products, and we introduce rules for logarithms

- *Differential and integral calculus*: In this section, we introduce basic notation for differentiation and integration

- *Analysis*: In this section, we introduce notation for describing limits, and order notation

- *Combinatorics*: In this section, we introduce notation for binomial coefficients

Some of this notation you may already be familiar with. For example, complex numbers, matrices, logarithms, and basic differential calculus you will have seen either in high school or in the first year of an undergraduate degree in a numerate subject. Other topics, such as order notation, you may have encountered as part of a university degree course on mathematical analysis or algorithm complexity, or it may be new to you. For the most part, the notation we recap in this chapter you will have seen before. You can skip this chapter if you want to and if you are already familiar and comfortable with the symbols and notation recapped here. You can easily come back later or read those sections that contain notation that is new to you.

We should emphasize that this chapter is a recap. It is brief. It is not meant to be an exhaustive and comprehensive review. We focus on presenting a few main facts, but also on trying to give a feel for why the notation may be useful and how it is likely to be used.

Finally, we will encounter new notation, terminology, and symbols as we progress through the book when we are discussing specific topics. We will introduce this new notation and terminology as and when we need it.

Technical requirements

As this chapter solely recaps some of the mathematical notation we will use in later chapters, there are no code examples given and hence no technical requirements for this particular chapter.

For later chapters, you will be able to find code examples at the GitHub repository: `https://github.com/PacktPublishing/15-Math-Concepts-Every-Data-Scientist-Should-Know`

Number systems

In this section, we introduce notation for describing sets of numbers. We will focus on the real numbers and the complex numbers.

Notation for numbers and fields

As this is a book about data science, we will be dealing with numbers. So, it will be worthwhile recapping the notation we use to refer to the most common sets of numbers.

Most of the numbers we will deal with in this book will be *real* numbers, such as 4.6, 1, or -2.3. We can think of them as "living" on the real number line shown in *Figure 1.1*. The real number line is a one-dimensional continuous structure. There are an infinite number of real numbers. We denote the set of all real numbers by the symbol $\mathbb{R}$.

The Real Number Line

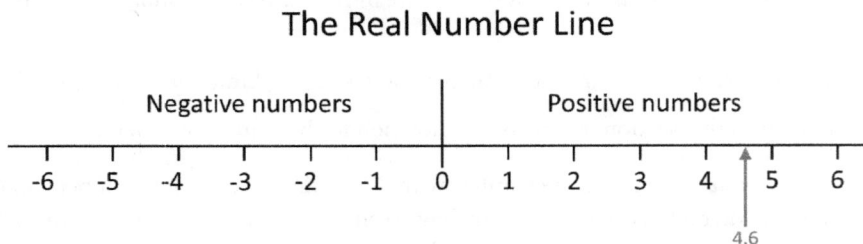

Figure 1.1: The real number line

Obviously, there will be situations where we want to restrict our datasets to, say, just integer-valued numbers. This would be the case if we were analyzing count data, such as the number of items of a particular product on an e-commerce site sold on a particular day. The integer numbers, ..., -2, -1, 0,

1, 2, …, are a subset of the real numbers, and we denote them by the symbol $\mathbb{Z}$. Despite them being a subset of the real numbers, there are still an infinite number of integers.

For the e-commerce count data that we mentioned earlier, the integer value would always be positive. If we restrict ourselves to strictly positive integers, 1, 2, 3, …, and so on, then we have the *natural* or counting numbers. These we denote by the symbol $\mathbb{Z}^+$, clearly meaning positive integers. The fact that these strictly positive integers are the natural numbers means we also denote them using the symbol $\mathbb{N}$.

As well as real numbers, we will occasionally deal with *complex numbers*. As the name suggests, complex numbers have more structure to them than real numbers. The complex numbers don't live on the real number line and so are not a subset of the real numbers, but instead, they have a two-dimensional structure, which we'll explain in a moment. We denote the set of complex numbers by the symbol $\mathbb{C}$.

Sometimes, there are very specific occasions when we may want to refer to other subsets of the real numbers. Other common symbols you may encounter are $\mathbb{Q}$, for the set of rational numbers, and $\mathbb{Z}_2$ for the two-element set {0,1}. The latter you may encounter when we talk about modeling binary discrete target variables or working with binary features.

Numbers such as 4.6 are specific instances of a real number. When we are talking about algorithms or code, we will want to talk about variables, in which case we use a symbol such as x to represent a number, which could take on a range of different values depending on what we do with it. But what could that range be? When we are documenting an algorithm, we may want to tell the reader that x will always be a real number. We do that by writing $x \in \mathbb{R}$, which is mathematical language for "x is in the set of real numbers," or more succinctly, "x is real."

Likewise, if we wanted to say x was always a positive integer, then we would write $x \in \mathbb{Z}^+$. Or, if we wanted to say x was a complex number, we would write $x \in \mathbb{C}$.

When we have several variables that all have similar properties or that may be related in some way – for example, they represent different features of a data point in a training set – then we use subscripts to denote the different variables. For example, we would use x_1, x_2, x_3 to represent three features of a dataset. Just as with the single variable x, if we want to say that those three features will always contain real numbers, then we would write $x_1, x_2, x_3 \in \mathbb{R}$.

Complex numbers

If the real numbers live on the one-dimensional structure that is the real number line, this raises the question of whether we can have numbers that live in a two-dimensional space. Complex numbers are such numbers. A complex number, z, has two components or parts. These are a real part, x, and an imaginary part, y, with both x, y being real numbers. The real and imaginary parts are combined, and we write the complex number z as follows:

$$z = x + iy$$

Eq. 1

The symbol i has a special meaning. It is in fact the square root of -1, so that $i^2 = -1$. We can think of the pair of numbers, (x, y), as picking out a point in a 2D plane. That plane is the **complex plane**, sometimes also called the **Argand plane**. *Figure 1.2* shows the point z in the complex plane:

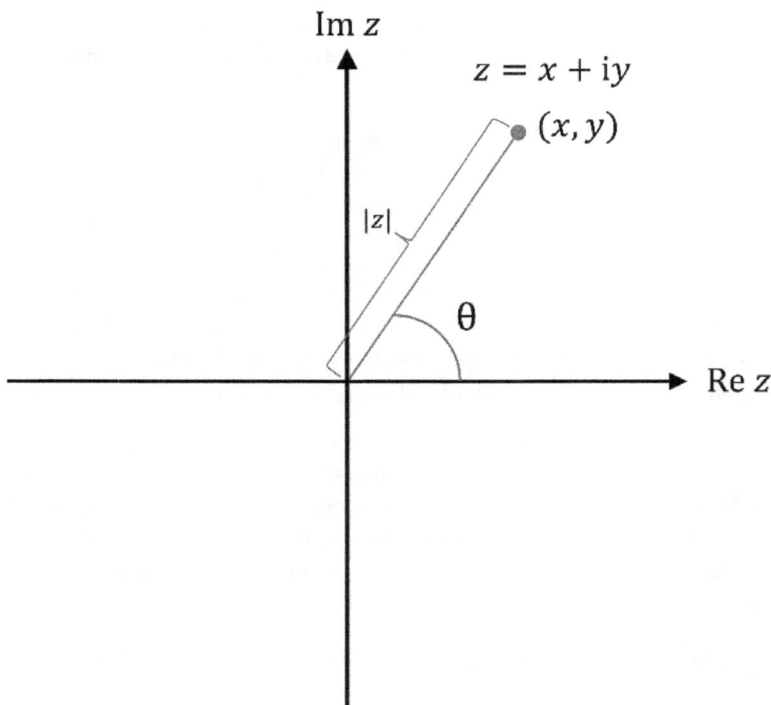

Figure 1.2: The complex number plane

The position of z along the x-axis is given by the real part of z, while the position of z along the y-axis is given by the imaginary part of z. We also use Re z to denote the real part of z, and Im z to denote the imaginary part of z, so that we have the following:

$$\text{Re } z = x \;, \; \text{Im } z = y$$

Eq. 2

Consequently, we have used Re z and Im z to label the axes of the complex plane in *Figure 1.2*.

A number that has $y = 0$ sits entirely on the x-axis and is a purely real number. Likewise, a complex number that has $x = 0$ sits entirely on the y-axis and is a purely imaginary number.

Just as with other 2D planes, we can represent a point in the complex plane not just with Cartesian coordinates (x, y) but with polar coordinates as well. This is also illustrated in *Figure 1.2*. A quick bit of high-school trigonometry gives us the following:

$$z = |z| \times (\cos \theta + i \sin \theta)$$

Eq. 3

The symbol $|z|$ denotes the modulus of z and is the same as the distance of the point z from the origin in *Figure 1.2*. Looking at *Figure 1.2* and using Pythagoras' theorem, we can calculate $|z|$ using the following:

$$|z|^2 = x^2 + y^2 = (\text{Re } z)^2 + (\text{Im } z)^2$$

Eq. 4

The angle θ is conventionally measured in a counterclockwise direction and in radians, so that a point on the positive y-axis would have $\theta = \pi/2$ (remember 2π radians = 360°). Euler's formula is as follows:

$$e^{i\theta} = \cos\theta + i\sin\theta$$

Eq. 5

This means we can also write z in the following form:

$$z = |z| e^{i\theta}$$

Eq. 6

This last form for writing a complex number will be useful when we introduce Fourier transforms, which are used to represent functions as a sum of sine and cosine waves. In fact, this is our main reason for introducing complex numbers.

One important concept relating to the complex number z is that of its complex conjugate. The complex conjugate of z we will denote by $\bar{z}$. Sometimes, the symbol z^* is used instead. The complex conjugate $\bar{z}$ is related to z by flipping the sign of the imaginary part of z. So, if $z = x + iy$, then $\bar{z} = x - iy$. In *Figure 1.3*, this is shown by simply reflecting z in the x-axis. A useful relation that follows is the following:

$$z\bar{z} = x^2 + y^2 = |z|^2$$

Eq. 7

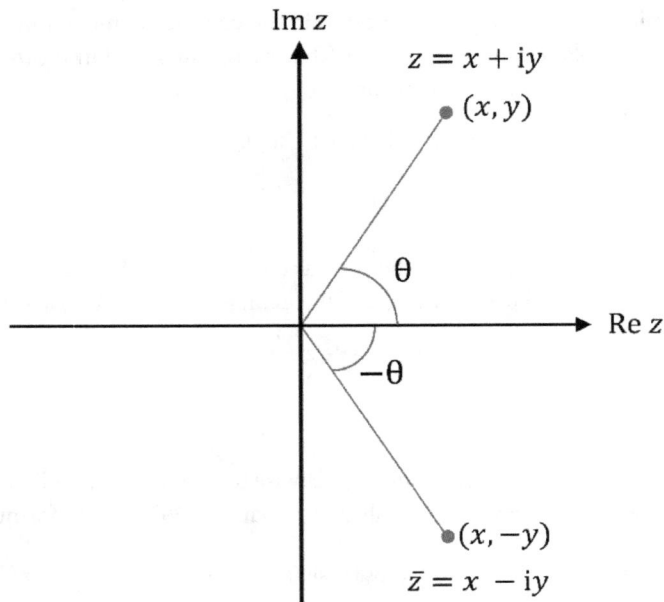

Figure 1.3: The complex conjugate

The integers, real numbers, and complex numbers represent the overwhelming majority of the numbers we will meet throughout this book, so this is a good place to end our recap of number systems.

Let's summarize what we learned.

What we learned

In this section, we have learned the following:

- The notation $\mathbb{R}$, for describing the real numbers
- The notation $\mathbb{Z}$, for describing the integer numbers
- The notations $\mathbb{Z}^+$ and $\mathbb{N}$, for describing the strictly positive integers, also known as the natural numbers
- The notation $\mathbb{Z}_2$, for describing the binary set $\{0,1\}$
- The notation $\mathbb{C}$, for describing the complex numbers
- How complex numbers have a real and an imaginary part
- How a complex number z can also be described in terms of a modulus, $|z|$, and a phase, θ
- How to calculate the complex conjugate $\bar{z}$ of a complex number z

In the next section, having learned how to describe both real and complex numbers, we move on to how to describe collections of numbers (vectors) and how to describe mathematical objects (matrices) that transform those vectors.

Linear algebra

In this section, we introduce notation to describe vectors and matrices, which are key mathematical objects that we will encounter again and again throughout this book.

Vectors

In many circumstances, we will want to represent a set of numbers together. For example, the numbers 7.3 and 1.2 might represent the values of two features that correspond to a data point in a training set. We often group these numbers together in brackets and write them as (7.3, 1.2) or [7.3, 1.2]. Because of the similarity to the way we write spatial coordinates, we tend to call a collection of numbers that are held together a vector. A vector can be two-dimensional, as in the example just given, or d-dimensional, meaning it contains d components, and so might look like $(x_1, x_2, \ldots, x_d)$.

We can write a vector in two ways. We can write it as a row vector, going across the page, such as the following vector:

$$(x_1, x_2, \ldots, x_d) = \text{a } d\text{-dimensional row vector}$$

Eq. 8

Alternatively, we can write it as a column vector going down the page, such as the following vector:

$$\begin{pmatrix} x_1 \\ x_2 \\ \vdots \\ x_d \end{pmatrix} = \text{a } d\text{-dimensional column vector}$$

Eq. 9

We can convert between a row vector and a column vector (and vice versa) using the transpose operator, denoted by a T superscript. So, the transpose of a row vector is a column vector. See the following example:

$$(x_1, x_2, \ldots, x_d)^\top = \begin{pmatrix} x_1 \\ x_2 \\ \vdots \\ x_d \end{pmatrix}$$

Eq. 10

And vice-versa in the following example:

$$\begin{pmatrix} x_1 \\ x_2 \\ \vdots \\ x_d \end{pmatrix}^\top = (x_1, x_2, \ldots, x_d)$$

Eq. 11

Symbolically, we often write a vector using a boldface font – for example, y would mean a vector. Sometimes, we use an underline to denote a vector, so you may also see $\underline{y}$. Throughout this book, I will use an underline to denote a vector. This will make it clear when I am talking about a vector.

Matrices

Usually, we will want to transform a vector more than just transposing it. Linear transformations of vectors can be done with matrices. We will cover such transformations in *Chapter 3*, but for now, we will just show how we write a matrix. A matrix is a two-dimensional array. For example, the following array is a matrix:

$$\underline{\underline{M}} = \begin{pmatrix} 7 & 3 & 2 & 5 \\ 1 & -2 & -1 & 6 \\ 1 & -9 & 14 & 0 \end{pmatrix}$$

Eq. 12

We have used a double underline to denote the matrix $\underline{\underline{M}}$. Note that a matrix has a double underline because it is a two-dimensional structure, while we use a single underline for a vector, which is a one-dimensional structure.

Because a matrix is a two-dimensional structure, we use two numbers to describe its size: the number of rows and the number of columns. If a matrix has R rows and C columns, we describe it as an R x C matrix. The matrix $\underline{\underline{M}}$ in *Eq. 12* is a 3 x 4 matrix.

We pick out individual parts of a matrix by referring to a *matrix element*. The symbol M_{ij} or $\underline{\underline{M}}_{ij}$ refers to the number that is in the position of the ith row and j^{th} column. So, for the matrix in *Eq. 12*, $M_{24} = 6$.

The matrix elements in the previous example are all integers. This need not be the case. A matrix element could be any real number. It can also be a complex number. If all the matrix elements are real, we say it is a real matrix, while if any of the matrix elements are complex, then we say the matrix is complex.

That short recap on notation for vectors and matrices is enough for now. We will meet vectors and matrices again in *Chapter 3*, but for now, let's summarize what we have learned about them.

What we learned

In this section, we have learned about the following:

- How to represent a vector as a collection of multiple components (numbers)

- Row vectors and column vectors and how they are related to each other via the transpose operator

- How a matrix is a two-dimensional collection of components (numbers) and how the notation M_{ij} is used to pick out individual components or matrix elements

In the next section, now we have learned about various notations for individual numbers and collections of them, we move on to notation for performing operations on them. We start with the simplest operations – adding numbers together, multiplying numbers together, and taking logarithms.

Sums, products, and logarithms

In this section, we introduce notation for doing the most basic operations we can do with numbers, namely adding them together or multiplying them together. We'll then introduce notation for working with logarithms.

Sums and the Σ notation

When we want to add several numbers together, we can use the summation, or Σ, notation. For example, if we want to represent the addition of the numbers x_1, x_2, x_3, x_4, x_5, we use the Σ notation to write this as follows:

$$\sum_{i=1}^{i=5} x_i$$

Eq. 13

This notation is shorthand for writing $x_1 + x_2 + x_3 + x_4 + x_5$. This essentially defines what the Σ notation represents – that is, the following:

$$\sum_{i=1}^{i=5} x_i = x_1 + x_2 + x_3 + x_4 + x_5$$

Eq. 14

In the **left-hand side** (**LHS**) of *Eq. 14*, the integer indexing variable, *i*, takes the values between 1 (indicated beneath the Σ symbol) and 5 (indicated above the Σ symbol) and we interpret the LHS as "take all the numbers x_i for the values of *i* indicated by the Σ symbol and add them together."

You may wonder whether the shorthand notation on the LHS of *Eq. 14* is of any use. After all, the **right-hand side** (**RHS**) isn't very long. However, when we want to represent the adding up of lots of numbers, then the Σ notation really comes into its own. For example, if we want to add up the numbers $x_1, x_2, \ldots$ up to x_{100}, then we use the Σ notation to write this compactly, as follows:

$$\sum_{i=1}^{i=N} x_i$$

Eq. 15

Sometimes, we will use the Σ notation to add together a set of numbers where the size of the set (the number of numbers being added together) is variable. For example, see the following notation:

$$\sum_{i=1}^{i=N} x_i$$

Eq. 16

This means "add together the N numbers, $x_1, x_2, \ldots, x_N$." Clearly, we would get a different result for different choices of N. This means the expression given in *Eq. 16* is a function of N.

Sometimes, you may see variants of the expression in the previous equation. Sometimes, a person may omit the upper value of i or both the lower and upper values in the Σ notation because it is taken as understood what the values should naturally be. For example, you may see the following:

$$\sum_{i} x_i$$

Eq. 17

This usually means "add up all values of x_i in the problem we are analyzing." Similarly, the expressions $\sum_{i=1}^{i=N} x_i$ and $\sum_{i=1}^{N} x_i$ mean the same thing.

Finally, note that when writing sums using the Σ notation, we haven't said where the values of x_i come from. We could in fact use the Σ notation to add up the values we get after we have applied a function f to the values x_i. In this case, we would write the following:

$$\sum_{i=1}^{i=N} f(x_i) = f(x_1) + f(x_2) + \cdots + f(x_N)$$

Eq. 18

The LHS of *Eq. 18* is the Σ notation way of writing the RHS. The example in *Eq. 19* makes this clearer. If we set $N = 5$ so we had five numbers, x_1, x_2, x_3, x_4, x_5, and we want to apply the sine function to these five numbers and add them up, then we would write the following:

$$\sum_{i=1}^{i=5} \sin(x_i)$$

Eq. 19

Finally, it is worth pointing out that we can also use the Σ notation to add numbers that are simple functions of the index variable i. For example, using the Σ notation, we can write the sum of the first 100 squares as follows:

$$\sum_{i=1}^{100} i^2$$

Eq. 20

This is obviously shorthand notation for $1^2 + 2^2 + 3^2 + \ldots + 99^2 + 100^2$.

Products and the Π notation

Having introduced the Σ notation and explained it at length, we can now introduce the complimentary idea of a concise, shorthand notation for multiplying lots of numbers together. We do this with the Π or product notation. If we want to multiply x_1, x_2, x_3, x_4, x_5 together, we can write this as follows:

$$\prod_{i=1}^{i=5} x_i$$

Eq. 21

As with the Σ notation, we can use the Π notation more generally. For example, we can write $\prod_{i=1}^{i=N} x_i$ as shorthand for $x_1 \times x_2 \times \ldots \times x_N$. Again, we can use the product notation as shorthand for multiplying function values together, as follows:

$$\prod_{i=1}^{i=N} f(x_i) = f(x_1) \times f(x_2) \times \cdots \times f(x_N)$$

Eq. 22

Logarithms

Logarithms are extremely useful for describing how quickly a quantity or function grows. In particular, the logarithm tells us the exponent that describes the rate of growth of a quantity or function. Let's make that more explicit. The logarithm to base a of the number a^x is x. Mathematically, we write this as follows:

$$\log_a(a^x) = x$$

Eq. 23

The symbol $\log_a$ is shorthand for taking the logarithm to base a. This shorthand is so common that even in the text, I will use the word log when I mean logarithm. It is also not uncommon to omit the brackets in the previous equation and write $\log_a a^x = x$. The most common bases we use for taking logarithms are base e, base 10, and base 2. Of these, base e is so commonly used that we use a different symbol, ln, when taking the log. So, in effect, this means $\ln = \log_e$. This symbol means the **natural**

logarithm or **natural log** to denote the fact that taking the log to base e is the most natural or common thing to do. Because taking the natural log is so common or natural, most mathematicians don't really consider taking the log to any other base, and so by default, we use the symbol log to mean ln. Watch out for this. If you see the symbol log without a base specified, then it either means the base is not important – for example, the proof of the mathematical statement does not depend upon the base – or base e is implicitly meant. This is also the case in most computer programming languages. Applying the operator log will return the natural logarithm. For example, in Python, if we use the `numpy.log(y)` NumPy function, we will get the natural logarithm of y returned.

We can see from *Eq. 23* that the logarithm does in fact tell us the exponent (in base a) of the number we are taking the log of. So, if $\log_a b = c$, then $b = a^c$. Because taking the log effectively gives us an exponent value, the logarithm of a number is typically much smaller than the number itself. More importantly, it also means that the logarithm function is monotonic, so that $\log(x)$ increases as x increases. The word "monotonic" means "of one tone" or "of one direction," and so it means either only going up (monotonically increasing) or only going down (monotonically decreasing). This is shown in *Figure 1.4*, which shows the natural logarithm function $\ln(x)$, from which we can see the value of $\ln(x)$ increasing as x gets bigger:

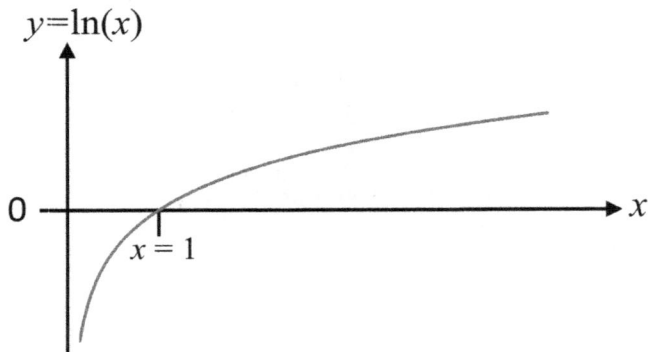

Figure 1.4: Graph of the natural logarithm function

An important consequence of the monotonically increasing nature of the logarithm function is that if we have a function $f(x)$ and we want to find the value of x where $f(x)$ has its highest (or maximum) value, then that maximal value of x, let's call it x^*, is also the point where $\log(f(x))$ has its maximal value. In mathematical notation, we can write this fact as follows:

$$\text{If } f(x) \leq f(x^*) \text{ when } x \neq x^* \text{ then } \log f(x) \leq \log f(x^*) \text{ when } x \neq x^*$$

Eq. 24

We will refer to this again in a moment.

There are well-known rules for taking logarithms of reciprocals, products, and ratios. These are (for any base):

$$\log_a\left(\tfrac{1}{y}\right) = -\log_a(y)$$

Eq. 25

And the following:

$$\log_a(xy) = \log_a(x) + \log_a(y)$$

Eq. 26

Combining these two rules, we get the rule for taking the log of a ratio:

$$\log_a\left(\tfrac{x}{y}\right) = \log_a(x) + \log_a\left(\tfrac{1}{y}\right) = \log_a(x) - \log_a(y)$$

Eq. 27

The rule for taking the log of a product is particularly useful when we have a product formed from many numbers. Using the Σ and Π notations we introduced earlier, we can write the following:

$$\log_a\left(\prod_{i=1}^{i=N} x_i\right) = \log_a(x_1 x_2 \ldots x_N) = \log_a(x_1) + \log_a(x_2) + \ldots + \log_a(x_N) = \sum_{i=2}^{i=N} \log_a(x_i)$$

Eq. 28

This, in conjunction with the fact that taking the log is a monotonic transformation, will be very useful to us when we start to use the concept of maximum likelihood to build probabilistic models in *Chapter 5*.

We will make lots of use of sums, products, and logarithms throughout this book, but we have all the notation we need to work with them, so let's summarize what we have learned about that notation.

What we learned

In this section, we have learned about the following:

- The Σ notation for adding lots of numbers together
- The Π notation for multiplying lots of numbers together
- How we can also use the Σ and Π notations when we have a function, $f(x)$, applied to our numbers, $x_1, \cdots, x_N$
- How a logarithm function transforms a number into an exponent
- How a logarithm function is defined with respect to a specific base

- How base e is the most commonly used base for taking logarithms, and the corresponding logarithm function is called the natural logarithm and is often denoted as $\ln(x)$

- How to take logarithms of products of numbers and ratios of numbers

In the next section, we will stick with functions applied to numbers, but we'll learn how to describe how fast a function $y(x)$ changes as we change x, using the notation of differential calculus. We will also learn how to describe the area under a function using the notation of integral calculus.

Differential and integral calculus

In this section, we won't go into the fundamentals of differential calculus, but instead just recap some basic results and notation. Therefore, we are assuming you already have some basic familiarity with differentiation and integration.

Differentiation

Let's start with what the derivative of a function or curve $y(x)$ intuitively represents. An example curve is shown in *Figure 1.5*. The derivative of this function is denoted by the following symbol:

$$\frac{dy}{dx}$$

Eq. 29

The derivative of $y(x)$ is itself a function of x. The numerical value of the derivative evaluated at a particular value of x, let's say at $x = x_0$, is the gradient (or slope) of the tangent to the curve $y(x)$ at $x = x_0$. As such, we can think of the derivative as defining the local gradient value of the curve. This is illustrated in *Figure 1.5* as well:

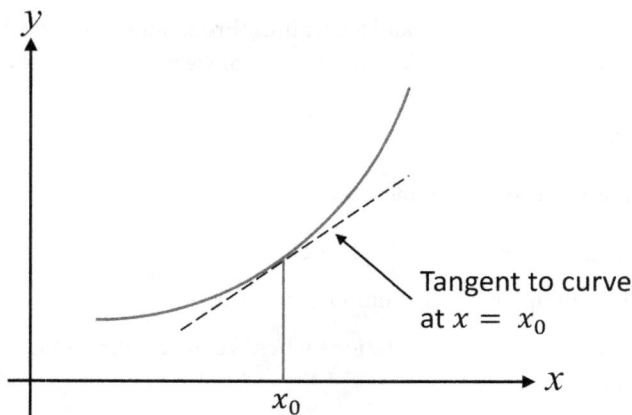

Figure 1.5: The derivative as the gradient of the tangent to the curve

Sometimes, when we want to be explicit about a particular point on the curve where we are evaluating the derivative, then we will use the notation $\frac{dy}{dx}\big|_{x=x_0}$, or more simply, $\frac{dy}{dx}\big|_{x_0}$. This means the derivative function is evaluated at $x = x_0$, while the notation $\frac{dy}{dx}$ tends to get used to mean the derivative function in general.

Clearly, if $\frac{dy}{dx}$ is a function of x, then we can ctalculate its derivative just like we could calculate the derivative of the function $y(x)$. Using the derivative notation, this second derivative is written as follows:

$$\frac{d}{dx}\frac{dy}{dx} = \frac{d^2y}{dx^2}$$

Eq. 30

What does this second derivative represent? Well, it is the gradient of the curve of $\frac{dy}{dx}$ – that is, the rate of change of the rate of change of y. Similarly, we can calculate, if we wish, higher derivatives of y – for example, the third derivative denoted by $\frac{d^3y}{dx^3}$, or the fourth derivative denoted by $\frac{d^4y}{dx^4}$.

Due to the history of how differential calculus was developed, we have a second, commonly used notation for the first and second derivatives etc. of a function. In this second notation form, the first derivative of the function $y(x)$ is written as $y'(x)$, or simply y'. So, $y' \equiv \frac{dy}{dx}$. As you might have guessed, the second and third derivatives are written as y'' and y''' in this new notation. Obviously, when we get to very high order derivatives – for example, fifth, sixth, and so on – writing all those apostrophes in the superscript becomes a bit clunky, so we also use the notation $y^{(n)}$ to denote the nth derivative, with $y^{(0)}$ meaning the function $y(x)$ itself. Consequently, the notations $\frac{dy}{dx}, y', y^{(1)}$ all mean the same thing – the first derivative – and we often will use them interchangeably in the same document, proof, or explanation.

Now, let's recap how to calculate the derivative of some common functions we're likely to encounter. The derivative of a linear function $ax + b$ is straightforward and intuitive – it is the gradient of that linear function. So, more explicitly, the derivative of a constant is zero and we have the following:

$$\frac{dax}{dx} = a$$

Eq. 31

The derivative of a power x^n is calculated as follows:

$$\frac{dx^n}{dx} = nx^{n-1}$$

Eq. 32

The derivative of the exponential function e^x is itself, as follows:

$$\frac{de^x}{dx} = e^x$$

Eq. 33

The derivative of the natural logarithm function ln$x$ is calculated as follows:

$$\frac{d\ln x}{dx} = \frac{1}{x}$$

Eq. 34

For calculating the derivative of more complicated functions, we typically make use of the "chain rule," which is a rule for how to calculate the derivative of a composite function. If we have a composite function $y(x) = g(h(x))$, then the chain rule says the following:

$$y' = g'(h(x)) \times h'(x)$$

Eq. 35

Let's illustrate that with an explicit example. Imagine I have the function $y(x) = \ln(2x + 3x^3)$. Using the chain rule and the results for the derivatives of the natural logarithm function and for a power of x, we have the following:

$$\frac{dy}{dx} = \frac{1}{2x + 3x^3} \times (2 + 9x^2) = \frac{2 + 9x^2}{2x + 3x^3}$$

Eq. 36

Similarly, if we have the function $y(x) = e^{bx}$, then the chain rule and the derivatives of exponential and linear functions combine to give us the following:

$$\frac{de^{bx}}{dx} = be^x$$

Eq. 37

Finally, we come to the situation where our function may depend upon more than just one variable. For example, what if our function y is a function of x and z so that $y = y(x, z)$? What does the derivative mean or represent now? One way to look at it is to ask, what if we just kept z constant and calculated the derivative with respect to x? To do this, we just apply the simple rules we have outlined previously. However, that calculation gives us the gradient in only the x direction of the function. It gives us only partial information about how the function is changing, so we call this the *partial derivative* of y with respect to x . We use a slightly different symbol to denote a partial derivative, namely $\frac{\partial y}{\partial x}$. Just as we can calculate a partial derivative of y with respect to x, we can calculate a partial derivative of y with respect to z by holding x constant and applying the rules of one-dimensional differential calculus. As you may have guessed, we use the symbol $\frac{\partial y}{\partial z}$ for this partial derivative with respect to z.

Finding maxima and minima

A common task we will want to carry out is to find the maximum value or the minimum value of a function. For example, we may want to find the model parameter values that have the smallest (minimum) error on a training dataset. In these cases, even if we can't find the absolute best (global minimum error) parameters, a local minimum may still be useful to us, as this would represent the best model parameters in a limited but relevant region of the model parameter space.

Finding the maximum value of a function and finding the minimum value of a function are closely related tasks because if we find the maximum value of the function $-y(x)$, then we have found the minimum value of the function $y(x)$. So, from now on, we will largely discuss only how to find the maximum value of a function.

To help find the maximum value of a function, we make use of the differential calculus we have just recapped. If we look at *Figure 1.6*, we can see that at the maximum point of the function shown, the gradient is zero:

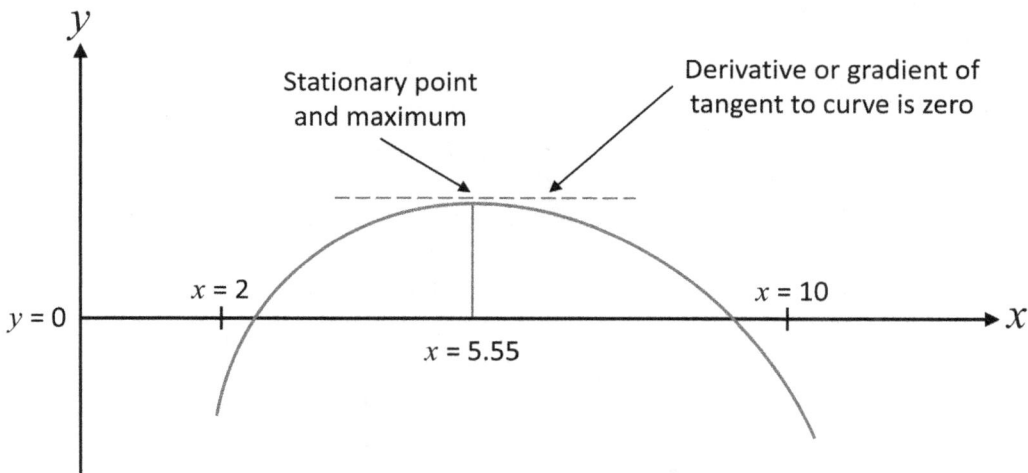

Stationary point and maximum

Derivative or gradient of tangent to curve is zero

$x = 2$

$x = 10$

$y = 0$

$x = 5.55$

Figure 1.6: A function with a single maximum

Now, since the gradient is given by the function $\frac{dy}{dx}$, we just have to find the point where $\frac{dy}{dx} = 0$. That is, we solve the equation $\frac{dy}{dx} = 0$ for values of x. Clearly, from *Figure 1.6*, in this example, there is only one solution to this equation in the region between $x = 2$ and $x = 10$, and it occurs at $x = 5.55$. For other functions, we may have multiple solutions to the equation $\frac{dy}{dx} = 0$. Solutions to this equation are called *stationary points*. Why? Well, the gradient is zero, so the function isn't changing much in the region around the stationary point, so the function is effectively "stationary."

The maximum value of a function is not always a stationary point. Look at *Figure 1.7*:

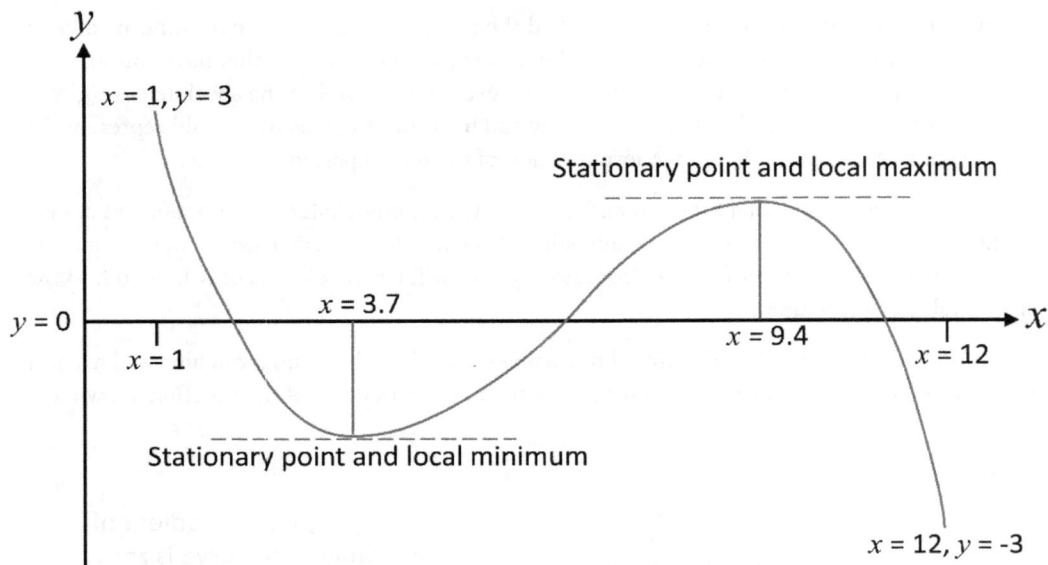

Figure 1.7: A function with a minimum and a maximum

There is a stationary point at $x = 9.4$. It is clearly not the highest value of the function. It is a maximum of the function, but only in the small region around $x = 9.4$, so we refer to it as being a local maximum. It is not the global maximum of the function. However, this local maximum may still be useful to us.

Notice in *Figure 1.7* that we have another stationary point at $x = 3.7$. It is a local minimum, but not the global minimum. So, if we are interested in finding maxima of a function, even if they are just local maxima, how can we do this if finding solutions to $\frac{dy}{dx} = 0$ can't distinguish between a maximum and a minimum? Well, take a closer look at *Figure 1.7*. To the left of the (local) maximum at $x = 9.4$, the gradient is positive, while to the right of the maximum, the gradient is negative. So, the gradient is decreasing in the region around this stationary point. That is, the second derivative is negative at $x = 9.4$. Conversely, if we look at the gradient in the region around the local minimum at $x = 3.7$, the gradient is increasing, and the second derivative is positive at $x = 3.7$. This means the second derivative gives us a means of distinguishing maxima from minima. Putting this together, we have the following:

$$\text{If } \frac{dy}{dx} = 0 \text{ and } \frac{d^2y}{dx^2} < 0 \Rightarrow \text{A stationary point and a maximum.}$$

Eq. 38

$$\text{If } \frac{dy}{dx} = 0 \text{ and } \frac{d^2y}{dx^2} > 0 \Rightarrow \text{A stationary point and a minimum.}$$

Eq. 39

Finally, let's return to our actual goal – to find the maximal (highest) value of a function. Let's say we are interested only in the region between $x = 1$ and $x = 12$. Just from a visual inspection of *Figure 1.7*, we can see the global maximum is at the boundary at $x = 1$ and is not even a stationary point. This emphasizes that while a stationary point with $\frac{d^2y}{dx^2} < 0$ is at least a local maximum, it may not be a global maximum – and a global maximum may not be a stationary point.

Integration

Integral calculus is the counterpart to differential calculus, in that if we know that a function $y(x)$ is the derivative of another function, say, $h(x)$, then we can easily calculate the integral of $y(x)$.

But what does the integral of $y(x)$ represent? We write the integral as follows:

$$\int_a^b y\,dx$$

Eq. 40

The similarity in shape between the integral symbol $\int$ and the Σ symbol used to denote summation is no coincidence. Integration is derived as a limit of a summation of lots of small contributions. We won't go into that derivation here, other than to say the integral in the previous equation represents a summation of lots of values of the function $y(x)$ evaluated at different points between a and b. The integral gives the area between a and b under the curve given by $y(x)$, as shown in *Figure 1.8*:

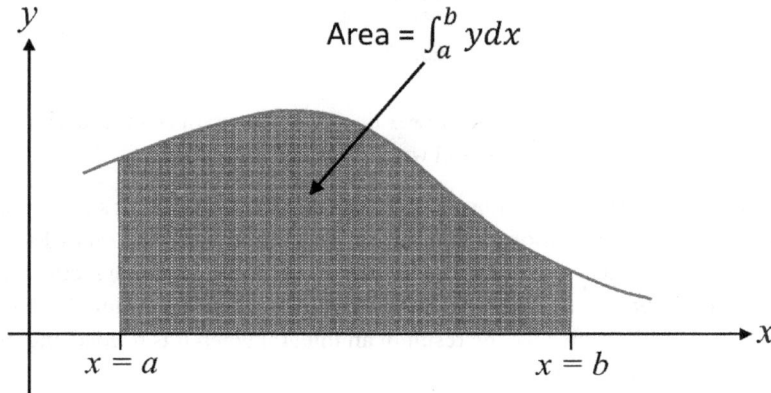

Figure 1.8: The integral as the area under the curve

There are numerous situations where we want to calculate such a quantity. For example, we may need to calculate the total probability of a particular event happening over a range of values of a predictive feature. In this case, we would have to add up lots of small probability contributions that change as the value of the predictive feature changes.

Unfortunately, calculating integrals is less of a routine process compared to differentiating a function. Calculating integrals can be more of an art. Where we can spot that the function being integrated is the derivative of another function, we can easily calculate the integral. So, for example, if we know that $y(x) = h'(x)$, then,

$$\int_a^b y\,dx = h(b) - h(a)$$

Eq. 41

This equation is called the *fundamental theorem of calculus* because it links integral calculus to differential calculus.

Frequently, you will see an integral written without explicit limits, such as $\int y\,dx$. The result is a function, which is defined up to a constant. For the previous example, where $y(x) = h'(x)$ the integral without explicit limits would be written as follows:

$$\int y\,dx = h(x) + \text{Constant}$$

Eq. 42

Once we have the function on the RHS, we can plug in explicit limits to get the result of the integral with limits. In this case, we would get the following:

$$\int_a^b y\,dx = [h(b) + \text{Constant}] - [h(a) + \text{Constant}]$$

Eq. 43

Again, notice that we *subtract* the value at the lower limit. The constants cancel each other out and we recover the original definition of the integral with limits, as in *Eq. 41*.

Often, we are not so lucky, and we won't know that $y(x)$ is an exact derivative of some other function. So, calculating an integral can rely more upon having seen the integral before or looking up the integral in a big book of integrals – see point 1 in the *Notes and further reading* section at the end of the chapter. Consequently, we are not going to dwell upon techniques for evaluating integrals, and throughout the book, we will simply give the result of an integral when it is needed and only explain briefly, where possible, how it was calculated.

However, there are some integrals that we will encounter again and again throughout this book, so we'll recap them here.

From our recap of differentiation, we know that $\frac{dx^n}{dx} = nx^{n-1}$. By making use of the fundamental theorem of calculus, we can work out the following:

$$\int x^n\,dx = \frac{x^{n+1}}{n+1} + \text{Constant}$$

Eq. 44

The other integral we will make use of a lot is related to the normal or Gaussian probability distribution, and we will encounter it a lot when we start building probabilistic predictive models. The integral result we refer to is as follows:

$$\int_{-\infty}^{\infty} e^{-\frac{1}{2}\left(\frac{x-\mu}{\sigma}\right)^2} dx = \sqrt{2\pi\sigma^2}$$

Eq. 45

This integral result is needed when defining the probability density function of the normal distribution. As the normal (a.k.a. Gaussian) distribution is one of the most common probability distributions you will encounter, this integral result is extremely useful.

Differentiation and integration are big mathematical topics. We have only been able to give a very short recap of the key bits of notation. Let's summarize what we learned about those key bits of notation.

What we learned

In this section, we have learned about the following:

- What the derivative of a function represents
- The different notations for the derivative of a function
- How to differentiate basic and composite functions
- How to calculate the partial derivatives of a function of many variables
- How stationary points of a function can be located by finding where its first derivative is zero
- How stationary points correspond to either a maximum or minimum of the function
- How maxima and minima of a function can be distinguished using its second derivative
- Integral calculus and what an integral of a function represents
- The integrals of some basic and important functions

In the next section, we move on to how to describe the behavior of a function using the notation of limits and order notation. We will then show how this notation can be used to describe commonly used approximations of functions.

Analysis

In this section, we recap the notation that is used in analyzing and describing the behavior of functions. This includes notation for describing limits, notation for describing the relative ordering of functions, and notation for describing standard approximations of functions.

Limits

When we are talking about limits, we are talking about the mathematical behavior of some quantity, often a function, as some other quantity approaches a particular value, often infinity. Let's make that more concrete. Consider the function $f(x) = \frac{1}{x}$. As x gets bigger and bigger, then clearly, the value of $f(x)$ gets smaller and smaller, until eventually, as x becomes infinitely large, $f(x)$ becomes zero. We say that $f(x)$ approaches its limit of 0 as x approaches ∞. Mathematically, we write this as follows:

$$\lim_{x \to \infty} f(x) = 0$$

Eq. 46

The word *lim* denotes the fact that we are talking about a limit, while the symbols $x \to \infty$ describe what limit we are talking about. The RHS of *Eq. 46* gives the actual limiting value.

Sometimes, the limiting value may depend on whether we are approaching the limit from above or below. For example, consider the function $f(x) = \frac{1}{1 + \exp(-\frac{1}{x})}$. As x approaches 0 from above (from positive numbers), $f(x)$ tends to 1, while as x approaches 0 from below (from negative numbers), $f(x)$ tends to 0. There is not a single value for the limit but two different limit values, depending on whether we are approaching 0 from above or below. In mathematical notation, we have to distinguish those two possibilities. We do this by being more specific in how we say we are approaching $x = 0$. To denote approaching $x = 0$ from above (from the positive side), we use the notation $\lim_{x \to 0^+}$ or the notation $\lim_{x \downarrow 0}$. Personally, I prefer the first notation and that is what I will use in this book if I need to be specific about how I'm approaching a limit. Similarly, we denote approaching $x = 0$ from below using the notation $\lim_{x \to 0^-}$ or $\lim_{x \uparrow 0}$. With these notations, we would write the two different limiting values as follows:

$$\lim_{x \to 0^+} \frac{1}{1 + \exp(-\frac{1}{x})} = \lim_{x \downarrow 0} \frac{1}{1 + \exp(-\frac{1}{x})} = 1$$

Eq. 47

And as follows:

$$\lim_{x \to 0^-} \frac{1}{1 + \exp(-\frac{1}{x})} = \lim_{x \uparrow 0} \frac{1}{1 + \exp(-\frac{1}{x})} = 0$$

Eq. 48

We should also mention that it is often the case that a quantity does not approach a well-defined limiting value. In this case, we sometimes say "the limit does not exist." For example, if our function was $f(x) = x^2$, then as x approached infinity, so would the value of $f(x)$. We would say the function $f(x)$ does not have a finite limit as $x \to \infty$. Instead, we might denote this behavior by writing $f(x) \to \infty$ as $x \to \infty$.

Order notation

Order notation does precisely what it says. It helps us put contributions to an expression, a formula, or a quantity in order of size. This is useful when we are wanting to approximate something or get a ballpark feel for the size of something. Order notation helps us identify what the biggest contributions are or helps us identify when a contribution is so small that we can safely ignore it. Let's introduce order notation first and then we can see some examples of how it can help us.

Big-O notation

The most common order notation that you will have encountered before is Landau's O-notation, named after the German mathematician Edmund Landau. It is also attributed to Paul Bachmann, so it is sometimes called Bachmann-Landau O-notation. You may have heard it called *big-O notation* because it uses the symbol O.

Formally, we say the following for some positive constants M and δ:

$$f(x) = O(g(x)) \text{ as } x \to a \text{ if } |f(x)| \leq Mg(x) \text{ when } 0 < |x - a| < \delta$$

Eq. 49

Note the use of the limit notation here, $x \to a$. Frequently, the limit we consider is $x \to \infty$, and so the notation $0 < |x - a| < \delta$ effectively means once x is sufficiently big enough – that is, once $x > x_0$ for some value x_0.

Okay, this definition can be a bit dry and impenetrable, so let's give an example to make this more concrete. If we say the following:

$$f(x) - e^{3x} = O(x^{-1}) \text{ as } x \to \infty$$

Eq. 50

What we are saying is that as x gets bigger and bigger, the function $f(x)$ gets closer and closer to the function e^{3x}, with the difference decreasing at least as fast as x^{-1}. So, if the difference between $f(x)$ and e^{3x} is, say, 0.1 at $x = 50$, then at $x = 100$, the difference will be at least halved and so will be 0.05 or smaller. Practically, what this tells us is that when x is large, we can approximate $f(x)$ by e^{3x}, with the error decreasing proportionally to x^{-1} or faster.

Providing a means to symbolically quantify the closeness of an approximation is one of the main uses of big-O notation, so much so that we typically write *Eq. 50* in a more convenient form:

$$f(x) = e^{3x} + O(x^{-1}) \text{ as } x \to \infty$$

Eq. 51

This emphasizes the fact that the RHS approximates the function $f(x)$ and that is our aim.

The thing inside the O does not have to be a function. It could be a constant, in which case we would write $O(1)$, since the definition of the big-O notation already contains a constant M. For example, we might have the case that a function $h(x)$ can be approximated as follows:

$$h(x) = x^3 + O(1) \text{ as } x \to \infty$$

Eq. 52

This means that the difference between $h(x)$ and x^3 is not bigger than a constant amount as x gets bigger and bigger. This is still an approximation worth using because the fractional error decreases as x gets bigger.

Sometimes, we may just use the big-O notation to describe the first part (the leading order) of the approximation. So, for the previous example, we can also write the following:

$$h(x) = O(x^3) \text{ as } x \to \infty$$

Eq. 53

Going back to the formal definition of big-O notation, *Eq. 53* tells us that as x gets very big, there is a constant M such that $h(x) \leq Mx^3$. We say that $h(x)$ scales as x^3 as x approaches infinity.

This use of big-O notation is useful for quantifying the complexity of an algorithm. For example, if we say the number of floating-point operations (and therefore runtime) needed for inverting a square matrix of size N is $O(N^3)$, then this indicates that compared to inverting a matrix of size $N = 10$, the computational time taken to invert a matrix of size $N = 1000$ will be $100^3 = 1$ million times longer. When planning and designing algorithms, knowing these aspects of the complexity of an algorithm is vitally important – they can make the difference between an algorithm being usable and being useless, and it is better to know this before trying to code and deploy the algorithm.

Little-o notation

Note that in the formal definition of the big-O notation, $|f(x)| \leq Mg(x)$ as $x \to \infty$, which means we can find a value M such that $Mg(x)$ bounds the absolute value of $f(x)$. Sometimes, we can make even stronger statements about the behavior of $f(x)$, such as saying $\varepsilon g(x)$ bounds $|f(x)|$ for any positive value of ε once x is close enough to the limit we are considering. This means $g(x)$ is always much bigger than $|f(x)|$ once we get close enough to our limit, while big-O notation only tells us that if we scale up $g(x)$ by some positive number M does it become bigger than $|f(x)|$.

We write these stronger statements using *little-o* notation. Formally, we say, $f(x) = o(g(x))$ as $x \to \infty$, if there exists a positive value x_0 such that for every value $\varepsilon > 0$ we have $|f(x)| < \varepsilon g(x)$ for all values $x > x_0$.

The statement $f(x) = o(g(x))$ as $x \to \infty$ is a much stronger statement than the statement $f(x) = O(g(x))$ as $x \to \infty$. In other words, $f(x) = O(g(x))$ is often used loosely to mean $f(x)$ behaves similarly to $g(x)$, while $f(x) = o(g(x))$ is used to mean that $f(x)$ is negligible compared to $g(x)$.

Taylor series expansions

Our next recap of mathematical notation regards function expansions. Specifically, we will focus on the Taylor series expansion of a function about a particular point. The usefulness to us of the Taylor expansion is that close to the expansion point, the higher-order terms in the expansion get smaller and smaller. This means that after a while, the higher-order terms can be ignored, and if we truncate the expansion, we have a useful approximation to the original function in the region around the expansion point.

So, how do we construct the Taylor expansion of a function $f(x)$? We use the differential calculus we recapped earlier. Specifically, we can expand $f(x)$ about the point x_0 by writing the following:

$$f(x) = f(x_0) + (x - x_0)\frac{df}{dx}\bigg|_{x=x_0} + \frac{1}{2}(x - x_0)^2\frac{d^2f}{dx^2}\bigg|_{x=x_0} + \dots$$

Eq. 54

The multiple dots refer to the fact that there are higher-order terms. Using the order notation we recapped earlier, we can write this more rigorously as follows:

$$f(x) = f(x_0) + (x - x_0)\frac{df}{dx}\bigg|_{x=x_0} + \frac{1}{2}(x - x_0)^2\frac{d^2f}{dx^2}\bigg|_{x=x_0} + O\big((x - x_0)^3\big) \text{ as } x \to x_0$$

Eq. 55

We could continue the expansion to even higher orders, so using the Σ notation, we can write the following:

$$f(x) = \sum_{k=0}^{\infty}\frac{(x - x_0)^k}{k!}\frac{d^kf}{dx^k}\bigg|_{x=x_0}$$

Eq. 56

While the sum on the RHS of the previous equation technically represents the function $f(x)$, numerically, it may not be that helpful as the sum may not converge – for example, if $|(x - x_0)|$ is larger than the radius of convergence of the series, in which case, the RHS is essentially infinite in value. More frequently, we truncate the expansion after a few terms, such as at second-order or third-order, and we have a low-order finite approximation to $f(x)$ which will be more and more accurate as x gets closer to the expansion point x_0.

We refer to the series on the RHS of the previous equation as an expansion, essentially because we are using our knowledge about the behavior of the function at $x = x_0$, namely our knowledge of its value $f(x_0)$ and its derivatives, to calculate the function away from x_0. So, we are expanding our knowledge of $f(x)$ from x_0 to a region around x_0.

When the expansion point $x_0 = 0$, the series is also called a **Maclaurin series** and it gives a power-series representation of the function $f(x)$. Specifically, when $x_0 = 0$ we have the following:

$$f(x) = \sum_{k=0}^{\infty} \frac{1}{k!} f^{(k)}(0) x^k$$

Eq. 57

Note that here, we have used the alternative notation for the kth derivative of $f(x)$. Some of the power-series representations or expansions you may have learned in high school are, in fact, Maclaurin (and hence Taylor) series expansions. Take the following examples:

$$e^x = 1 + \frac{x}{1!} + \frac{x^2}{2!} + \frac{x^3}{3!} + \frac{x^4}{4!} + \dots$$

Eq. 58

$$\sin(x) = x - \frac{x^3}{3!} + \frac{x^5}{5!} - \frac{x^7}{7!} + \dots$$

Eq. 59

$$\cos(x) = 1 - \frac{x^2}{2!} + \frac{x^4}{4!} - \frac{x^6}{6!} + \dots$$

Eq. 60

$$\ln(1 + x) = x - \frac{x^2}{2} + \frac{x^3}{3} - \frac{x^4}{4} + \dots$$

Eq. 61

These are all Maclaurin series. The usefulness of a truncated series expansion is illustrated in *Figure 1.9*, where we plot the exact values of $\ln(1 + x)$ and its 4-term Maclaurin series approximation given in *Eq. 61*, for a range of small positive values of x. We can see that for $x < 0.5$, the exact and approximate values are almost indistinguishable. As x increases beyond 0.5, we see that the error made by the approximation increases. Consequently, for small values of x, the 4-term approximation, or even 1-term or 2-term approximations, can be useful both for numerical calculations and as approximate formulae:

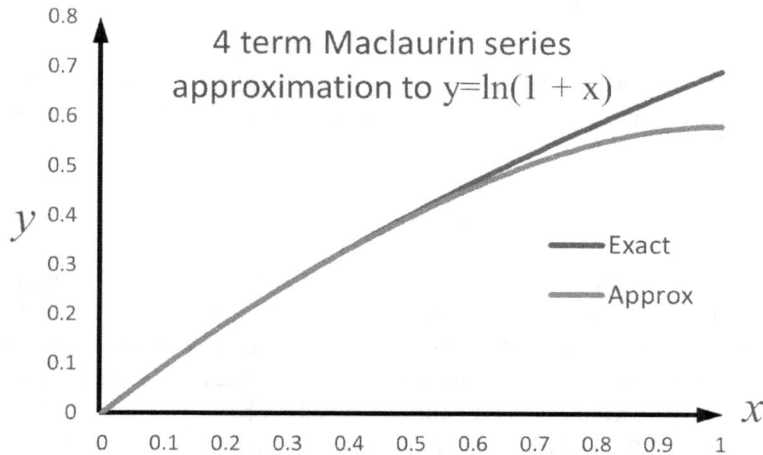

Figure 1.9: Approximation of ln(1+x) by a 4-term Maclaurin series

Analysis is another big math topic. Again, we have only covered a small number of sub-topics and the associated notation. Let's summarize what we have learned about the notation of those sub-topics.

What we learned

In this section, we have learned about the following:

- How limit notation can be used to describe the approach to particular locations on the real line, including $+\infty$ and $-\infty$

- How the limit behavior of a function can be different as we approach the limit point from below compared to when we approach the limit point from above

- The Landau big-O and little-o notations for describing how quickly a function $f(x)$ grows as we approach a particular limiting point of x

- How a function can be represented as a Taylor series expansion consisting of a sum of terms, each involving derivatives of increasing order

- How we can truncate the Taylor series expansion to give an approximation that involves just a few terms

- How the accuracy of the truncated Taylor series expansion is better the closer we are to the point about which we are expanding the function

In the final section of this chapter, we take a change in direction and look at binomial coefficients, which will be useful when we introduce the binomial distribution in *Chapter 2*.

Combinatorics

Our final section regards binomial coefficients. They are part of the mathematical field of combinatorics, but we will introduce them in the context of the binomial distribution, which we will meet multiple times in the book.

Binomial coefficients

Along with the normal or Gaussian distribution, the binomial distribution is one of the most common distributions we will encounter as data scientists. It is the distribution of the number of times, n, we observe a particular outcome in a set of N observations, where in each observation there are only two possibilities that can occur. Given we are interested only in the total number, n, of successful outcomes of a particular type, a large part of calculating the associated probability comes down to calculating how many ways we can distribute or arrange the n successes between the N observations. The answer is given by the *binomial coefficient* $\binom{N}{n}$. This is defined mathematically as follows:

$$\binom{N}{n} = \frac{N!}{n!(N-n)!}$$

Eq. 62

Here, $n!$ means n factorial and is defined as $n! = n \times (n-1) \times (n-2) \times \ldots \times 2 \times 1$. So, $\binom{N}{n}$ is calculated as follows:

$$\binom{N}{n} = \frac{N \times (N-1) \times (N-2) \times \ldots \times 2 \times 1}{[n \times (n-1) \times (n-2) \times \ldots \times 2 \times 1]\,[(N-n) \times (N-n-1) \times \ldots \times 2 \times 1]}$$
$$= \frac{N \times (N-1) \times \ldots \times (N-n+1)}{n \times (n-1) \times (n-2) \times \ldots \times 2 \times 1}$$

Eq. 63

It is also useful to remember that $0! = 1! = 1$.

What does the quantity given by $\binom{N}{n}$ represent? It is equal to the number of distinct ways you can choose a group of n identical items from a larger set of N identical items. Because of this, you may hear the binomial coefficient $\binom{N}{n}$ being referred to as "N choose n." Likewise, the symbol $^{N}C_{n}$ is also used, where the symbol C refers to the fact that we are "choosing" or "making choices." Throughout this book, I will use the $\binom{N}{n}$ symbol.

That was a very short section, but we will make use of the binomial coefficients in multiple places in this book, so it is important to introduce the notation for them. Let's summarize what we have learned about that notation.

What we learned

In this section, we have learned about the following:

- The $\binom{N}{n}$ notation for the binomial coefficients
- The binomial coefficient $\binom{N}{n}$ representing the number of ways we can choose n items from a pool of N identical items

Summary

We have completed our brief recap of the main notation and terminology that we will need for our tour of math concepts in data science. We are now ready for that tour, so let's begin.

In the next chapter, we will meet our first key math concept when we learn about random variables and probability distributions.

Notes and further reading

1. An example of one of the more well-known "big book of integrals" is *Table of Integrals, Series, and Products* by I.S. Gradshteyn and I.M. Ryzhik, 8th Edition, Daniel Zwillinger (Editor), Academic Press (Cambridge, Massachusetts, USA), 2014. This is the book of integrals that I use.

What you learned

In this section we learned about the following:

- ...

Summary

Notes and further reading

2

Random Variables and Probability Distributions

In this chapter, we are going to learn about randomness in data and how we can use probability distributions to describe and handle this randomness. Because of the importance of this chapter, we will deliberately spend a lot of time emphasizing in words the math concepts we are introducing. This makes this chapter a long one. Take your time and digest its contents fully. Doing so will pay dividends because so many of the other math concepts we introduce in this book are underpinned by what we introduce in this chapter. We will do this in five separate topics, each building upon the previous topic. Those topics are the following:

- *All data is random*: In this section, we learn how randomness in data arises and why it is important to understand it

- *Random variables and probability distributions*: In this section, we learn the basics of random variables and probability distributions, why they are useful for describing randomness in data, how to summarize their characteristics, how to transform them, as well as the details of some commonly occurring distributions you will encounter as a data scientist

- *Sampling from distributions*: In this section, we learn how datasets are created or sampled from probability distributions, and how to generate our own samples

- *Understanding statistical estimators*: In this section, we learn how samples of data differ from the distribution from which the data was generated, and how to use our new knowledge of probability distributions to make accurate inferences from samples of data

- *The Central Limit Theorem* (*CLT*): In this section, we learn why and how the normal distribution is one of the most common distributions we will encounter as a data scientist

Technical requirements

All code examples given in this chapter (and additional examples) can be found at the GitHub repository, https://github.com/PacktPublishing/15-Math-Concepts-Every-Data-Scientist-Should-Know/tree/main/Chapter02. To run the Jupyter notebooks, you will need a full Python installation, including the following packages:

- `numpy` (>=1.24.3)
- `scipy` (>=1.11.1)
- `scikit-learn` (>=1.3.0)
- `matplotlib` (>=3.7.2)

All data is random

If you read only one chapter in this book, read this one. Why? Well, because it explains the most important math concept in data science – all data is random. Or, more precisely, all data contains a random component.

Is this really the case? Let's explain. To start, we must explain what we mean by random. I'm not going to give some dry technical definition here, expressed in mathematical symbols. I'm going to give a technical, but intuitive definition: *random means non-predictable*.

What do we mean by that? Precisely what it says. If something is random, it can't be computed or calculated in advance.

A little example

I have an old ship's barometer that belonged to my father (he was a ship's captain). The barometer is damaged and a bit temperamental, so the measurement is imprecise. This means the measured atmospheric pressure is not the same as the actual atmospheric pressure but deviates from it, possibly by as much as +/- 40 mbar. We can capture the idea of that deviation in the following schematic equation:

Measured atmospheric pressure = True atmospheric pressure + Measurement 'Noise'

Eq. 1

I have referred to this as a schematic equation because it expresses a key concept. The data we have, the measured pressure, deviates from what we'd like to know – the true pressure – due to the addition of the random measurement noise. So, the data (the observation) contains a random component.

Sometimes, we also refer to measurement noise as measurement error or just error. In fact, *error* is a word used a lot to describe the random component in data. Sometimes we will also use the word *stochastic* instead of random, so you may see the terms *random error, random component, stochastic error*, and *stochastic component* used interchangeably.

You may say, "*But that measurement error is just because you're using an old barometer. If you were just using a modern digital barometer, the error in the pressure measurement would be inconsequential.*" True, but measurement error would still be there because we have to use some proxy physical process by which the force of the atmosphere is transferred to some other object whose change we can measure. Philosophically, what matters is that the error is there. One of our jobs as data scientists is to assess/quantify the scale of the error and determine how consequential it is.

One thing we should emphasize is that the use of the word *error* does not imply a mistake has been made. It simply implies that there is a difference between the measurement and the true value. In our preceding example, the use of the word *error* means there is a difference between the measured pressure and the true (and unobserved) pressure. That difference is natural, and we understand why it comes about. No mistake has been made. Measurement error is a natural consequence of the fact that we measure things via proxy physical processes; for example, the atmosphere exerting pressure indirectly on a metal spring.

> **Pro tip**
>
> On this point about what we mean, as data scientists, when we use the word *error*, be careful. Don't use the word *error* when talking to non-data science stakeholders, or be very careful using it. A non-data scientist may hear *mistake* when you use the word *error*. So, instead, I tend to explain using words such as *deviation* or *difference*.

Systematic variation can be learned – random variation can't

Sometimes I'm interested in just how inaccurate the old ship's barometer is, so I also take a measurement using a small digital barometer I have. The measurement from the digital barometer I consider to be accurate – the measurement error is much smaller – and so the reading from the digital barometer I interpret as being the true atmospheric pressure. The following scatterplot shows how the two measurements relate to each other:

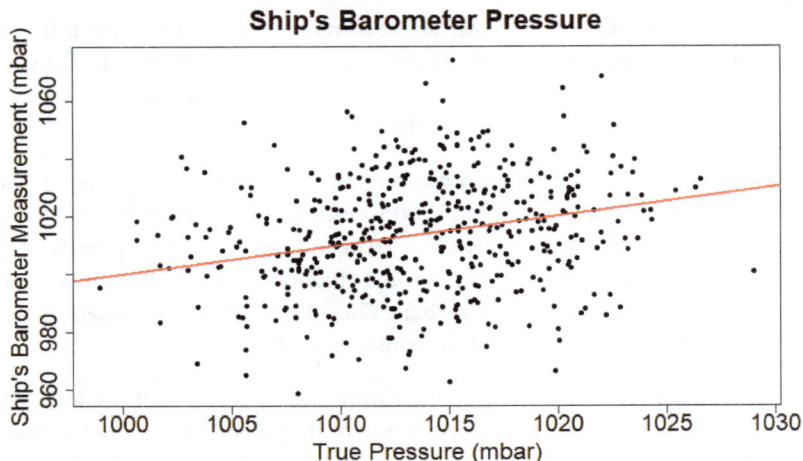

Figure 2.1: Scatterplot showing the accuracy of my old ship's barometer

The solid red line is the 1:1 line and so shows that on average the ship's barometer reading gives the correct atmospheric pressure, but the vertical scatter about that line clearly highlights the random nature of the measurement from the ship's barometer. What it also highlights is that if I wanted to predict what the ship's barometer measurement would be, I wouldn't be able to do so, even if I knew what the true pressure was. How far above or below the line any particular individual ship's barometer measurement is is determined by the random measurement error. We can't predict this in advance because, by definition, it is random. No amount of clever mathematics will change this. No **machine learning** (**ML**) algorithm will be able to predict what the next measurement from my ship's barometer will be, no matter how much training data is given to that algorithm.

What the solid line on the scatterplot does show, though, is that there is something about the ship's barometer measurement that can be predicted – its average behavior. The average ship's barometer measurement is the same as the true pressure, and so it varies systematically with the true pressure. It is this average or systematic behavior that an ML algorithm could learn and would learn with increasing accuracy as we used an increasing amount of training data. An alternative schematic way of writing *Eq. 1* would be the following:

Observed Data = Systematic component + Random component

Eq. 2

It is the systematic component that data science algorithms can learn and not the random component. In this simple example, the random component is additive, and without loss of generality, we can assume its average is zero. Consequently, the systematic component here is equal to the average value of the observed data, or more correctly, the average value of the observed data is equal to the systematic contribution to the data. This is not always the case, as we explain next.

Random variation is not just measurement error

"But," I can hear you say, "what about something where there is no measurement error, where it is highly unlikely that we measured something incorrectly?" Consider something such as what kind of drink I'm drinking now, as I write this at 11:15 a.m. on a Saturday morning. Surely, I can't be mistaken about what kind of drink, tea or coffee, that I have in my cup? No, but what type of drink I choose to put in the cup can still have a large random component to it. Let me illustrate.

In the morning when I'm writing, I like to have a hot drink around 11:00 a.m. Overall, I choose tea 40% of the time and coffee 60% of the time. There is some good logic as to which drink I will choose. Largely, which type of drink depends on whether I'm already getting tired by 11:00 a.m., and that depends on how well I've slept. How well I've slept depends on whether I've had my cat trying to sit on my head during the night while I've been trying to sleep. Whether the cat tries to sit on my head depends on whether the cat wants to be inside the house or outside hunting. This chain of connections is summarized in the following diagram:

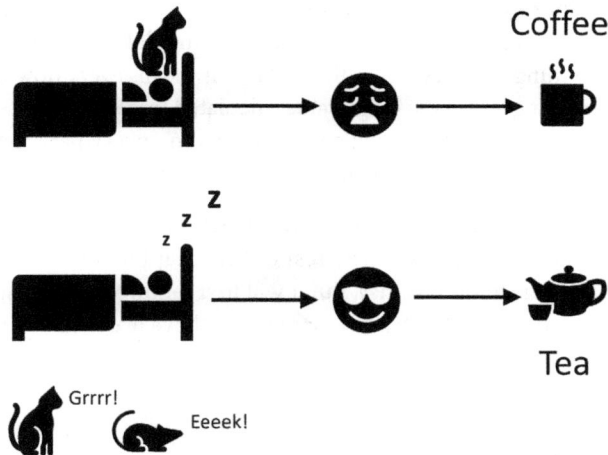

Figure 2.2: How my cat's random behavior makes my choice of morning drink effectively random

Despite there being a strong deterministic element to how I make my choice of drink in the morning, there is a large random variability in the outcome because my choice is now largely determined by the whim of my cat's behavior the night before.

This example highlights three important things:

- The random component within data is not necessarily additive. Random variation does not always present itself in the form of *Eq. 2*. Instead, random variation can be variation in the chosen outcome. The potential outcomes may have well-defined long-run frequencies of occurrence, but any single instance of an outcome is unpredictable – that is, random – because in this example, on any particular night, I don't know what my cat is going to do.

- What we consider to be random can be a matter of modeling choice. I've said my morning drink is essentially determined by what my cat did the night before. Now, you may think that my cat's behavior could potentially be predicted. Personally, I doubt it, but let's say that in theory, it could. However, whatever equations are determining my cat's behavior, they must be so complex as to give the appearance of something almost random because I have yet to work my cat out. What this illustrates is that sometimes we have variation in data that is systematic – meaning it is deterministic in nature and could in principle be predicted in advance – but that systematic variation is so complex, or its deterministic causes hidden from us, that we choose to view that variation as effectively being random and we model it as such.

- Wherever human decisions are involved, we have random variability in outcomes, and that variability can be marked. When modeling datasets of human decisions, understanding this is important.

What are the consequences of data being random?

Okay – so I may have convinced you that all data contains some random component. So what? Why is this important for data science? Unfortunately, if our starting data has some random variation within it, then so does everything we derive from that data. This means every downstream quantity we calculate. And I mean everything – every average value calculated from a dataset, every loss-function value we calculate from a dataset when training an ML algorithm, every parameter value of a **deep learning (DL) neural network (NN)** trained using a dataset, every single ML metric we calculate.

If the random variation within data affects every single calculation we do in data science, we'd better learn how to handle this random variation. The rest of this chapter is devoted to giving you the mathematical tools and language to do just that. But it will first require learning some new concepts, so this is a good place to finish this section and recap what we have learned.

What we learned

In this section, we have learned the following:

- All data we will work with as data scientists has a random component to it

- Because of this randomness in data, all quantities derived from data also contain a random component

- Only the non-random or systematic variation in a dataset is learnable by a data science algorithm

- Sometimes, parts of the systematic variation in a dataset will be hidden from us – we will be unaware that the variation is systematic, and so because of our ignorance we will treat it as random

- Sometimes, parts of the systematic variation in a dataset will be so complex that we deliberately choose to treat it as effectively random

Having learned how randomness appears in all datasets we work with as data scientists, in the next section, we move on to how we describe and quantify that randomness.

Random variables and probability distributions

We start this section by introducing a new concept that is necessary to describe the randomness we find in data.

A new concept – random variables

In computer code, when we want to use a variable, we type something such as $x=5$. In many programming languages, we may change the value of the variable x. We even use the word *variable* to indicate that its value may change. However, those changes are caused by us or by code we have written, and so typically they happen in a deterministic way; that is, we compute when the changes should happen, and we can compute the new value of the variable.

For data that contains a random component, we need a new concept. Remember – random means non-predictable. When we record, observe, or capture the value of that variable, its value is not pre-determined. Instead, it could take on a number of values. The new concept we need is that of a **random variable**. A random variable is a variable, but its value when measured can be one of many different potential outcomes, each occurring with different probabilities. For example, the hot drink I will choose tomorrow morning is a random variable with two possible outcomes (tea or coffee) with probabilities of 0.4 and 0.6, respectively.

Listing the possible outcomes of a random variable and their associated probabilities gives us all the information we need to work with that random variable. We call this set of probabilities a **probability distribution**. Graphically, we can display this as a simple chart. For example, the bar chart in *Figure 2.3* shows the two possible outcomes for the tea/coffee example and their associated probabilities:

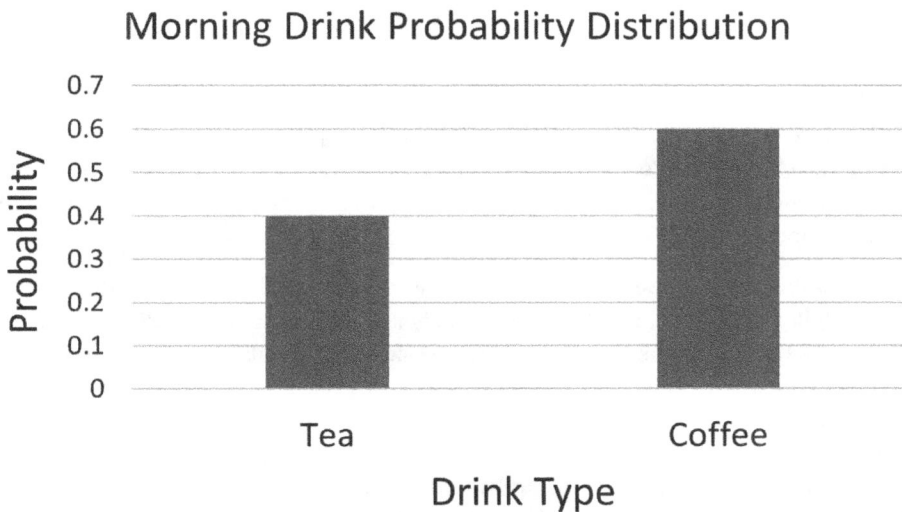

Figure 2.3: The probability distribution for my morning drink choice

We could equally have listed the probability distribution as a table, such as we have in *Table 2.1*:

Drink	Probability
Tea	0.4
Coffee	0.6

Table 2.1: My morning drink probability distribution displayed in table form

Where we have many outcomes, the graphical visualization is more intuitive. We can quickly pick out which outcomes have the highest probability, and we can see where the probability becomes very small, so outcomes occur very infrequently. In many situations, the possible outcomes of a random variable will have a natural ordering; for example, if I was considering the number of items bought from an e-commerce website in a day. Ten items bought in a day is clearly larger than five items bought. The number of items bought in a day probably has an upper limit, but some random variables can have an infinite number of potential outcomes.

One of the most obvious characteristics of the probability distribution shown in *Figure 2.3* or *Table 2.1* is that the probabilities for the two possible outcomes add up to 1. This is true for any probability distribution. The probabilities across all the possible outcomes always sum to 1. This is because the sum of the probabilities represents the probability of getting any outcome, and we have absolute certainty that we will get some outcome. If a probability distribution has outcomes x and associated probabilities $p(x)$, we write this condition as follows:

$$\sum_x p(x) = 1$$

Eq. 3

Some types of probability distributions are so commonly used that we i) give them a name so that we can conveniently refer to them using shorthand rather than having to list all the outcomes and probabilities and ii) because we want to study them in depth and share our findings with other scientists who may also be working with the same distribution. Giving something a name facilitates easier communication about that thing.

The probability distribution where we have just two possible outcomes is called the **Bernoulli distribution** (named after Jacob Bernoulli from the famous Bernoulli dynasty of Swiss mathematicians). Without loss of generality, we call those outcomes 1 and 0 because the choice of what labels we map or associate to the 1 and 0 outcomes is a matter of individual choice, and so doesn't affect the mathematical properties of the Bernoulli distribution. The two outcomes are also sometimes called "success" and "failure," reflecting the fact that one of the outcomes may be more beneficial or preferrable. For the Bernoulli distribution, we also only need to know the probability associated with the 1 (or success) outcome. Let's denote that probability by p. The fact that the probabilities of all possible outcomes must sum to 1 means that the probability of the 0 (or failure) outcome is $1 - p$. For the tea/coffee example, because the probability of me choosing a tea drink was 0.4, then the probability of me choosing coffee was automatically 0.6. So, a Bernoulli distribution is entirely specified once we know the value of p, and so we write a Bernoulli distribution using a shorthand notation as *Bernoulli(p)*. If I see the notation *Bernoulli(p)*, I know we are talking about a probability distribution with two possible outcomes, 1 and 0, with the probability of 1 being p and the probability of 0 being $1 - p$.

Now, when we have a random variable X that follows a Bernoulli distribution, we write the following:

$$X \sim Bernoulli(p)$$

Eq. 4

Let's unpack that notation. Firstly, the use of ~ (the tilde symbol) and the presence of a named distribution on the right-hand side of *Eq. 4* tells us that X is a random variable. We have used a capital letter for the random variable. This is common practice, to distinguish it from an ordinary variable that we would more commonly represent by a symbol such as x. However, it is not unusual to see lowercase symbols used to represent random variables as well, and so the main giveaway that something is a random variable is the presence of the tilde symbol (~). The use of ~ means that the random variable on the left-hand side of *Eq. 4* follows the probability distribution on the right-hand side. Overall, we read *Eq. 4* as meaning X is distributed as *Bernoulli(p)*. This immediately tells us that X has two possible outcomes, 1 and 0, with associated probabilities of p, $1 - p$. What the outcomes 1 and 0 map to in terms of more useful human interpretable labels is usually explained elsewhere in the documentation that had the equation in it.

For my morning drink example, and using the encoding 1=Coffee, 0=Tea, I can write the following,

$$Drink \sim Bernoulli(0.6)$$

Eq. 5

Having learned about how random variables and their associated probability distributions are natural math concepts to describe the randomness we find in data, we are now going to learn how to quantify and summarize probability distributions in a more intuitive way.

Summarizing probability distributions

The probability distribution associated with a random variable tells us everything we need to know about that random variable. It tells us what the probability of any particular outcome is. Imagine we wanted to communicate information about the random variable to a colleague. Obviously, the most complete way would be to communicate all the individual probabilities for all the individual outcomes. If we have a large or infinite number of outcomes, that is a lot of probabilities we must communicate to our colleague. We could send our colleague the bar chart showing the probability distribution. However, while that bar chart is a great visual way of communicating the distribution, it doesn't easily communicate all the numerical values of the probabilities to our colleague. But what if our colleague doesn't want all the probabilities? What if they just want to get a quantitative feel for what the distribution is about? Is there a single number or a couple of numbers that conveniently summarize most of the information contained in the probability distribution? The answer is yes, and we will now introduce the most important ones.

The mean of a distribution

The **mean** of a distribution is a single number that we calculate from the full probability distribution. The mean is used to give us an idea about the average value of the distribution; that is, the typical value we would expect to see if we drew lots of values from the same distribution. However, be aware that the mean does not always do a good job of this. We will see why later. For now, let's see how the mean is calculated.

The mean of a probability distribution that has outcomes x and associated probabilities $p(x)$ is defined as follows:

$$\text{Mean} = \sum_x xp(x)$$

Eq. 6

Here, the summation is over all the possible outcomes. Let's make that a bit more concrete. Imagine I have the probability distribution shown in *Figure 2.4*:

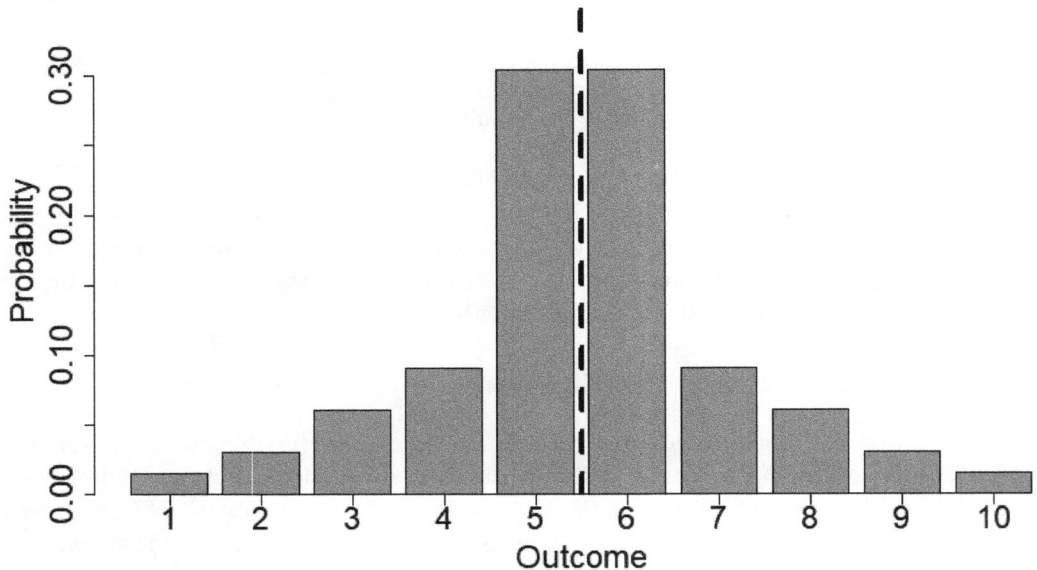

Figure 2.4: Example probability distribution

The outcomes are the integer values 1, 2, …, 10 (perhaps resulting from throwing a strangely shaped 10-sided die). The mean is then the sum of each of those integer values multiplied by the corresponding probability shown in *Figure 2.4*. The mean value turns out to be 5.5 and is shown by the vertical dashed line in *Figure 2.4*. Is the mean value useful to us? Yes – in this case, it is very useful. You can see from the shape of the probability distribution in *Figure 2.4* that the most probable outcomes are 5 and 6, with other outcomes having a lower probability and hence occurring less frequently. When drawing a single value from this distribution, we would expect to get 5 or 6 a lot of the time. For this reason, the mean is what's called an **expectation value** or **expected value**. The mean of the random variable X is called the expectation value of X or the expected value of X. We have a special symbol for an expectation value. We use the symbol $\mathbb{E}(X)$ for the expectation value of X. So, $\mathbb{E}(X)$ means the same as calculate the mean of X. That is shown in the following formula:

$$\mathbb{E}(X) = \sum_x x\, p(x)$$

Eq. 7

From the notation $\mathbb{E}(X)$, you'll see that we sort of consider $\mathbb{E}$ to be like a function, in that it is applied to the thing inside the brackets. It is actually an operator, but we won't go into that distinction here. You can think of $\mathbb{E}$ as a function that gets applied to random variables. That means we can also apply $\mathbb{E}$ in a composite way. Say we wanted to calculate the mean value of taking the exponential of any outcome x of the random variable X. Taking the exponential of a random variable is just another random variable – remember what we said about anything derived or computed from something random being also random. We'll denote this newly derived random variable as e^X to signify that it is a random variable obtained from taking the exponential of the original random variable X. The expectation value of this newly derived random variable e^X is calculated as follows:

$$\mathbb{E}(e^X) = \sum_x e^x\, p(x)$$

Eq. 8

To calculate the expectation value of this newly derived random variable, we have simply applied the same function (the exponential function) to the outcome values x. As you might guess, for a general function f we calculate the expectation value of $f(X)$ by calculating the following:

$$\mathbb{E}(f(X)) = \sum_x f(x) p(x)$$

Eq. 9

The variance of a distribution

Looking at *Figure 2.4*, we would think from the shape of the distribution that it was obvious that the mean of X was representative of the typical value we would get when we draw a value from the distribution, because of how tightly concentrated the distribution is around its highest probability outcomes at $x = 5$ and $x = 6$. What would happen if we had a distribution that was more widely spread? Take a look at *Figure 2.5*:

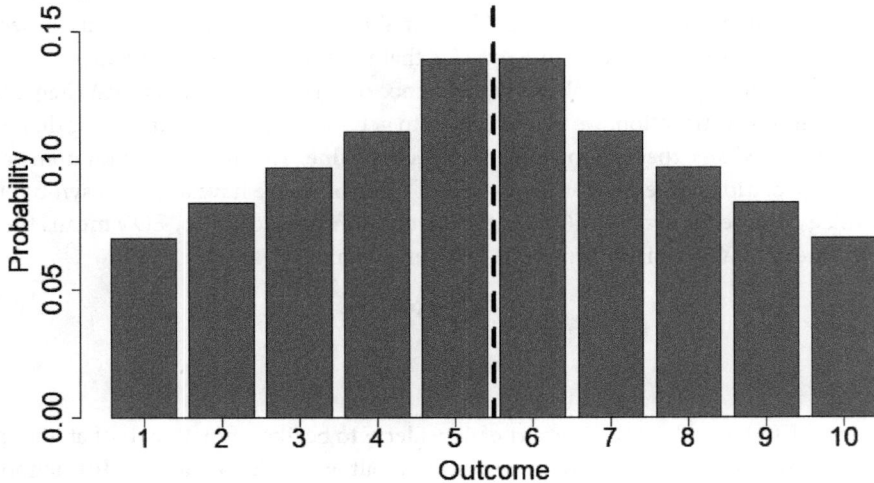

Figure 2.5: Another example probability distribution

The mean of this distribution is also 5.5, but clearly, values that are very different from 5.5 are now more likely compared to when we were drawing values from the distribution in *Figure 2.4*. So, for *Figure 2.5*, the mean of X is the average value we would get when we draw lots of values from the distribution in *Figure 2.5*, but as a single number, the mean value of 5.5 is not typical as there is a lot spread of values around 5.5. Clearly, knowing **only** the mean isn't a good way of summarizing the distribution in *Figure 2.5*. Whether that spread is small or large is important to know. How can we quantify that spread?

Just as the mean quantifies the average value drawn from the distribution, we can also calculate the average difference from that mean value. If the mean of a distribution is μ and we look at a specific outcome value x, then the deviation of the outcome from the mean is $x - \mu$. We could then just take the average of this deviation over all possible outcomes. This would give us the following:

$$\sum_x (x - \mu)p(x)$$

Eq. 10

A quick bit of algebra will tell you that this is always zero, so it isn't a very good way of measuring the spread of a distribution. The reason for this is that for small values of x, the deviation $x - \mu$ is negative, while for large values of x the deviation is positive, and overall, the positive and negative values exactly cancel each other out. We can stop this cancellation by using the squared deviation $(x - \mu)^2$ instead and calculate the average of this squared deviation because the squared deviation is always positive or zero. This is calculated using the following formula:

$$\text{Variance} = \sum_{x}(x - \mu)^2 p(x)$$

Eq. 11

We call this average squared deviation the **variance** of the distribution. Now, you're probably asking how this gives us a measure of the spread of a distribution if we have calculated the average squared deviation. The answer is it doesn't. To get a measure of spread, we take the inverse operation at the end; that is, we take the square root of the variance. The square root of the variance is called the **standard deviation**. So, putting that together, we have the following:

$$\text{Standard Deviation} = \sqrt{\text{Variance}} = \left[\sum_{x}(x - \mu)^2 p(x)\right]^{\frac{1}{2}}$$

Eq. 12

In general, we have the following:

$$\text{Variance} = \text{Standard Deviation}^2$$

Eq. 13

We use the symbol σ for the standard deviation, and so from the preceding equation, we can also use σ^2 to denote the variance. Sometimes, you will also see Var used as shorthand for the variance. For example, we might write $\text{Var}(X)$ to indicate the variance of the random variable X.

What does the standard deviation tell us? Well, the clue is in the name. It is the standard, typical, or expected size of deviation from the mean that we should expect when we draw a number from the distribution. If we draw a number from the distribution, we should not be surprised if it differs from the mean by something comparable to the standard deviation; for example, by as much as 1σ or 2σ.

We've said that the standard deviation of a distribution is the size of the typical deviation we should **expect** when drawing a number from the distribution. So, you might ask, can the standard deviation be written as an **expectation value** of some random variable? Not quite. But the variance can. If we look at *Eq. 11*, we can see that it is a weighted average of the squared deviation. So, we can write the variance calculation in *Eq. 11* as follows:

$$\text{Variance}(X) = \mathbb{E}((X-\mu)^2) = \mathbb{E}((X-\mathbb{E}(X))^2)$$

Eq. 14

The second part of the formula on the right-hand side in *Eq. 14* just comes from replacing $\mu = \mathbb{E}(X)$. The formula in *Eq. 14* tells us that the variance of the random variable X is the same as the expectation value of the random variable $(X-\mu)^2$. The left-hand side of the preceding equation shows that calculating the variance of a distribution is a function or operation that we apply to a random variable, so you might ask whether there is a special symbol we use when calculating the variance of a distribution, just like we use the symbol $\mathbb{E}$ when calculating the expectation value of a random variable. Well, there is. We have already used it. It is $\text{Var}(X)$. But you may also see $\mathbb{V}(X)$ used to represent the operation of evaluating the variance of the random variable X. However, in my personal experience, $\mathbb{V}(X)$ is a lot less commonly used compared to $\text{Var}(X)$.

Other characteristics of a distribution

When summarizing a distribution, it is common to give just the mean and standard deviation, or equivalently the mean and the variance. Are these two numbers enough to summarize a distribution? The answer is no. Sometimes, we may want to quote higher-order moments about the mean μ. For example, the third moment looks like this:

$$\mathbb{E}((X-\mu)^3) = \sum_x (x-\mu)^3 \, p(x)$$

Eq. 15

This can tell us about how lop-sided or asymmetric a distribution is. Using this third **central moment**, we can calculate the skewness of a distribution:

$$\text{Skewness} = \mu_3 = \mathbb{E}\left(\left(\frac{(X-\mu)}{\sigma}\right)^3\right)$$

Eq. 16

The graph on the left-hand side of *Figure 2.6* has a skewness of zero, while the middle graph has a positive skewness, and the right-hand graph has a negative skewness. The dashed line in each of the plots shows the position of the mean of each distribution, so we can see how the skewness reflects the shift in probability mass from one side of the mean to the other:

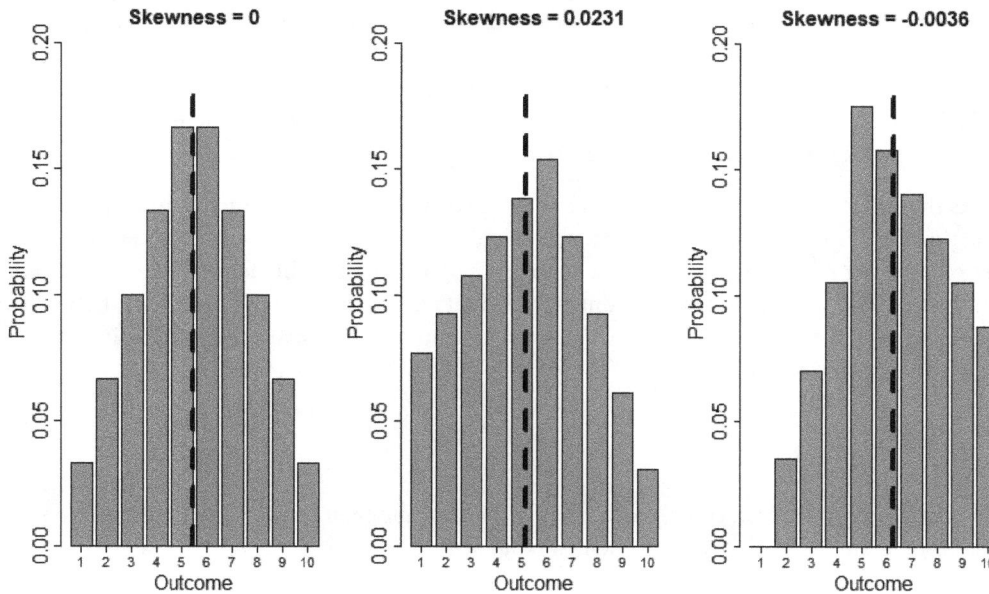

Figure 2.6: Example probability distributions with different skewness values

The symbol μ_3 is typically used for the skewness. Likewise, we can use the fourth central moment to calculate what is called the **kurtosis** of a distribution. We won't define the kurtosis here, other than to say that the kurtosis measures relatively how fat or thin a distribution is. If you want to convey to another data scientist characteristics such as how lop-sided or thin a distribution is, it is often easier just to show a plot of the distribution. So, most data scientists tend to calculate only the mean and standard deviation of a distribution and then show a plot if they want to communicate additional details.

Having learned how to summarize and characterize a probability distribution for random variables that have discrete outcomes, we are going to learn how to extend this to continuous-valued random variables.

Continuous distributions

The sharp-eyed among you will have spotted that all the examples of probability distributions that we have given so far in this section are of discrete outcomes – outcomes that are clearly distinct, such as tea and coffee, or the integer numbers 1, 2, 3…and so on. Surely some outcomes can be continuous. The ship's barometer example we started the chapter with is an example of a continuous quantity. So, yes – an outcome can be continuous, but there are some subtleties with continuous-valued outcomes that we will now explain.

The first subtlety stems from the number of possible outcomes. Imagine I have a random variable X that is a real number and can take any value between $-\infty$ and $+\infty$. It has an infinite number of possible outcomes. Given we have said the sum of the probabilities across all outcomes must be 1, then what happens when we sum an infinite number of probabilities? Doesn't this mean the probability of any particular outcome is zero? Even the answer to this question is slightly complicated, but just asking this question does highlight that the concept of a probability distribution needs modification when we are dealing with continuous-valued outcomes. Instead of talking about probability distributions, we talk about **probability density functions (PDFs)**.

A PDF, as the name suggests, tells me the density of probability in the region around a particular outcome point x. With physical densities, we can calculate the amount of material; for example, the number of atoms or molecules, present in a volume V, by multiplying the density ρ by the volume. Similarly, with a PDF, to calculate the amount of probability in a small interval of width dx between x and $x + dx$, we simply multiply the PDF by the interval width dx. This gives us the following:

$$\text{Prob}(x < X < x + dx) = pdf(x)\ dx$$

Eq. 17

So, for a continuous-valued outcome, we talk not of the probability of having a particular outcome x but of the probability of having an outcome in a range.

Note that we have used here the differential calculus concept of a small interval dx. This is necessary because, unlike with physical densities and physical material where the density may be constant over a significant sized volume, the probability density can vary markedly as we change the outcome value x.

Secondly, note that if we decrease the size of the interval dx down to zero so that we are looking at a single outcome point and not an interval, then the right-rand-side of *Eq. 17* becomes zero – just as we explained previously. So, the probability of a point outcome is zero, but the probability density at that point can be nonzero.

Just in case you're wondering if the aforementioned line of reasoning is dependent on the range of possible outcomes being $-\infty$ to $+\infty$, the answer is no. If the range of outcomes is in a finite interval, say between -10 and 10, we still have an infinite number of outcomes, and so everything we have said before still holds. It just means the PDF is zero outside of that interval -10 to 10.

Now that we have explained the subtle difference between a probability distribution and a PDF, it is worth highlighting the following:

- Most statisticians and data scientists will use the term *probability distribution* when they mean PDF, with the assumption that the reader will implicitly know what is really meant.

- Given an infinite number of possible outcomes and probabilities referring to small intervals, all the results we have given about expectation values for discrete random variables also hold for continuous random variables, with the simple replacement of the probability distribution $p(x)$ by the density function $f(x)$, and the summation symbol Σ by the integration symbol $\int$. We will recap those results now, with those replacements.

If we have a continuous-valued random variable X with outcomes denoted by x and a PDF $f(x)$, then we have the following results:

$$\int_{-\infty}^{+\infty} f(x)\ dx\ = 1 \qquad \text{Normalization}$$

Eq. 18

$$\mathbb{E}(X)\ =\ \mu\ =\ \int_{-\infty}^{+\infty} xf(x)\ dx \qquad \text{Mean}$$

Eq. 19

$$\mathbb{E}((X-\mu)^2) = \sigma^2 = \int_{-\infty}^{+\infty} (x-\mu)^2 f(x)\ dx \qquad \text{Variance}$$

Eq. 20

Now we have learned about single random variables, whether discrete or continuous, we are next going to learn how to transform and combine multiple random variables. This will enable us to describe and handle the impact of randomness in data when we apply transformations and aggregations to data.

Transforming and combining random variables

Now we have learned some properties of random variables, we are going to learn about what happens when we combine random variables.

Why is this important? Well, often, we will want to modify or aggregate our data. If each observation in a dataset is a random variable, then what does that mean for the total of all the observations? We know the total is also a random variable, but what are its properties?

As a simple example, consider whether an individual shopper buys an item from a website. We can model that individual purchase decision as a Bernoulli random variable. But what about the total number of items sold in a day, bought by the 1,000 visitors to the website? How do we model that total?

Linear transformations

We'll start with something simpler. What happens when we just linearly transform a random variable? If we have a random variable X, we can create a new random variable $Y\ =\ aX + b$. What does this transformation mean? Well, if the random variable X has possible outcomes x, then it means that the random variable Y has possible outcomes $y\ =\ ax + b$. In terms of a dataset, it means that if we have a particular value – say, 10 – for an outcome of X, then the corresponding outcome value of Y is just $a \times 10 + b$.

Given the linear transformation $Y = aX + b$, how does the mean of Y relate to the mean of X? From *Eq. 7*, we find that for discrete random variables, the following applies:

$$\mathbb{E}(Y) = \mathbb{E}(aX + b) = \sum_x (ax + b)\, p(x) = a\sum_x x\, p(x) + b\sum_x p(x) = a\mathbb{E}(X) + b$$

Eq. 21

So, the mean of Y is simply related to the mean of X by the same linear transformation that we applied to the random variable. An identical result holds for continuous-valued random variables – as you might have guessed, we simply replace $\sum$ with $\int$ and the probability $p(x)$ with the PDF $f(x)$ in *Eq. 21*.

What happens to the variance? A similar simple calculation shows that for both discrete and continuous random variables, if $Y = aX + b$, then the following applies:

$$\mathrm{Var}(Y) = a^2\,\mathrm{Var}(X)$$

Eq. 22

Non-linear transformations

What happens if we apply a more general non-linear transformation to a random variable? This would be the case if we were to apply a non-linear transformation to our dataset. Let's apply the function $h(x)$ to the values x in our dataset. We model this by saying we have a random variable $Y = h(X)$. For discrete outcomes, to calculate the mean of Y we just calculate $\sum_y y\, p(y)$, so we just need to know the probability, $p(y)$, of each outcome. But since for each value of x we know the corresponding value of y and we know the probabilities $p(x)$, then calculating the mean of Y is easy:

$$\mathbb{E}(Y) = \sum_y y\, p(y) = \sum_x h(x) p(x)$$

Eq. 23

The formal proof of the right-hand side of the formula in Eq. 23 is more nuanced. In the way we have simply written the right-hand side, we are making use of "the law of the unconscious statistician." For the purposes of this book, we will take the right-hand side of *Eq. 23* as the definition of $\mathbb{E}(Y)$ when Y is a transformation of another random variable.

For the case where the transformation $h(x)$ is linear, it is easy to see we get the same result as in *Eq. 21*.

Now, for continuous random variables, we have a similar result:

$$\mathbb{E}(Y) = \int_{-\infty}^{+\infty} h(x) f(x)\, dx$$

Eq. 24

To write $\mathbb{E}(Y)$ as an integral over outcomes y we would have to perform a change of variable under the integration sign. For simplicity, we'll assume the transformation $h(x)$ is monotonic. If we do this, we get the following:

$$\mathbb{E}(Y) = \int_{-\infty}^{+\infty} h(x)f(x)dx = \int_{f(-\infty)}^{f(+\infty)} yf(x = h^{-1}(y))\frac{dx}{dy}\,dy$$

Eq. 25

Looking at what is inside the integral in the last expression on the right-hand side of *Eq. 25*, we can see we have an effective PDF for the random variable Y. That effective PDF is the following:

$$f(h^{-1}(y))\left|\frac{dx}{dy}\right| = f(h^{-1}(y))\frac{1}{|h'(h^{-1}(y))|}$$

Eq. 26

We can make some immediate comments on the preceding result:

- Firstly, the obvious – if we apply a transformation to a continuous random variable, the probability density changes.

- We can calculate the PDF of the new transformed random variable using the expression in *Eq. 26*, but only because the function $h^{-1}(y)$ exists and can be calculated. This is because the transformation $y = h(x)$ is monotonic and so invertible, meaning that a single value of y can only have come from a single possible value of x. If the transformation $y = h(x)$ is not monotonic it is still possible to determine a PDF for y by dividing $h(x)$ into individual monotonic segments, but the resulting expression is more complicated.

- The expression in *Eq. 26* looks complicated, but underneath it is a very simple principle. Probability is about counting how many outcomes of a particular type we see. If we count how many outcomes we see in an interval between x and $x + dx$, then that number is simply the density times the volume of the interval (using our physical analogy). That number is also the same (it is invariant) whether we count using x as our measure of volume or whether we count using y as our measure of volume. It is like counting the number of molecules in a given volume – it is the same number of molecules whether we measure volume in millimeters or centimeters. So, for probability densities, we have the following:

Probability Density of Y × Volume of interval for Y = Probability Density of X × Volume of interval for X

Eq. 27

Or, more exactly, we have the following:

$$\left|f_Y(y)dy\right| = \left|f_X(x)dx\right|$$

Eq. 28

Here, we have been more explicit and used $f_X(x)$ to denote the PDF of the random variable X, evaluated at outcome value x, and $f_Y(y)$ to denote the PDF of the random variable Y, evaluated at outcome value y.

- If you just then plug the transformation $y = h(x)$ into *Eq. 28*, you will get the same result as in *Eq. 26*. Personally, I prefer this method of working out how a PDF transforms because I can never remember the complex formula in *Eq. 26*, and because the result in *Eq. 28* is based on a simple principle that I can remember – it doesn't matter how you count how many things are in a given volume; the number is always the same.

Combining random variables

Now, let's get back to our original question of how to combine random variables. Imagine I have two random variables X_1 and X_2 that have outcomes x_1 and x_2 respectively. What is the mean of the random variable $Y = X_1 + X_2$? You can probably guess, but let's work things out properly:

$$\mathbb{E}(Y) = \sum_{x_1, x_2}(x_1 + x_2)\, p(x_1, x_2) = \sum_{x_1} x_1 \sum_{x_2} p(x_1, x_2) + \sum_{x_2} x_2 \sum_{x_1} p(x_1, x_2)$$

Eq. 29

Now, $p(x_1, x_2)$ is the probability of seeing outcomes x_1 and x_2 together, or jointly, and so is called the **joint probability** or **joint distribution** of X_1 and X_2. One of the properties of the joint distribution is that if we sum over all the possible values of x_2 we get $p(x_1)$, or in other words, we get the following:

$$\sum_{x_2} p(x_1, x_2) = p(x_1) \quad \text{and similarly} \quad \sum_{x_1} p(x_1, x_2) = p(x_2)$$

Eq. 30

Plugging this into *Eq. 29*, we get the following:

$$\mathbb{E}(X_1 + X_2) = \sum_{x_1} x_1 p(x_1) + \sum_{x_2} x_2 p(x_2) = \mathbb{E}(X_1) + \mathbb{E}(X_2)$$

Eq. 31

Hopefully, this is what you will have guessed. The same result holds for continuous-valued random variables, again just changing summation for integration.

We can iteratively apply the result in *Eq. 31* to adding together many random variables to get the following:

$$\mathbb{E}(X_1 + X_2 + \cdots + X_N) = \mathbb{E}(X_1) + \mathbb{E}(X_2) + \cdots + \mathbb{E}(X_N)$$

Eq. 32

It is also a simple extension to include linear transformations of those N random variables to give the following:

$$\mathbb{E}\left(a_1 X_1 + a_2 X_2 + \cdots + a_N X_N\right) = a_1 \mathbb{E}(X_1) + a_2 \mathbb{E}(X_2) + \cdots + a_N \mathbb{E}(X_N)$$

Eq. 33

Here, the weights $a_1, a_2, \cdots , a_N$ are fixed numbers.

Okay – but what about the variance of $X_1 + X_2 + \cdots + X_N$? A lengthy, but similar calculation to that in *Eq. 29* shows that if $X_1, X_2, \cdots , X_N$ are independent of each other, then the following applies:

$$\mathrm{Var}(X_1 + X_2 + \cdots + X_N) = \mathrm{Var}(X_1) + \mathrm{Var}(X_2) + \cdots + \mathrm{Var}(X_N)$$

Eq. 34

Likewise, if the different random variables are independent of each other, we have the following:

$$\mathrm{Var}\left(a_1 X_1 + a_2 X_2 + \cdots + a_N X_N\right) = a_1^2 \mathrm{Var}(X_1) + a_2^2 \mathrm{Var}(X_2) + \cdots + a_N^2 \mathrm{Var}(X_N)$$

Eq. 35

Now we have learned the basics of random variables and probability distributions, we will describe in detail some of the most common distributions you will encounter as a data scientist. These are the distributions you will make extensive use of in your career as a data scientist. We divide our descriptions into two obvious categories: descriptions of discrete-valued distributions and descriptions of continuous-valued distributions.

Named distributions

There are a few distributions that we encounter a lot as data scientists. These are the distributions that have characteristics that match the sort of data we encounter in real-world datasets, so it is unsurprising that we should use these common distributions when analyzing real data or when building predictive models. For that reason, it is worth understanding these distributions in a little more detail. We will go into that detail now.

Discrete distributions

We'll start with some of the most important named discrete distributions.

The Bernoulli distribution

We already met the Bernoulli distribution when we first introduced random variables. However, we can calculate some of its key properties with what we have learned since then. You'll recall that a Bernoulli random variable has two outcomes, 0 and 1. If $X \sim$ *Bernoulli(p)* and we draw an observation, then $\text{Prob}(X = 0) = (1 - p)$, $\text{Prob}(X = 1) = p$. The mean of the Bernoulli distribution is then calculated as follows:

$$\mu = \sum_x x\, p(x) = [0 \times (1 - p)] + [1 \times p] = p$$

Eq. 36

Similarly, the variance is calculated as follows:

$$\sigma^2 = \sum_x (x - \mu)^2\, p(x) = [(0 - p)^2 \times (1 - p)] + [(1 - p)^2 \times p] = p(1 - p)$$

Eq. 37

Note the symmetry in the expression for the variance. If we swapped p for $(1 - p)$ and vice versa, we would get the same expression. This is because the choice of how we encode outcomes is arbitrary; that is, it is our subjective choice whether we use 1/0 or 0/1 to represent success/failure, and so certain properties of the distribution will be invariant to our choice of encoding. Switching from an encoding of 1/0 to 0/1 would obviously lead to us swapping the values p and $(1 - p)$.

The binomial distribution

Obviously, a Bernoulli random variable can be used to model data where we have two possible outcomes, such as whether a user clicked a button on an e-commerce website or whether a shopper bought a particular product. However, in a real-world situation, we are more interested in how many items we sell in total or how many website users in total click the button. Each individual choice to click or choice to buy is a Bernoulli random variable, but what is the distribution of their sum?

Imagine we have N shoppers, each deciding to buy or not, so we can model each shopper choice as a Bernoulli random variable, X_i, $i = 1, \cdots, N$. For simplicity, we will assume that the shoppers are all similar so that the probability that any shopper makes a purchase is the same; that is, p. This means $X_i \sim Bernoulli(p)$ for all values of $i = 1, 2, \cdots, N$. Each of these random variables has the same distribution. Note this doesn't mean they are all the same random variable, nor does it mean that all shoppers are making the same choice. It just means the distribution of possible outcomes is the same for each shopper. We consider this to be a reasonable assumption for our problem because our shoppers are all similar in other characteristics, and so we might expect them to have a similar level of preference for the shopping item we are modeling. When we have a set of random variables that all follow the same distribution, we say they are **identically distributed (i.d.)**. If the random variables are also **independent** of each other, (in this case, that means the choice of one shopper does not have an influence on the choice of another shopper), then we say the random variables are **independently and identically distributed**, or **i.i.d.** for short.

Now, what we are interested in is the total number of items sold. This is $S = X_1 + X_2 + \cdots + X_N$. What is the distribution of S? Well, let's use n to represent the actual number of units sold and consider one particular set of outcomes that would give us $n = 4$ items sold when we had $N = 10$ shoppers. The sequence in *Figure 2.7* shows one such set of outcomes:

Outcome: Yes Yes   No   No   No   Yes   No   Yes   No   No

Probability: $p \times p \times (1 - p) \times (1 - p) \times (1 - p) \times p \times (1 - p) \times p \times (1 - p) \times (1 - p)$

Figure 2.7: Outcomes from a sequence of 10 Bernoulli trials with their corresponding probabilities below

Here, we have used *Yes* to indicate a shopper purchased the item and *No* to indicate that they didn't. What is the probability of this outcome? Since the random variables are independent of each other, we can multiply their individual outcome probabilities together, as shown in the lower row of *Figure 2.7*. So, the probability of this particular pattern is $p^4(1 - p)^6$. Now, you may have spotted that the reason we got a factor of p^4 in that result was because four shoppers bought the item, but it didn't matter which four. So, another but different pattern with four shoppers buying the item would also have a probability of occurring of $p^4(1 - p)^6$. To find the total probability of $n = 4$ items sold in a set of ten shoppers, we just need to find how many patterns of ten shoppers we can have where there are four shoppers who buy. In other words, how many ways can we distribute the $n = 4$ successes among $N = 10$ attempts (trials)? This is the binomial coefficient $\binom{10}{4}$ that we recapped in *Chapter 1*. So, the overall probability of four items being bought by ten shoppers is $\binom{10}{4} p^4(1 - p)^6$. It is straightforward to generalize to N shoppers and n items bought, to get the probability of n items bought as follows:

$$\text{Prob}(S = n) = \binom{N}{n} p^n (1 - p)^{N-n}$$

Eq. 38

This is the **binomial distribution**. It is the distribution of the sum of N i.i.d. Bernoulli random variables (or trials) each with success probability p. Notice that the probability $\text{Prob}(n)$ depends upon two quantities or parameters: N and p. When we write that a random variable has a binomial distribution, we write *Binomial*(N, p) or *Binom*(N, p). In our example, the total number of items sold, S, has a binomial distribution, so we would write $S \sim Binomial(N, p)$. In a modern e-commerce setting where we potentially have tens of millions of visitors to a website, N can be very, very big.

Now we have derived the probabilities of the binomial distribution, we can calculate some of its characteristics, such as its mean and variance:

$$\text{Mean} = \mu = \sum_{n=0}^{N} \binom{N}{n} n p^n (1-p)^{N-n} = Np$$

Eq. 39

$$\text{Variance} = \sigma^2 = \sum_{n=0}^{N} \binom{N}{n} (n - Np)^2 p^n (1-p)^{N-n} = Np(1-p)$$

Eq. 40

So, to summarize, for a binomial distribution the mean is Np and the variance is $Np(1-p)$. These results can be derived very simply. Since we know what the mean and variance of a *Bernoulli(p)* random variable is, and our total $S = X_1 + X_2 + \cdots + X_N$ is just a sum of i.i.d. *Bernoulli(p)* random variables, we can just use the results in *Eq. 32* and *Eq. 34* to derive them – see if you can do this for yourself.

The Poisson distribution

For the binomial distribution, we were interested in the distribution of the total number of successes when performing N independent trials. Another situation where we may be interested in looking at the distribution of counts of something is when we want to count how many occurrences of something we get within, say, a given time interval or within a given area. For example, continuing our e-commerce example, we may be interested in the number of items sold within an hour. If we make the simplest possible assumption that the average rate of sale is constant, then the distribution of the number of actual items sold is a **Poisson** distribution. The Poisson distribution is a discrete distribution – it tells us how counts (that is, integer values) are distributed. The possible outcome, k, can be $0, 1, \ldots, \infty$. The probabilities of those outcomes are given by the following formula:

$$\text{Prob}(k) = \frac{\lambda^k e^{-\lambda}}{k!}$$

Eq. 41

Here, λ is the mean outcome value, so it is the constant rate that formed our starting assumption. There are two things to highlight about λ. Firstly, λ can be a non-integer even though the Poisson distribution is a distribution of integer counts. For example, if $\lambda = 3.4$ in our e-commerce example, it means we sell on average 3.4 items per hour. The actual number of items sold in any hour period will be an integer; for example, 2, 3, or 4 items. The second thing to highlight about λ is that it is the only parameter that is in *Eq. 41*. This implies that the variance of the Poisson distribution will be some function of λ. Likewise, other characteristics such as the skewness and the kurtosis. The variance of the Poisson distribution is easily calculated and turns out to also be λ. To recap, that means that if $X \sim Poisson(\lambda)$, then the following applies:

$$\mathbb{E}(X) = \lambda \quad \text{and} \quad \text{Var}(X) = \lambda$$

Eq. 42

An immediate implication of this is that if the mean of a Poisson random variable is increased, then so does its variance and so does its standard deviation. As the bulk of a Poisson distribution is shifted to the right, then it also spreads out. *Figure 2.8* shows two examples of a Poisson distribution. The left-hand plot is for $\lambda = 2.5$, while the right-hand plot is for $\lambda = 7.5$. We can see from the right-hand plot that more of the distribution is at higher values, but it is also more spread out compared to the left-hand plot:

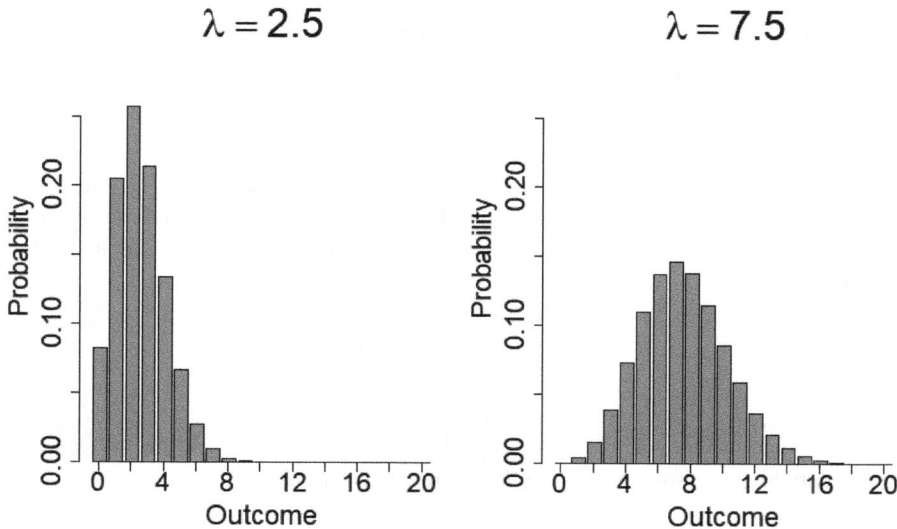

Figure 2.8: Two examples of a Poisson distribution with different means

Continuous distributions

Now, we'll introduce some of the most important named continuous distributions.

The uniform distribution

The uniform distribution is a continuous distribution. It has a minimum possible outcome value (let's call that a) and a maximum possible outcome value (let's call that b). By definition, the probability of getting an outcome smaller than a or an outcome larger than b is zero. As the name suggests, in between a and b the probability of getting any outcome is the same; that is, the probability is uniform between a and b. Consequently, the uniform distribution is a two-parameter distribution, meaning once we know the two parameters a and b, we know everything there is to know about it. If a random variable X is distributed according to a uniform distribution between a and b, we write $X \sim Uniform(a, b)$ or using a shorthand notation, $X \sim U(a, b)$. The PDF, $f(x)$, is given by the following:

$$f(x) = \frac{1}{b - a} \quad \text{for} \quad a \leq x \leq b$$

Eq. 43

This follows simply from the need to have the probabilities of all possible outcomes add up to 1. The mean and variance are also easily calculated using high school math:

$$\mathbb{E}(X) = \tfrac{1}{2}(a+b) \quad \text{Var}(X) = \tfrac{1}{12}(b-a)^2$$

Eq. 44

The uniform distribution may look like it is a bit boring, but in practical terms, it forms the building block when we want to generate example random data from other distributions, so it is a distribution worth knowing about.

The Gaussian distribution

The Gaussian distribution is a two-parameter continuous distribution, so its density function is characterized by two parameters – in this case, its mean μ and its variance σ^2. The formula for its PDF is given next:

$$f(x) = \frac{1}{\sqrt{2\pi\sigma^2}} \exp\left(-\tfrac{1}{2}\left(\frac{x-\mu}{\sigma}\right)^2\right)$$

Eq. 45

A plot of the density function for $\mu = 0, \sigma = 1$ is shown in *Figure 2.9*:

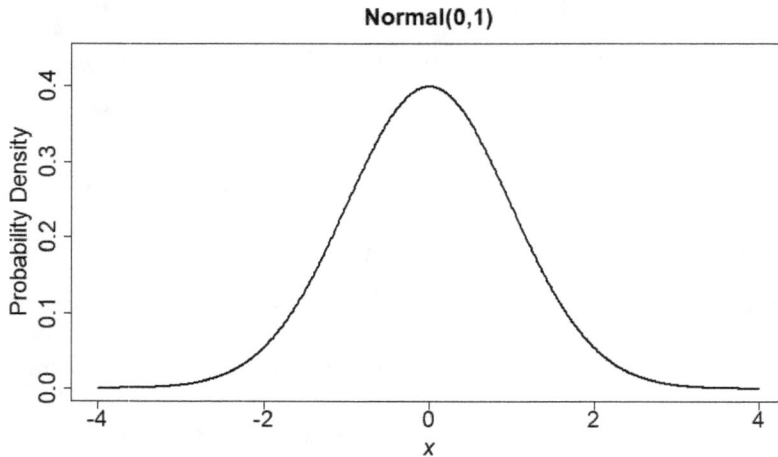

Figure 2.9: The probability density of the standard normal distribution

As the shape of the density function resembles a bell, the Gaussian distribution is sometimes called the bell-curve distribution. However, it is more commonly referred to as the normal distribution, so much so that if X is a Gaussian random variable, we write $X \sim Normal(\mu, \sigma^2)$. This notation also highlights that the distribution depends on the two parameters μ and σ^2. A warning – sometimes you will also see some authors write $X \sim Normal(\mu, \sigma)$, meaning that the value they are giving as the second argument is the standard deviation, not the variance. However, whether you are given σ or σ^2, the density is always given by the formula in *Eq. 45*.

Why are we interested in the Gaussian distribution? The Gaussian distribution is probably the most common distribution you will encounter as a data scientist. Many natural processes produce data that follow a Gaussian distribution, and so many datasets you will analyze will be best modeled as Gaussian random variables. There are some very natural reasons for this, which we will touch upon at the end of this chapter.

From the rules applying to linear transformations of random variables, we know that if we have Gaussian random variable $X \sim Normal(\mu, \sigma^2)$ and we construct a new random variable $Y = \frac{X-\mu}{\sigma}$, then $Y \sim Normal$ $(0, 1)$. What this means is that we can understand the properties of any Gaussian random variable by understanding the behavior of the distribution $Normal(0, 1)$. Given the central importance of the Gaussian distribution, this makes the distribution $Normal(0, 1)$ a key distribution in its own right, and it has its own special name – it is called the **standard** normal distribution, reflecting the fact it has standardized values for the mean and variance. The values $\mu = 0$ and $\sigma^2 = 1$ also have their own special names. We refer to these values as **zero mean and unit variance**, with unit meaning a value of 1 here. So, the standard normal distribution is a normal distribution with zero mean and unit variance.

Phew! That was a long section. But it will be worth it. Data underpins everything in data science, and all data contains a random component, so learning the basics about how we use math to describe and handle randomness will pay big dividends. For now, though, let's do a short recap of what we learned in this section.

What we learned

In this section, we have learned the following:

- Random variables are the natural math concept to describe randomness. Probability distributions give the probabilities of possible outcomes of a discrete random variable, while PDFs perform a similar role for continuous random variables.

- Probability distributions and random variables can be characterized by their mean and variance, and other characteristics such as their skewness and kurtosis.

- How to transform and combine random variables and how to calculate the mean and variance of transformed quantities.

- The details of the most important discrete and continuous probability distributions we will encounter in data science.

Now we have learned about the basics of random variables and probability distributions, we are going to learn in the next section how datasets are generated from probability distributions.

Sampling from distributions

So far, we've learned a lot about random variables, probability distributions, and how to calculate some of the key characteristics of a distribution such as its mean and variance, and we've learned about some commonly occurring distributions. But so far, it doesn't feel like we've learned much about data. We'll now change that.

How datasets relate to random variables and probability distributions

We said at the beginning of this chapter that all data is random. This means when data is captured or generated, we are drawing or **sampling** values from some underlying probability distribution. This is illustrated schematically in *Figure 2.10*:

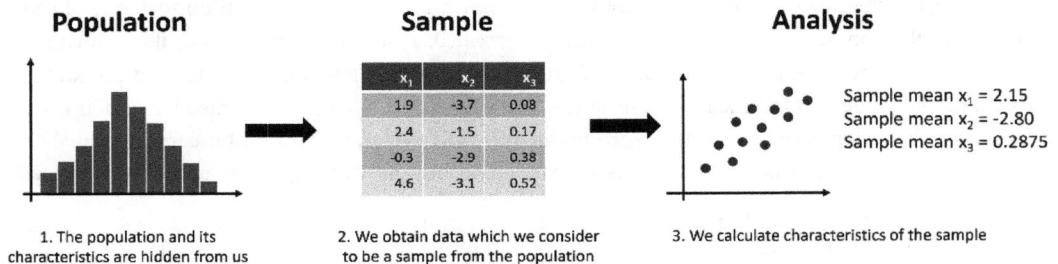

Figure 2.10: Diagram illustrating how real data is generated as samples from a population

A sample is finite. It represents a snapshot or subset of the entirety of possible outcomes; for example, a subset of all users who might visit a website. But from a business perspective, it is the behavior of the collection of all users I want to understand. When we analyze a dataset, what we really want to understand are the characteristics of the underlying distribution that we think the data has come from. We call this underlying distribution the **population** distribution.

Unfortunately, the underlying population distribution from which a dataset is generated is usually hidden from us. Instead, we use the sampled data as a proxy for the underlying population and use the summary characteristics of the sample as proxies for their population counterparts. Due to randomness, no sample is a perfect copy of the underlying population distribution from which it was taken. Consequently, different samples (datasets) taken from the same population can have different characteristics. This is called **sampling variation** and can lead us to different conclusions if we don't know how to correctly account for this randomness inherent in any sample. Alternatively, an individual dataset can give misleading conclusions about the underlying population if we analyze it naively and aren't aware of the random nature of the data. This is why we have spent a lot of effort on understanding randomness, random variables, and distributions and why we consider this to be the most important chapter in the book.

Take my e-commerce example. I want to understand what the **click-through-rate** (**CTR**) of a particular advertisement design is. I want to understand what proportion of the 10 million visitors I get to the website each year will click the ad. So, I track 20 visitors to the website, see whether they click or not, and find 12 of them did. That CTR of 12/20 = 60% may be good enough for my business model to succeed, so I go away with the intention of using that ad design for all website visitors. However, the CTR of all 10 million visitors can be very, very different from the CTR of 60% of those 20 people I tracked. Those 20 people are human beings, and humans have their own quirks and idiosyncrasies and so have a degree of randomness to their click behavior. Those 20 people we tracked are a sample of the 10 million – a very small sample. And we have seen that when we take a sample – when we draw each of their click behaviors from a Bernoulli click/no-click distribution – then we could get anywhere between 0 and 20 clicks. The true CTR of the 10 million annual visitors – the number I'm basing my business decisions on – might be 40%, but I could still see a CTR in that sample of 20 people that is anywhere between 0% and 100%. Suddenly, I don't feel so confident that the sample of 20 people is helping me make good business decisions. What should we do?

This is where math comes to our rescue. Our inferences about the CTR of all 10 million website visitors have some randomness or uncertainty within them. Using our knowledge of distributions, we can quantify that uncertainty and so control the level of risk associated with our inferences and decisions. We are taking the situation shown in *Figure 2.10* and closing the loop, as illustrated in *Figure 2.11*:

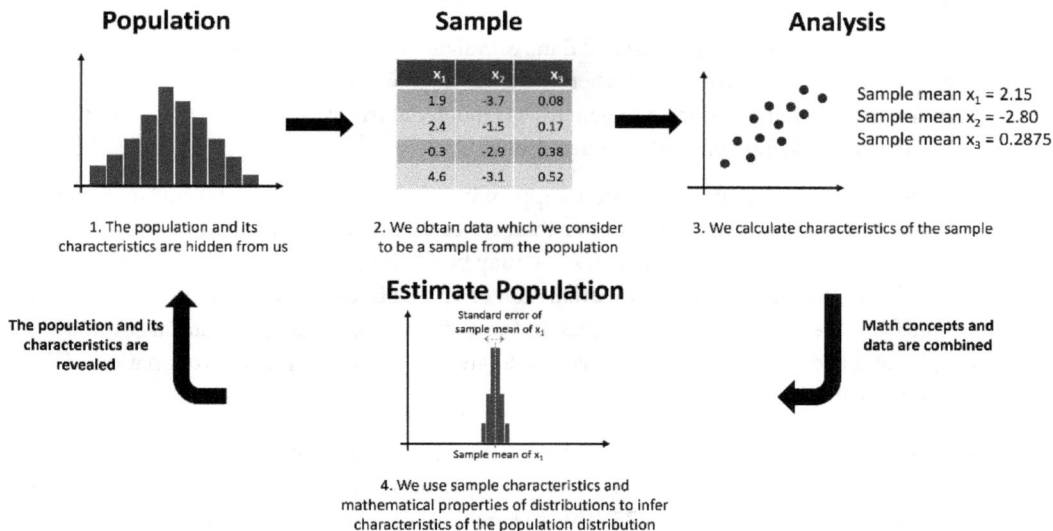

Figure 2.11: Diagram illustrating how we close the loop and use knowledge of probability distributions to go from samples back to making inferences about the underlying population

How big is the population from which a dataset is sampled?

In theoretical situations, we can usually consider the population to be of infinite size, meaning that I can obtain finite-sized sample datasets from it as big as I want. In real-world situations, the population may also be of finite size due to genuine constraints. For example, in our CTR example, we have 10 million visitors a year to the website, so the largest sample I could study in a year would consist of 10 million people. Often, even if the true population is of finite size, it is so big compared to the typical dataset (sample) we will study that we can consider the population to be effectively infinite and ignore the finite size implications of the population. However, I say this to point out that sometimes, we can't ignore the finite size of the population from which we are drawing our samples, and so it is always worth thinking about this aspect; that is, how big is our sample and how big do we think the population is from which the sample was obtained.

Now we have seen that a dataset can be considered as a sample from an underlying probability distribution, we will learn how to generate our own samples using snippets of Python code. We will also learn why it is useful to be able to generate our own samples.

How to sample

Why is being able to generate a sample useful? There are two main reasons:

- We can use sampling to create simulated data. Simulated data can be used to test data science algorithms and is particularly useful where we don't have any existing real ground-truth data. We can use the simulation process to generate as much data as we want and where we specify what the true underlying parameter values should be.

- Sometimes it is easier to use sampling to approximate a statistical calculation, such as the calculation of an expectation value, rather than try to calculate the expectation value exactly using rigorous mathematics. Sometimes, we may be dealing with complex data generation processes for which exact mathematical evaluation of a mean is just not possible, but it is possible to sample from the data generation process. In these situations, we can generate large numbers of samples and use a simple numerical average to give us a very accurate approximation of the population mean.

Let's say we have a binomial distribution $N = 20, p = 0.6$. Now, given that information, we can use the formula in *Eq. 38* to calculate the probability of, say, getting a count outcome of $n = 9$ or fewer successes. We find this probability is 0.1275. This means that if I were to repeatedly sample from a Binomial(20, 0.6) distribution, I would expect about 12.75% of the time I would get a number 9 or less. In fact, I can do this calculation for all possible values of n. This calculation is easy to do – it just involves cumulatively adding up the probabilities from the smallest value of the outcome to the largest value of the outcome. This is illustrated in *Figure 2.12*. The resulting curve is called the **cumulative probability distribution** (CDF) for obvious reasons. The CDF for any distribution always monotonically increases from 0 to a maximum of 1:

CDF of Binomial(20,0.6)

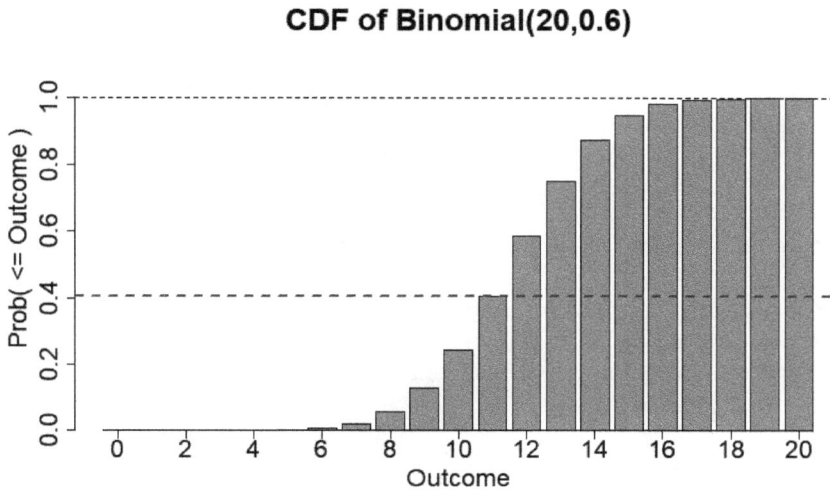

Figure 2.12: The CDF of the Binomial(20,0.6) distribution

Now, let's say I draw a number from the uniform distribution $U(0,1)$. The number I get from that uniform distribution effectively splits the interval $[0,1]$ into two portions. Or equivalently, it splits the range 0%-100% into two regions. Let's say the number I got was 0.41, so I've split the 0%-100% range into two portions: one containing a total probability of 0.41 (or 41%) and the other containing a total probability of 0.59 (or 59%). What I can now do is ask what value of n in my binomial distribution example would give me that same split. Given the cumulative distribution shown in *Figure 2.12*, that is easy to do, and we find we get the same 41%-59% split at $n = 11$. This is illustrated by the blue horizontal dashed line in the middle of *Figure 2.12*. So, what we now have is a method for going from generating uniform random numbers to giving numbers drawn from the actual distribution I'm interested in – the Binomial(20,0.6) distribution in this example. Most computer programming languages will provide you with an easy way to generate random numbers from $U(0,1)$. The Python code snippet shown next shows how to use the numpy uniform random number generator to do this and shows how to generate 1,000 Binomial(20, 0.6) random numbers.

Generating your own random numbers code example

A fuller version of the following code is given in the Code_Examples_Chap2.ipynb Jupyter notebook in the GitHub repository:

```
import numpy as np

# Set parameters of the Binomial distribution
# We will use 20 trials and a success probability of 0.6
N = 20
p = 0.6
```

```
# Calculate the cumulative probability distribution
# of the Binomial
cumulative_probs = cumulative_binomial_probability(N=N,p=p)

# Set how many random numbers we want to generate
# and initialize an empty array to hold them
n_random = 1000
random_binomial = np.empty(n_random, dtype=np.int8)

# Loop to generate the random Binomial numbers
for i in range(n_random):
    # Draw a random number from U(0,1)
    r_tmp = np.random.rand(1)

    # See where the cumulative probability distribution first exceeds
    # the uniform random number we just created
    random_binomial[i] = np.where(
        cumulative_probs - r_tmp > 0.0)[0][0]
```

This method can be applied to any discrete distribution. A modified version can be used to sample from continuous distributions, but it can be a bit harder.

Okay – now I'm going to let you into a little secret. You don't have to do any of this. The specialist numerical modules or packages of most modern general-purpose programming languages come with in-built methods for sampling from the most common distributions you are likely to encounter (and some more obscure ones as well). We'll now take a brief tour of some of those methods with some bits of example Python code.

Sampling from numpy distributions code example

First, here's how to sample from a binomial distribution using numpy:

```
# Generate Binomial random numbers directly using numpy

import numpy as np

N = 20
p = 0.6
n_random = 1000

random_binomial2 = np.random.binomial(n=N, p=p, size=n_random)
```

Second, here's how to sample from a Poisson distribution using numpy:

```python
# Using numpy to sample from a Poisson distribution
# We set the mean number of counts to 5.2
# This is specified by the 'lam' argument of the
# numpy function.

import numpy as np

n_random = 10000
random_poisson = np.random.poisson(lam=5.2, size=n_random)
```

Finally, here's how to sample from a Gaussian distribution using numpy:

```python
# Using numpy to sample from a Gaussian distribution
# We set the population mean=2.0 and the
# standard deviation=1.5.

import numpy as np

mean = 2.0
sd = 1.5

# We first sample from the standard Gaussian
# distribution (zero mean and unit variance)
# and apply a simple linear transformation
n_random = 10000
random_gaussian1 = mean + (sd*np.random.randn(n_random))

# Alternatively, we can use the normal function of
# numpy.random without having to apply the linear
# transformation
random_gaussian2 = np.random.normal(
    loc=mean, scale=sd, size=n_random)
```

All the preceding code snippets can be found in full in the Code_Examples_Chap2.ipynb Jupyter notebook in the GitHub repository.

That code example is a good place to end this shorter section, so let's summarize what we have learned about generating our own random data.

What we learned

In this section, we have learned the following:

- How the datasets we work with as data scientists can be considered as samples from a distribution
- How to generate our own samples from any probability distribution
- Why and when generating our own samples from a probability distribution is useful

Having learned how datasets can be viewed as samples from a distribution, we will learn in the next section how to characterize and summarize a sample and how those summaries of a sample can be used to make accurate inferences about the underlying population distribution from which we think the data was drawn.

Understanding statistical estimators

When we were looking at various example probability distributions, we learned how to calculate their mean and variance. Now, you may ask: Is it possible to calculate the mean and variance of a sample (of a dataset)? The answer is yes. You have probably done this before in high school or college. So, you may be wondering how the mean and variance of a dataset are connected to the mean and variance of a population distribution. What we are going to do now is explain the following:

- How to calculate the mean and variance of a sample
- How they differ from the mean and variance of the population distribution from which the sample was generated
- How they are connected to the mean and variance of the population distribution from which the sample was generated
- How to use our understanding of the population distribution to make quantified inferences about it from the sample

Let's start with the first of those. Given a set of n numbers (a sample) $x_1, x_2, \cdots, x_n$, the **sample mean** is calculated as follows:

$$\text{Sample mean } = m = \frac{1}{n} \sum_{i=1}^{n} x_i \ .$$

Eq. 46

This is the formula you will have learned in high school. For example, if my sample consists of the five numbers 3.7, 1.2, 2.3, 4.1, 2.7, then the sample mean is the following:

$$\frac{1}{5} \times (3.7 + 1.2 + 2.3 + 4.1 + 2.7) \ = \ 2.8$$

Eq. 47

You will see that we have used a different symbol for the sample mean. We have used m and not μ. This is to distinguish it from a population mean. In fact, we have explicitly called it "the sample mean." We will show how m is related to μ later.

Similarly, we can calculate the **sample variance** as the average squared deviation of the data from its mean. Here, I'm going to define it as follows:

$$\text{Sample Variance} = s^2 = \frac{1}{n-1} \sum_{i=1}^{n} (x_i - m)^2$$

Eq. 48

For the example five numbers shown previously, the sample variance is the following:

$$\frac{1}{4} \times [(3.7 - 2.8)^2 + (1.2 - 2.8)^2 + (2.3 - 2.8)^2 + (4.1 - 2.8)^2 + (2.7 - 2.8)^2] = 1.33$$

Eq. 49

Again, we have used a different symbol, s^2, for the sample variance, and not σ^2 that we use for the population variance. As you may have guessed, the sample standard deviation is simply the square root of the sample variance, so the sample standard deviation is s.

Now, you will probably be wondering why we had $n - 1$ in the denominator on the right-hand side of *Eq. 48*. Why didn't we calculate the sample variance using a formula, like so?

$$\frac{1}{n} \sum_{i=1}^{n} (x_i - m)^2$$

Eq. 50

The short answer is that using a denominator of $n - 1$ means the resulting value of s^2 gives us a more accurate estimate of the true population variance σ^2 than if we had used a denominator of n. Remember – it is the underlying population distribution and its properties that we are ultimately interested in.

Consistency, bias, and efficiency

Now, let's give the long answer and explain in more detail. Those five numbers in the preceding example came from a normal distribution with a mean $\mu = 2.5$ and a variance $\sigma^2 = 1.5$. I used the code example given in the *How to sample* subsection to generate them. I also rounded to 1 decimal place for convenience of presentation. You can see that the sample mean m and sample variance s^2 are different from the mean μ and variance σ^2 of the distribution from which the five numbers were sampled. Ideally, we want m and s^2 to be good estimates of μ and σ^2, and so be close in numerical value to μ and σ^2. You would suspect that the reason m and s^2 are different from μ and σ^2 is the small sample size; that is, because we have used only five numbers in our sample. We would hope that m and s^2 get closer to μ and σ^2 as we increase the sample size n. *Figure 2.13* shows what happens to m and s^2 as we increase the number of samples we draw from the distribution *Normal(2.5, 1.5)*:

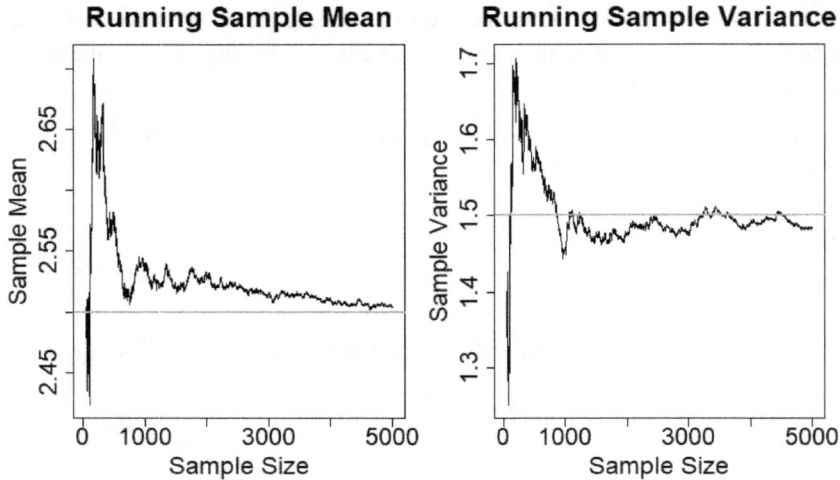

Figure 2.13: Running sample mean and sample variance for samples drawn from Normal(2.5, 1.5)

Here, we have generated 5,000 random values from *Normal*(2.5, 1.5) and then calculated the running sample mean; that is, the average of the first n numbers, for $n = 50, \ldots, 5000$. In the right-hand plot, we have also plotted the running sample variance. In the left-hand and right-hand plots, the horizontal gray lines show the values of the population mean and population variance, respectively. We can see that visually, the sample mean and sample variance appear to converge to their population values as the sample size increases. That is, we have the following:

$$m \to \mu \quad s^2 \to \sigma^2 \text{ as } n \to \infty$$

Eq. 51

Because of this, we say that m and s^2 are **consistent estimators**. They get better at estimating the things we want them to estimate as we use more and more data, until ultimately with an infinite amount of data they are consistent with (give the same value) as the things we want to calculate (μ and σ^2).

Excellent! Well not quite. Usually, we don't have an infinite amount of data in our real-world data science problems. So, while an estimator that is consistent is almost a minimum requirement, it still doesn't guarantee that when we have a finite-sized dataset, our estimator is based upon a good formula.

How could we even tell if a formula gives a good estimator at finite n? Remember what we said at the very start of this chapter that all data has a random component, and so all quantities derived from data also have a random component? Well, that means that the sample mean and sample variance have a random component and so are random variables. As m is a random variable, we accept that any individual instance of m, calculated from an individual dataset of size n, will differ from μ, but we would want the average value of m to be the same as μ if we were using a good estimator. That means, that if we repeat lots of times our little experiment of drawing five numbers from *Normal*(2.5, 1.5) and calculate a value of m for each of those experiments, we want the average value of m to be equal to μ. *Table 2.2* shows 10 such experiments, with the values of m and s^2 given for each of the 10 experiments. The bottom row of the table shows the average values of m and s^2 across the 10 experiments:

Experiment	Sample Mean	Sample Variance
1	2.2008	2.9962
2	2.3386	0.4683
3	1.8398	4.0131
4	0.4696	1.8640
5	2.7111	0.7492
6	1.8808	1.1175
7	2.9189	0.0231
8	2.5560	1.0802
9	3.8317	0.6668
10	2.4434	1.9100
Average	2.3191	1.4888

Table 2.2: Sample mean and sample variance values in 10 experiments
drawing samples of 5 numbers from Normal(2.5, 1.5)

You can see from *Table 2.2* that the average across the 10 experiments of the sample mean is reasonably close to the true population mean of $\mu = 2.5$. In fact, if we took the average over an infinite number of experiments, we would get an average value of 2.5 exactly. So, **on average**, the sample mean gives the same value as the population mean. We will prove that now.

Since the sample mean is a random variable, we can just calculate its expectation value. The sample mean is defined as follows:

$$m = \frac{1}{n}\sum_{i=1}^{n}x_i$$

Eq. 52

So, its expectation value is the following:

$$\mathbb{E}(m) = \mathbb{E}\left(\frac{1}{n}\sum_{i=1}^{n}x_i\right) = \frac{1}{n}\sum_{i=1}^{n}\mathbb{E}(x_i)$$

Eq. 53

The last part on the right-hand side of *Eq. 53* follows on from the rules of expectations of linear transformations of random variables. Now, if we have i.i.d. data, then since each random variable x_i is drawn from the same population distribution we have $\mathbb{E}(x_i) = \mu$ for all i. Plugging that result into *Eq. 53*, we then get the following:

$$\mathbb{E}(m) = \frac{n}{n}\mu = \mu$$

Eq. 54

So, on average, for i.i.d. data, the sample mean will equal the true underlying population mean we are trying to estimate. This is true even though our sample mean is based on a finite sample size! We say that the sample mean is an **unbiased estimator** of the population mean. Even if the different random variables x_i are not independent of each other but they still have the same mean – that is, we still have $\mathbb{E}(x_i) = \mu$ – then the sample mean m is still an unbiased estimator of the population mean μ.

Now, it turns out that for the sample variance, if we have $n - 1$ in the denominator of the definition of s^2, then s^2 is an unbiased estimator of the population variance σ^2. That is, $\mathbb{E}(s^2) = \sigma^2$ for any value of n. We can prove this as well using a similar proof to that in *Eq. 53*, but we will simply state it here and leave the proof as an exercise for the reader. If we had used n in the denominator of *Eq. 48* instead of $n - 1$, our definition of s^2 would not give us an unbiased estimator of σ^2.

Since the difference between having $n - 1$ or n in the denominator of *Eq. 48* vanishes as $n \to \infty$, it becomes clear that an estimator can be biased at finite n but still consistent; that is, if the bias at finite n vanishes as $n \to \infty$. So, bias and consistency are two different concepts – we will summarize them in a moment.

Tip

We have said that the sample variance (defined with a denominator of $n - 1$) is an unbiased estimator of the population variance. This means $\mathbb{E}(s^2) = \sigma^2$. However, this does not mean that the sample standard deviation is an unbiased estimator of the population standard deviation. We do not have $\mathbb{E}(s) = \sigma$. Why not? Well, $s = \sqrt{s^2}$ and taking the square root is a non-linear operation. Wherever we introduce a non-linear operation, we potentially introduce new biases when we take the expectation. So, i) be aware that applying a non-linear transformation to data can introduce biases in estimates calculated from that transformed data, ii) the sample standard deviation is not an unbiased estimate of the population standard deviation. This explains this seeming cryptic note in the documentation of the numpy.std function, "*The standard deviation computed in this function is the square root of the estimated variance, so even with ddof=1, it will not be an unbiased estimate of the standard deviation per se.*" This note in the documentation always seems to confuse people on first reading, but it is just saying s is not unbiased even if s^2 is.

Let's recap what we have learned so far:

- The sample mean and sample variance converge to their population equivalents as we use more data

- The sample mean and sample variance (with appropriate denominator) are unbiased estimates of their population counterparts at finite n

So, the sample mean and sample variance look like pretty good formulas to use, right? But in the real world, we can't do an infinite number of experiments. So, knowing that the sample mean is an unbiased estimator of the population mean isn't exactly helpful. For a finite-sized sample, we know that the sample mean will be in the right ballpark, but how good is m from any single dataset? Look at *Table 2.2*. In experiment number 4, the sample mean is quite a long way from the population value of 2.5. Since the sample mean m is a random variable, we know it will vary around its expectation value. In this case, it looks like for $n = 5$ the variations of m around the population mean of $\mu = 2.5$ can be large. Those variations are quantified by the standard deviation of m, the size of the typical or expected deviation of m from μ. Unsurprisingly, the variance of m decreases as n increases, meaning that the sample mean calculated from larger samples typically has smaller deviations from μ than the sample mean calculated from smaller samples. Ultimately, the variance of m becomes zero as $n \to \infty$. How quickly the variance of an estimate of a population quantity converges to its population equivalent is termed its **efficiency**. The efficiency tells us how well the formula makes use of the data given to it to construct its estimate of the population quantity. The efficiency of an estimator depends upon the details of the formula and the distribution from which we assume the data is coming. For i.i.d. data, the variance of the sample mean has a simple form:

$$\text{Var}(m) = \frac{\sigma^2}{n}$$

Eq. 55

So, the standard deviation of the sample mean (for i.i.d. data) is $\sigma/\sqrt{n}$. This is also referred to as the **standard error** of the sample mean – it is the typical size of error that the sample mean will make when we use it as an estimate of the population mean μ. Remember also here that σ is the population standard deviation of each individual data point. So, no matter how big σ is, we can still come up with a good estimate of the population mean if we have a sufficiently big sample size n, and the preceding formula helps us work out how big that sample size needs to be. This is one of the beauties of statistical analysis. No matter how big the noise in individual data points is, we can still recover an accurate estimate of the population mean by having enough datapoints.

To sum up: For any statistical estimator (that is, for any formula we use to estimate a population quantity), we ideally want three criteria to be met. Those criteria are the following:

- **Consistency** – An estimator converges to the (population) quantity it is attempting to estimate as the sample size $n \to \infty$. In simple terms, the estimator gets better and better as we give it more data. The sample mean and sample variance are consistent estimators.

- **Unbiasedness** – An unbiased estimator has an expectation value equal to the (population) quantity it is attempting to estimate. In simple terms, the estimate is accurate on average, even for finite sample sizes. The sample mean and sample variance (with a denominator of $n - 1$) are unbiased estimators.

- **Efficiency** – Efficiency measures how quickly the variance of an unbiased estimator converges to zero as the sample size $n \rightarrow \infty$. In simple terms, an efficient unbiased estimator makes good use of the data given to it. It turns out that the sample mean has the best efficiency of any linear estimator of the population mean.

Applying the concept of efficiency

Let's return to our CTR example. We had a sample size of $n = 20$. Our single sample had a sample mean of 0.6 (60%). We can model each user's decision to click as a *Bernoulli(p)* random variable. We know that the variance of each Bernoulli trial is $p(1 - p)$, and so the standard error of the sample mean is $\sqrt{p(1 - p)/20}$. But we don't know what the true value of p is, the true CTR. But we do have our estimate of the CTR, so we can plug our sample mean of 0.6 into the formula $\sqrt{p(1 - p)/20}$ to get an approximate value of the standard error. In this case, we find $\sqrt{0.6 \times 0.4/20} \approx 0.11$. So, we should not be surprised if the real (population) CTR was as low as 0.5 or 0.4. As we have already said, our sample of 20 website visitors does not give us confidence in our conclusions. However, we can easily work out how big our sample size would need to be for us to be confident in that estimated CTR of 0.6. Let's say I wanted the standard error to be smaller than 0.01. Again, if we assume our initial estimate of 0.6 is in the right ballpark, we can simply equate the formula in *Eq. 55* for the standard error of the sample mean with 0.01 and solve for the required sample size n. That is, we solve for the following:

$$\frac{p(1 - p)}{n} = 0.01^2$$

Eq. 56

This is done for n and where we plug in our sample mean of 0.6 as the value of p. Solving the preceding equation for n, we find the required sample size is the following:

$$n = \frac{0.6 \times 0.4}{0.01^2} = 2400$$

Eq. 57

So, we need a couple of thousand website visitors for our experiment to reach a sound business conclusion. With 10 million annual visitors this is still entirely doable, but now we are basing our method and decisions on some sound statistical analysis and data science rather than guesswork. The calculation of the required sample size we have just walked through is somewhat hand-waving in places. When we come to *Chapter 7*, we will learn how to do these calculations of the required sample size more rigorously. They are called **power calculations**. However, the principles and spirit of those rigorous power calculations are essentially the same as we have just outlined.

The empirical distribution function

We have now linked the properties and characteristics of samples of data to the properties and characteristics of the underlying population distributions from which we have assumed the data comes. We're now going to make the link between samples and distributions stronger by asking if we can view a sample as being some sort of distribution. The answer is yes. Enter the **empirical distribution function (EDF)** or, equivalently, the **empirical cumulative probability function (eCDF)**. We use the word *empirical* to emphasize that this distribution is based on real observations; that is, data.

If we had a sample of 30 data values $x_1, x_2, \cdots, x_{30}$ and we wanted to think of them as some sort of probability distribution, we would have to be able to plot them as a collection of bars as in *Figure 2.4*. We have 30 datapoints, so we can think of the distribution as 30 vertical bars, each centered on one of the datapoints. Since the total probability in a distribution must add up to 1, each bar has a weight 1/30. And since we have only seen values corresponding to the actual data values $x_1, x_2, \cdots, x_{30}$, those bars are infinitesimally thin – there is no probability of getting any possible value other than x_1 or x_2 or.....x_{30}. So, our empirical distribution is actually a series of infinitely narrow spikes. We have drawn that schematically in *Figure 2.14* for a set of 30 real x values:

A series of delta-functions

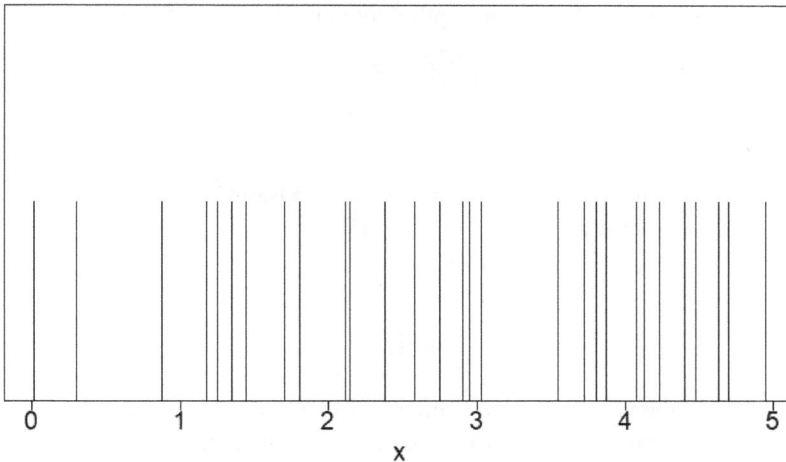

Figure 2.14: Representing a sample as a series of spikes

Figure 2.14 shows a schematic representation of the EDF, but how do we represent it mathematically? We'll build it up mathematically in stages. Mathematically, we want to represent the first spike as a function that picks out only the value x_1, which is 0.3 in this example. The mathematical function that does this is the Dirac delta function, or Dirac δ function, named after the famous physicist Paul

Dirac. The Dirac delta function, $\delta(x)$, is actually a distribution. It is sometimes referred to as being a generalized function, but it is easier to loosely think of it as a continuous probability distribution with the following properties:

$$\int_{c-\varepsilon}^{c+\varepsilon} \delta(x - c)\, dx = 1 \quad \text{for any } \varepsilon > 0$$

Eq. 58

$$\int_{c-\varepsilon}^{c+\varepsilon} \delta(x - c)g(x)dx = g(c) \quad \text{for any } \varepsilon > 0$$

Eq. 59

So, if we have a delta function, $\delta(x - x_1)$, centered on x_1, then using *Eq. 19* and *Eq. 20* for the mean and variance of a continuous probability distribution, we can think of $\delta(x - x_1)$ as a probability distribution with mean x_1 and variance zero. So, sampling from $\delta(x - x_1)$ would only ever give us the value x_1, which is what we want.

If we repeat the process by putting a Dirac delta function centered on each of our 30 datapoints x_1, x_2 , $\cdots$, x_{30}, we can think of our sample as being equivalent to the following PDF:

$$\frac{1}{30} \sum_{i=1}^{30} \delta(x - x_i)$$

Eq. 60

For a sample, $x_1, x_2, \cdots, x_n$, of n datapoints, the generalization is obvious – we can view our sample as being a distribution with density:

$$\frac{1}{n} \sum_{i=1}^{n} \delta(x - x_i)$$

Eq. 61

Why is this useful? Well, sometimes we have a metric or expression or concept that we know how to compute for continuous probability distributions, but we don't quite know how to calculate the equivalent or corresponding metric or formula for a sample of data. No problem! Simply plug *Eq. 61* into the formula for continuous probability distributions, simplify using *Eq. 59*, and you have an answer. You don't believe me? Plug the distribution in *Eq. 61* into *Eq. 19* and *Eq. 20* for the mean and variance of a continuous probability distribution and what you get out are the expressions in *Eq. 46* and *Eq. 48*, for the sample mean and sample variance (up to a factor of $\frac{n}{n-1}$). Give it a try.

From the EDF in *Eq. 61*, we can calculate the corresponding eCDF. The cumulative probability function *CDF(x)* is the probability of getting a value less than or equal to *x*. So for a continuous random variable with density function *f(x)*, the CDF is given by the following:

$$CDF(x) = \int_{-\infty}^{x} f(s)\,ds$$

Eq. 62

Plugging the empirical density function in *Eq. 61* into *Eq. 62*, we get the eCDF. Using *Eq. 58*, we can see that the eCDF goes through a series of steps, increasing by $\frac{1}{n}$ every time *x* passes one of the datapoints. The eCDF for the sample shown in *Figure 2.14* is shown in *Figure 2.15*:

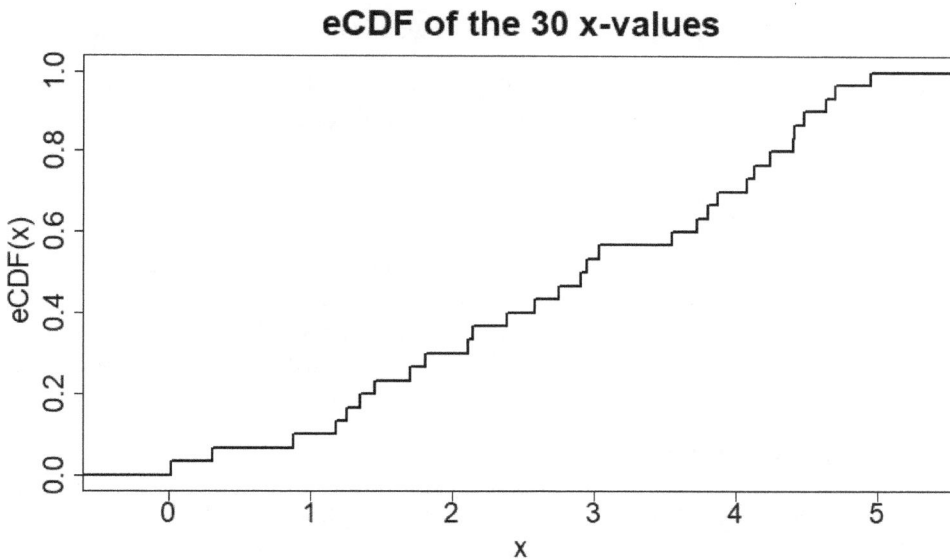

Figure 2.15: The eCDF for the series of 30 x values shown in Figure 2.14

The stepped nature of the eCDF is clearly visible in *Figure 2.15*. As the sample size *n* increases, the size of those steps decreases and the eCDF looks smoother and smoother. In fact, as $n \rightarrow \infty$, the EDF in *Eq. 61* and its associated eCDF converge to their population counterparts. So, just like we have used the sample mean and the sample variance as approximations of the population mean and population variance, we can use (when *n* is reasonably large) the empirical distribution in *Eq. 61* as an approximation of the population distribution.

Excellent! We are making good progress on understanding and taming randomness in data. Having seen how we can use the eCDF to approximate a population CDF, let's summarize what we have learned about statistical estimators and formulas in this section before we move on to our final section of the chapter.

What we learned

In this section, we have learned the following:

- How to calculate the sample mean and sample variance

- How the sample mean and sample variance converge to their population counterparts as we increase the sample size

- How the sample mean and sample variance are unbiased estimators of their population counterparts

- How to calculate the standard error (standard deviation) of the sample mean and how to use this to approximately calculate the sample size required to ensure the sample mean is a sufficiently accurate estimate of the population mean

- How to construct the EDF and use it as an approximation of the population distribution

Having learned how to relate the characteristics of a sample to the characteristics of the underlying population, in the next section we will learn why a particular distribution, the normal distribution, is the most common distribution from which data is generated.

The Central Limit Theorem

Earlier in the chapter, when we were introducing specific continuous-valued distributions, we described the Gaussian or normal distribution and we said that it was an extremely important distribution because it was an extremely common distribution. By this, we meant that many datasets you will encounter will effectively have been drawn from a normal distribution, or you will use a normal distribution to model those datasets. We will now explain why.

Sums of random variables

Lots of the quantities we analyze as data scientists are aggregations of other data. Aggregating observations over some dimension to simplify the data is a very natural thing to do.

For example, consider our e-commerce scenario where we are interested in how many items are sold. The number of items sold on any day of the year we might model as a binomial random variable, but what about for the whole year? Imagine we have a relatively niche website where we only get, say, 20 visitors a day who are thinking about buying a particular product, and the probability of any one of those visitors buying the product is 0.3. So, on any day t, we can model the total items sold X_t as $X_t \sim$ Binomial$(20, 0.3)$. We know from the properties of the binomial distribution that the mean number of items sold on any day is 20×0.3, while its variance is $20 \times 0.3 \times (1 - 0.3)$. The total number of items sold in a year is, simply, $X_1 + X_2 + \cdots + X_{365}$. This sum is a random variable itself. We will denote this sum by S. As S is a sum of random variables, we can use the rules for combining random variables that we explained earlier in the chapter, to find the following:

$$\mathbb{E}(S) = 365 \times 20 \times 0.3 = 2190$$

Eq. 63

We will also use the aforementioned rules to find the following:

$$\text{Var}(S) = 365 \times 20 \times 0.3 \times (1 - 0.3) = 1533$$

Eq. 64

We can see that we know the mean and variance of the yearly total sales S, but what about the shape of the distribution of S? There are many distributions of different shapes that could have a mean of 2190 and a variance of 1533.

This is where a very famous piece of math comes to our aid. The Central Limit Theorem (CLT) tells us that as we add more and more random variables together, their sum behaves more and more like a Gaussian random variable. Since we already know the mean and variance of the sum S, we can write a mathematical statement of the CLT. Specifically, if we have n i.i.d. random variables X_i, $i = 1, \cdots ,$ n, each of which has mean μ and variance σ^2, then if we define $S = X_1 + X_2 + \cdots + X_n$, the distribution of S has the following limiting behavior:

$$S \to Normal(n\mu, n\sigma^2) \quad \text{as } n \to \infty$$

Eq. 65

As the mean $n\mu$ and variance $n\sigma^2$ of the distribution on the right-hand side of *Eq. 65* become infinite as $n \to \infty$, we need to write *Eq. 65* in a slightly more rigorous way as follows:

$$\frac{S - n\mu}{\sigma\sqrt{n}} \to Normal(0,1) \quad \text{as n} \to \infty$$

Eq. 66

We won't go into the proof of the CLT as it involves some additional math techniques that we won't be covering in this book. What is more important is that you are aware of the CLT, specifically that when we sum up lots of random variables, the distribution of that sum will be very well approximated by a Gaussian random variable. Let's illustrate this with a code example.

CLT code example

For our first code example, we'll start slightly simpler. We'll add 20 *Uniform*(0,1) random variables together. We'll do this lots of times and plot the resulting histogram of those totals to see if it looks anything like a normal distribution. The code is shown next and is also in the Code_Examples_ Chap2.ipynb Jupyter notebook in the GitHub repository:

```
import numpy as np
import matplotlib.pyplot as plt


## Central Limit Theorem Uniform Example
n_random_var = 20 # The number of random variables we add together
```

```
n_simulations = 100000 # The number or experiments we will run

# Initialize an array to hold the results of the experiments
total_of_random_vars = np.zeros(n_simulations)

# Loop over the experiments
for i in range(n_simulations):
    # generate n_random_var Uniform(0, 1) values and sum them
    total_of_random_vars[i] = np.sum(np.random.rand(n_random_var))

# Plot the frequencies of the experiment totals as a histogram
plot = plt.hist(total_of_random_vars, bins=100, label='Sum')

# Construct expected frequencies according to Central Limit Theorem
# The mean of the sum of the random variables:
mean_CLT = n_random_var * 0.5
# The variance of the sum of the random variables:
var_CLT = n_random_var /12.0
# The standard deviation of the sum of the random variables:
std_CLT = np.sqrt(var_CLT)

# Store the x values at which we which to calculate the
# CLT approximation.
# Since we want to plot the CLT approximation on top of the
# histogram we
# will use the positions of the histogram bins as our x values
x_values = plot[1]

# Initialize an array to hold the results of the CLT approximation
clt_expected_frequency = np.zeros(len(x_values)-1)

# Loop over the x values
for i in range(len(x_values)-1):

    # Since we are comparing to the frequencies in the histogram we
    # need to calculate
    # the expect number of experiments whose sum is in between the
    # current and next
    # x value. This expected frequency is equal to the total number
    # of experiments
    # multiplied by the probability of an experiment having its sum
    # between the current
    # and next x value. This probability is the difference in
    # cumulative probability
```

```
# between those two x values. Since we are using the CLT, we are
# approximating the
# probability density function of an experiment sum by a Normal
# distribution with
# mean = number of random variables * mean of each random
# variable, and
# variance = number of random variables * variance of each random
# variable.
# To calculate the cumulative probability of a Normal
# distribution we use the
# scipy.norm.cdf function.
cumulative_prob1 = norm.cdf((x_values[i+1] - mean_CLT)/std_CLT)
cumulative_prob2 = norm.cdf((x_values[i] - mean_CLT)/std_CLT)
cumulative_prob_diff = cumulative_prob1 - cumulative_prob2
clt_expected_frequency[i] = n_simulations * cumulative_prob_diff

# Plot the CLT expected frequencies on top of the histogram
plt.plot(x_values[:-1], clt_expected_frequency, 'red',
         label='CLT Approx.')

# Add graph annotation and display
plt.legend(loc='upper right')
plt.title('CLT demonstration adding 20 Uniform(0, 1) \
          random variables')
plt.xlabel('Sum of random variables')
plt.ylabel('Frequency')
plt.show()
```

An example output from the code is shown in *Figure 2.16*:

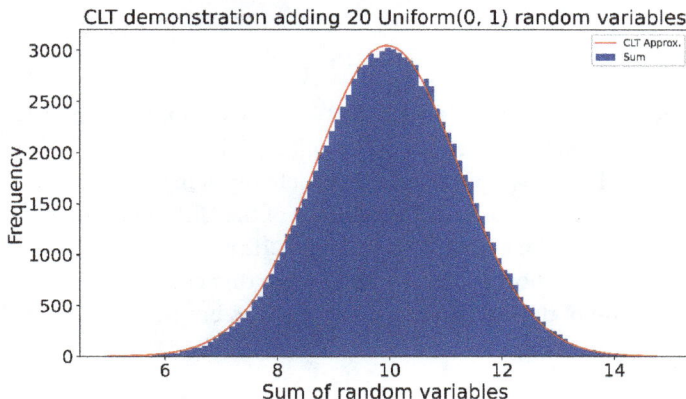

Figure 2.16: Example of the CLT when adding together the values of 20 Uniform(0,1) random variables

By comparing the red line of the CLT approximation to the histogram values, you can see how good the approximation of the CLT-based approximation is, even though we are only adding 20 random variables together in this example.

We have said that the CLT gives us an approximation here because technically, the CLT is an asymptotic result. *Eq. 66* tells us we only get a normal distribution when we add an infinite number of random variables together. What we have done is to take the CLT and say, well, when we add a finite number of random variables together, we will get something that is approximately normal. As you can see from this example, that approximation can be pretty good even when we are a long way from adding an infinite number of random values together. As you would expect, the more random variables we add together, the better the CLT approximation gets, and the approximation is better in the center of the distribution (around the mean) than in the tails (at the edges). How quickly the distribution of the sum of random values converges to a normal distribution as n increases depends not only on n (the number of random variables) but also on the details of the distribution of each random variable; for example, are they uniform random variables, binomial random variables, and so on.

The reason we chose adding *Uniform*(0,1) random variables to illustrate our first example of the CLT is partly because the shape of the density function of each of the random variables is flat, and so it is very different from the normal distribution that results from their sum. This emphasizes the fact that the resulting normal distribution shape is because we are adding lots of random variables together, not because of the properties or shape (within reason) of the individual random variables we are adding together. The normal distribution shape that results from adding lots of random variables together is a universal outcome.

We also chose to add *Uniform*(0,1) random variables to illustrate the CLT because *Uniform*(0,1) is a continuous distribution. Consequently, the sum of many *Uniform*(0,1) random variables will also be continuous, and when we compare the normal distribution of the CLT approximation, which is also continuous, it is possible to see a very good agreement between the histogram of the experimental sums and the CLT approximation, even when we are adding only a finite number of random values together.

CLT example with discrete variables

When our starting random variables are discrete, then so is their sum. Let's simulate our "items sold in a year" example, where we add together 365 *Binomial*(20, 0.3) random variables. We can do this using very similar code to that shown previously by simply replacing *Uniform*(0,1) with *Binomial*(20, 0.3) and updating the mean and variance calculations of the CLT approximation. Example code to do this simulation is given in the `Code_Examples_Chap2.ipynb` Jupyter notebook in the GitHub repository, but we only show example output from that code in *Figure 2.17*. The granular or discrete nature of the sum of the 365 discrete values is clear, but the CLT still provides us with a good approximation:

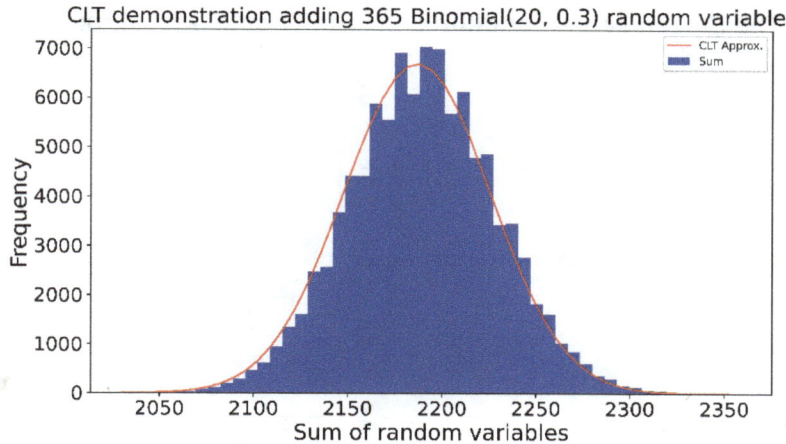

Figure 2.17: Example of the CLT when adding together the values
of 365 Binomial(20, 0.3) random variables

The CLT is incredibly important when we analyze a dataset because it tells us what probability density shape we should expect for a quantity that is a sum of many other things. But what if we are analyzing a quantity in a dataset that isn't an aggregation and we want to know its density shape? We will now show how to do this computationally, with some code examples.

Computational estimation of a PDF from data

Here, we will show you how to use functions from the `scikit-learn` package to calculate a computational estimate of a PDF. To do this, we use something called **kernel density estimation** (KDE). KDE works by approximating the PDF that underlies the data by a series of fixed-shape distributions, one placed at each data point. If $x_1, x_2, \cdots, x_n$ are the data values in our sample, the KDE estimates the PDF as $\hat{f}(x)$, with $\hat{f}(x)$ calculated from the following formula:

$$\hat{f}(x) = \frac{1}{nh} \sum_{i=1}^{n} K\left(\frac{x - x_i}{h}\right)$$

Eq. 67

The hat symbol on $\hat{f}(x)$ is used to denote the fact that $\hat{f}(x)$ is an **estimate** of the true PDF, $f(x)$. In *Eq. 67*, the function $K(x)$ is called the **kernel** function. It is also called a **window** function because, as we will see, it specifies the window or range over which data is smoothed. It is typically a function that is highest at $x = 0$ and falls away symmetrically to zero on either side, possibly within a finite distance from $x = 0$. *Figure 2.18* shows two example kernel functions; on the left is the Parzen window, which is a Gaussian, and on the right is the Tricube kernel function, which has finite support (it is zero outside of $|x| = 1$):

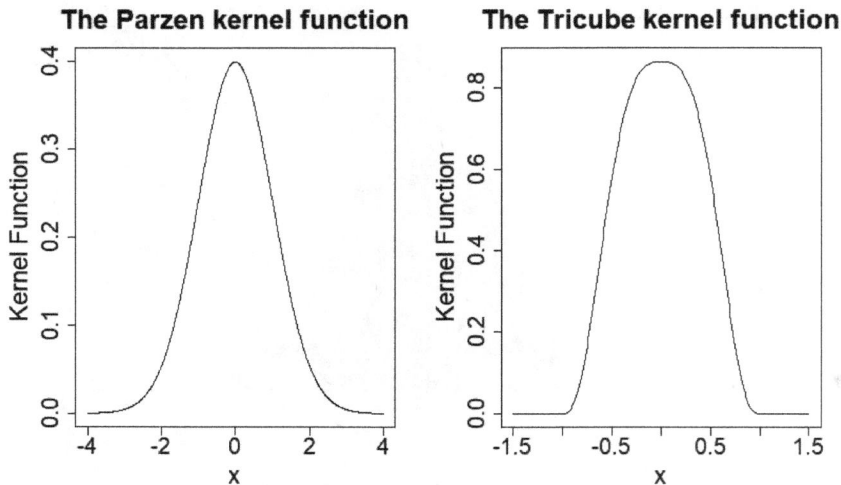

Figure 2.18: Two examples of commonly used kernel functions

Given each datapoint has a total weight $1/n$ when we calculate a sample average, we can see that the function K smooths out the impact of having a datapoint at x_i. How much it smooths out that impact is determined by, obviously, the shape of the function K, but also the parameter h, which modifies the width of the impact of the kernel smoothing. The parameter h is called the **kernel width**, or sometimes the **bandwidth**. A large value of h will mean each datapoint is smoothed out considerably, with the impact of each datapoint overlapping and the resulting approximation $\hat{f}(x)$ looking almost like a uniform distribution. A small value of h means each datapoint is smoothed out only a little and essentially is still a spike. With a small value of h, the resulting approximation $\hat{f}(x)$ doesn't look that much different from the EDF of the sample data.

So, how do we choose a value for h, or indeed a kernel function $K(x)$? We won't go into that here. There is a whole field of statistics devoted to how to automatically determine from the data a suitable value for h, and we will make use of one of them (Silverman's rule of thumb) now in a code example.

KDE code example

The following code constructs kernel density estimates from 30 values drawn from a gamma distribution. We do this for two different kernel functions – a Parzen kernel and an exponential kernel. This code example can be found in the `Code_Examples_Chap2.ipynb` Jupyter notebook in the GitHub repository:

```
## Kernel density estimation example
import numpy as np
from scipy.stats import norm, gamma, iqr
from sklearn.neighbors import KernelDensity
import matplotlib.pyplot as plt
```

```python
# Set the seed for the random number generator
np.random.seed(280)

# First we sample our x values
shape = 5.0
scale = 1.0
n_sample = 30 # The number of data points in our sample
x_sample = np.random.gamma(
    shape=shape, scale=scale, size=(n_sample,1))

# Calculate summary statistics of the sample
sample_std = np.std(x_sample)
sample_iqr = iqr(x_sample)

# Construct the kernel density function using a
# Parzen (Gaussian window)
# and using the default bandwidth (=1.0)
parzen_kde = KernelDensity(kernel='gaussian').fit(x_sample)

# Construct x values at which we want to calculate the
# kernel density estimate
x = np.linspace(0, 2.0*np.max(x_sample),200)

# Calculate the log of the kernel density estimate for the
# regular spaced points
log_density_parzen = parzen_kde.score_samples(x.reshape(200,1))
plt.plot(x, np.exp(log_density_parzen), label='KDE:Parzen')

# Repeat kernel density estimation using a different kernel function
# We use a exponential kernel. We will set the bandwidth using
# Silverman's rule of thumb
bandwidth = 0.9*np.min(
    [sample_std, sample_iqr/1.35]) * np.power(float(n_sample), -0.2)
exponential_kde = KernelDensity(
    kernel='exponential', bandwidth=bandwidth).fit(x_sample)
log_density_exponential = exponential_kde.score_samples(
    x.reshape(200,1))
plt.plot(x, np.exp(log_density_exponential), 'red',
    label='KDE:Exponential')

# Add the true density function
plt.plot(x, gamma.pdf(x, shape)/scale, 'green', label='True Density')
```

```
# Add graph annotation and display
plt.legend(loc='upper right')
plt.title('Kernel density estimates from gamma random values')
plt.xlabel('x')
plt.ylabel('Probability Density')
plt.show()
```

The output from the code is shown in *Figure 2.19*:

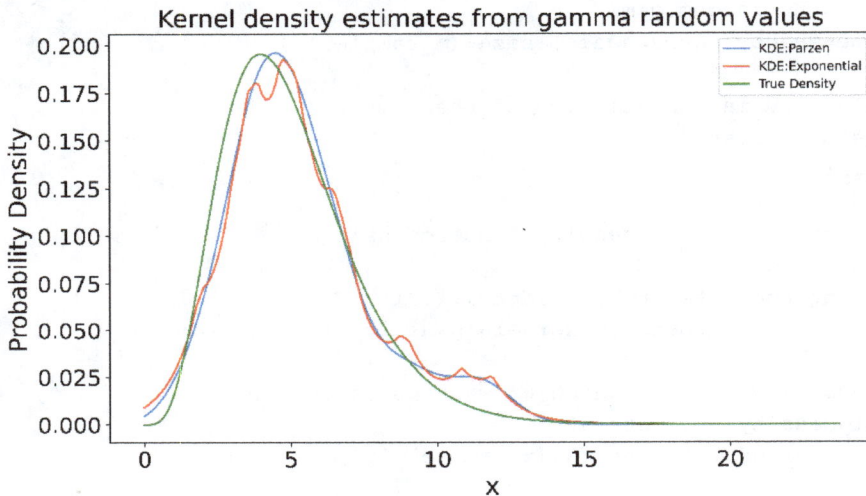

Figure 2.19: Example of KDE using different kernel functions

Figure 2.19 also shows the true PDF of the gamma distribution, and we can see that, even though we have only 30 datapoints in our sample, the kernel density estimates are close to the true density.

That practical example of how to estimate probability densities from data brings us neatly to the end of this section, so let's recap what we learned about KDE, and then wrap up by summarizing the chapter as a whole.

What we learned

In this section, we have learned the following:

- How adding together lots of i.i.d. random variables produces a random variable whose distribution is approximately normal.

- How many datasets we analyze will naturally contain variables that are themselves aggregations of many random values; for example, yearly total sales as the aggregation of 365 daily sales values.

- How the approximation by a normal distribution improves the more random values we are adding together. This is the CLT.

- How to use KDE to construct an estimate of a probability density from data even when we can't or don't want to make use of the CLT.

Summary

This chapter has been a long one. The effort will be worthwhile. Random variation is a component of any dataset, so knowing how to characterize and describe that random variation when analyzing data is a key skill for any data scientist. In this chapter, we have learned the following:

- How and why randomness arises in data

- How random variables are a natural concept to describe randomness in data

- Key aspects of random variables, such as their probability distributions, and how to use key metrics such as the mean and variance of a distribution to characterize a distribution

- How we can think of datasets as being samples drawn from an underlying distribution, and it is the underlying distribution we are really interested in understanding

- How to summarize a sample using the sample mean and sample variance

- How sample characteristics, such as the sample mean and sample variance, can be related back to the corresponding quantities of the underlying population distribution that we are interested in

- How the normal or Gaussian distribution is a commonly occurring distribution because it arises from the CLT, which tells us what happens when we add lots of data values together – a common task in data science

Randomness in data also flows through into any downstream calculation involving a dataset. Consequently, when constructing various data science algorithms, we need to take into account randomness within the data that those algorithms process. Unsurprisingly, many data science algorithms, such as **maximum likelihood estimation** (**MLE**) of model parameters and Bayesian probabilistic modeling, start by building from first principles upon the random nature of data. We will cover these concepts in *Chapter 5*. We can do so only because we have laid the solid foundations in this chapter.

Exercises

Next is a series of short exercises. They start easy and increase in difficulty. Answers to all the exercises are given in the `Answers_to_Exercises_Chap2.ipynb` Jupyter notebook in the GitHub repository. The exercises are a mix of code exercises and mathematical derivation exercises – this is designed to get you to start flexing your mathematical muscles. Give them a go and have fun:

1. Use the numpy package to sample 1,000 random values from a Beta distribution with $\alpha = 2$, $\beta = 5$, and plot a histogram of the resulting 1,000 values.

2. A mixture distribution is a distribution where the random variable has a certain specified probability of coming from one distribution, a certain specified probability of coming from a second distribution, a certain specified probability of coming from a third distribution, and so on. The number of different distributions is called the number of components in the mixture. We have a two-component mixture distribution, which we can write as follows:

$$A \sim Bernoulli(0.4)$$

$$z_1 \sim Normal(\mu = 1, \sigma^2 = 3.5) \quad z_2 \sim Laplace(\mu = 10, \lambda = 1.5)$$

$$X = A \times z_1 + (1 - A) \times z_2$$

Eq. 68

This means that with 40% probability X is drawn from a normal distribution with mean=1 and variance=3.5, and with 60% probability it is drawn from a Laplace distribution with mean=10 and scale=1.5.

Using `numpy.random.randn` and `numpy.random.laplace` functions, draw 3,000 values from this mixture distribution and plot a histogram of the resulting values.

3. In *Eq. 14*, we have given a definition of the variance of a random variable X as follows:

$$\text{Variance} = \mathbb{E}((X - \mathbb{E}(X))^2)$$

Eq. 69

An alternative way of writing this, which is sometimes useful computationally, is the following:

$$\text{Variance} = \mathbb{E}(X^2) - (\mathbb{E}(X))^2$$

Eq. 70

Derive this second way of writing the variance of X.

4. Create a dataset of 30 values sampled from the distribution $Normal(\mu = 2, \sigma^2 = 1.5)$. Using this data, create and plot kernel density estimates using a Parzen kernel, but with three different bandwidth values, $h = 0.1, h = 1.0$, and $h = 3.0$. Add the true probability density to the plot. What do you notice about the three different density estimates?

5. If we have i.i.d. data values $x_1, x_2, \cdots, x_n$ with $\mathbb{E}(x_i) = \mu$ and $Var(x_i) = \sigma^2$, prove that the expression in *Eq. 48* for the sample variance s^2 gives an unbiased estimator for the population variance σ^2.

3
Matrices and Linear Algebra

In this chapter, we are going to focus on linear algebra, specifically matrices and vectors. Vectors are the natural way to represent much of the data you will encounter as a data scientist, and matrices are the natural way to represent things that we do to that data, that is, transformations of the data.

Like the previous chapter, linear algebra is an absolute core part of the math behind data science, and so it is hugely beneficial to understand some of the intuition behind it. That is what this chapter aims to do, by covering the following topics:

- *Inner and outer products of vectors*: We will learn about the basic building block operations that we can apply to vectors.

- *Matrices as transformations*: We will learn about the basic operations involving matrices and what they represent.

- *Matrix decompositions*: We will learn key methods (eigen-decomposition and **Singular Value Decomposition (SVD)**) for representing matrices that make them simpler to work with.

- *Matrix properties*: We will learn about properties of matrices such as the trace and determinant of a square matrix.

- *Matrix factorization and dimensionality reduction*: We will learn how to apply matrix decomposition techniques to data, to identify structure within the data and simplify the data. We will do this using **Principal Component Analysis (PCA)** and **Non-Negative Matrix Factorization (NMF)** algorithms.

Technical requirements

All code examples provided in this chapter (and additional ones) can be found in the GitHub repository: `https://github.com/PacktPublishing/15-Math-Concepts-Every-Data-Scientist-Should-Know/tree/main/Chapter03`. To run the Jupyter notebooks, you will need a full Python installation including the following packages:

- `pandas` (>=2.0.3)
- `numpy` (>=1.24.3)
- `scikit-learn` (>=1.3.0)
- `matplotlib` (>=3.7.2)

Inner and outer products of vectors

As we have said, vectors are the natural way to store data points that have many features. You probably already have some experience with manipulating vectors or arrays, for example, by performing element-wise calculations on them. But we want to do more than that. We want to combine vectors. This section introduces the two most basic but important operations we can apply to two vectors – the **inner product** and the **outer product**.

Inner product of two vectors

To calculate the **inner product** between two vectors, $\underline{a}$ and $\underline{b}$, they need to be of equal length. In this and all subsequent calculations, we will assume that our vectors $\underline{a}$ and $\underline{b}$ are real-valued. If those vectors are d-dimensional and have components $a_1, a_2, \cdots, a_d \in \mathbb{R}$ and $b_1, b_2, \cdots, b_d \in \mathbb{R}$, then the inner product between them is denoted by the symbol $\underline{a} \cdot \underline{b}$ and is defined as follows:

$$\underline{a} \cdot \underline{b} = \sum_{i=1}^{d} a_i b_i$$

Eq. 1

Because of the dot between $\underline{a}$ and $\underline{b}$ in the left-hand side of Eq. 1, the inner product is also called the **dot-product** of $\underline{a}$ and $\underline{b}$. The result of the inner product calculation is a number, that is, a scalar, so the inner product is also sometimes called the **scalar product** of $\underline{a}$ and $\underline{b}$.

Let's look at what happens if we take the inner product between a vector $\underline{a}$ and itself. What we get is the following:

$$\underline{a} \cdot \underline{a} = \sum_{i=1}^{d} a_i^2$$

Eq. 2

You'll probably recognize the right-hand side of Eq. 2 as the d-dimensional version of Pythagoras' theorem written in terms of the components $a_1, a_2, \ldots, a_d$, and so the right-hand side of Eq. 2 gives the square of the length of the vector a. The length of a vector a is also denoted using the symbol $|a|$, and so we have the following:

$$a \cdot a = |a|^2$$

Eq. 3

We'll also give another way of calculating the inner product between a and b. Since we have two vectors, a and b, we can always find a plane in which those two vectors lie. Let θ be the angle between the two vectors in that plane. This is illustrated in *Figure 3.1*.

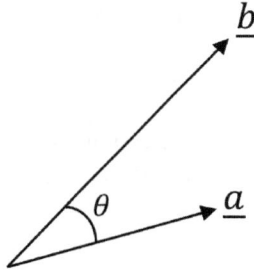

Figure 3.1: Two vectors in a 2D plane

In terms of θ, the inner product between a and b can be calculated using the formula in Eq. 4:

$$a \cdot b = |a|\,|b|\cos\theta$$

Eq. 4

If we re-arrange Eq. 4, we can write the following:

$$\frac{a}{|a|} \cdot \frac{b}{|b|} = \cos\theta$$

Eq. 5

Now $a/|a|$ is a vector in the same direction as a but with length 1. Likewise, $b/|b|$ is a vector in the same direction as b but with length 1. This tells us that $\cos\theta$ is the inner product between a unit-length vector in direction a and a unit-length vector in direction b. When a and b point in the same direction, $\theta = 0°$, and so $\cos\theta = 1$. Conversely, if a and b are orthogonal to each other, then $\theta = 90°$, and so $\cos\theta = 0$. When a and b point in directly opposite directions, then $\theta = 180°$ and $\cos\theta = -1$. From this, we can see that the inner product on the left-hand side of Eq. 5 is a measure of similarity between the two unit-length vectors. In other words, the inner product on the left-hand side of Eq. 5 is a measure of similarity between a and b that corrects for the fact that a and b may not be of unit lengths and may be of different lengths to each other.

Because of this, we tend to think of the inner product $\underline{a} \cdot \underline{b}$ itself as a measure of similarity between the vectors $\underline{a}$ and $\underline{b}$. The higher the inner product between $\underline{a}$ and $\underline{b}$ is, the more aligned they are, and the more they point in the same direction. We can make this idea more exact by introducing the concept of **projecting** one vector onto another. When we **project** $\underline{a}$ onto $\underline{b}$ we are measuring how much of the vector $\underline{a}$ lies along the direction given by the vector $\underline{b}$. This projection can be calculated using $|\underline{a}|$ and $\cos\theta$ and is given by the following:

$$\text{Projection of } \underline{a} \text{ onto } \underline{b} = |\underline{a}|\cos\theta$$

Eq. 6

Similarly, the projection of $\underline{b}$ onto $\underline{a}$ is calculated as follows:

$$\text{Projection of } \underline{b} \text{ onto } \underline{a} = |b|\cos\theta$$

Eq. 7

This tells us the amount of $\underline{b}$ that lies along the direction given by $\underline{a}$. When $\underline{a}$ and $\underline{b}$ are both of unit length, so that $|\underline{b}| = |\underline{b}| = 1$, then the projection of $\underline{a}$ onto $\underline{b}$ is the same as the projection of $\underline{b}$ onto $\underline{a}$, and is equal to the inner product $\underline{a} \cdot \underline{b}$.

Outer product of two vectors

If the inner product between two vectors leads to a scalar (a zero-dimensional object), a natural question is whether we can create a two-dimensional object from two real vectors. That is, can we create a matrix? The answer is yes, and the most obvious way is to create a matrix $\underline{A}$ with matrix elements $A_{ij} = a_i b_j$. A matrix formed in this way is called the **outer product** of the vectors $\underline{a}$ and $\underline{b}$. We'll typically use $\underline{a} \otimes \underline{b}$ to denote this operation of forming a matrix from $\underline{a}$ and $\underline{b}$.

For the outer product, the vectors $\underline{a}$ and $\underline{b}$ do not have to be of the same dimension. This contrasts with the inner product, which can only be calculated between vectors with the same dimensions. If we have an M-dimensional vector $\underline{a}$ with components $a_1, a_2, ..., a_M \in \mathbb{R}$, and an N-dimensional vector $\underline{b}$ with components $b_1, b_2, ..., b_N \in \mathbb{R}$, then the outer product, $\underline{a} \otimes \underline{b}$, is an $M \times N$ matrix $\underline{A}$ given by the following:

$$\underline{A} = \begin{pmatrix} a_1 b_1 & a_1 b_2 & \cdots & a_1 b_N \\ a_2 b_1 & a_2 b_2 & \cdots & a_2 b_N \\ \vdots & \vdots & \ddots & \vdots \\ a_M b_1 & a_M b_2 & \cdots & a_M b_N \end{pmatrix}$$

Eq. 8

Because the vectors $\underline{a}$ and $\underline{b}$ don't have to have the same dimensions to form the outer product $\underline{a} \otimes \underline{b}$, it means we can also create a second matrix from $\underline{a}$ and $\underline{b}$ by changing their order in the outer product, to calculate $\underline{b} \otimes \underline{a}$. This matrix will be an $N \times M$ matrix $\underline{B}$ given by the following:

$$\underline{B} = \begin{pmatrix} b_1 a_1 & b_1 a_2 & \cdots & b_1 a_M \\ b_2 a_1 & b_2 a_2 & \cdots & b_2 a_M \\ \vdots & \vdots & \ddots & \vdots \\ b_N a_1 & b_N a_2 & \cdots & b_N a_M \end{pmatrix}$$

Eq. 9

So, the order in which we specify the vectors matters when calculating an outer product. Contrast this with the inner product, where $\underline{a} \cdot \underline{b} = \underline{b} \cdot \underline{a}$ (provided $\underline{a}$ and $\underline{b}$ have the same dimensions), and so the order is irrelevant.

Inner and outer product code example

Let's look at how we calculate inner and outer products in code. Fortunately, the numpy Python package has in-built functions for both. This code example can be found in the Code_Examples_Chap3. ipynb Jupyter notebook in the GitHub repository. First, we'll calculate the inner product between two vectors:

```
import numpy as np
# Create the two vectors
a = np.array([1.0, -4.0, 0.5])
b = np.array([0.0, 3.5, 2.0])

# Calculate the inner product
# The answer should be = (1*0)+(-4.0*3.5)+(0.5*2.0) = 0-14+1 = -13
np.inner(a,b)
```

The following is a code example for calculating the outer product between the same two vectors:

```
# Calculate the outer product
# The answer should be a 3x3 array of the form
# [[1*0, 1*3.5, 1*2], [-4*0, -4*3.5, -4*2], [0.5*0, 0.5*3.5, 0.5*2]]
# which gives [[0, 3.5, 2], [0, -14, -8], [0, 1.75, 1]]
np.outer(a,b)
```

That simple code example is a good place to conclude this section on vector operations, so we'll summarize what we learned.

What we learned

In this section, we have learned about the following:

- How to calculate the inner product between two vectors, and that it outputs a scalar value
- How the inner product gives us a measure of similarity between two vectors
- How to calculate the outer product between two vectors, and that it outputs a matrix

Having learned about basic operations involving pairs of vectors, we will next learn about basic operations involving two or more matrices. We will also learn what those matrices represent.

Matrices as transformations

Matrices are typically applied to vectors and other matrices through the process of matrix multiplication. However, the dry mechanics of matrix multiplication tend to hide what a matrix really represents and what matrix multiplication does. We aim to shed light on what matrices really are in this chapter. We'll start by covering the basics of matrix multiplication and then show how matrices represent transformations.

Matrix multiplication

If we have a matrix $\underline{A}$ of size $M \times K$, and a matrix $\underline{B}$ of size $K \times N$, then we can multiply those two matrices together to get a new matrix $\underline{C} = \underline{A}\,\underline{B}$, which is of size $M \times N$. The matrix element C_{ij} is calculated as follows:

$$C_{ij} = \sum_{k=1}^{K} A_{ik} B_{kj}$$

Eq. 10

In this example, we are multiplying a $K \times N$ matrix by an $M \times K$ matrix. Schematically, we can write this as follows:

$$\underset{\text{Matrix } \underline{C}}{M \times N} = \underset{\text{Matrix } \underline{A}}{M \times K} \times \underset{\text{Matrix } \underline{B}}{K \times N}$$

Eq. 11

From this, it is clear that the "inner" dimensions in this example match, both being K. To multiply two matrices together, the inner dimensions must match, when we write the multiplication out in this schematic way. If the dimensions do not match, we cannot multiply the matrices together. For example, we cannot multiply a 10×4 matrix by a 6×7 matrix.

The inner product as matrix multiplication

In fact, if we think of a d-dimensional column vector $\underline{a}$ as a $d \times 1$ matrix, and its transpose, $\underline{a}^\top$, which is a d-dimensional row vector, as a $1 \times d$ matrix, then multiplying those two matrices together as $\underline{a}^\top \underline{a}$ gives a 1×1 result, or in other words, a scalar. The value of that scalar is calculated using the right-hand side of Eq. 10 for matrix multiplication and gives the same calculation as Eq. 2 for the inner product. In other words, we have the following:

$$\underline{a}^\top \underline{a} = \underline{a} \cdot \underline{a}$$

Eq. 12

More generally, if we have two vectors $\underline{a}$ and $\underline{b}$ of the same length, then $\underline{a}^\top \underline{b} = \underline{a} \cdot \underline{b}$. Let's look at the last formula more schematically, as shown in *Figure 3.2*:

$$\begin{pmatrix} a_1 & a_2 & \dots & a_d \end{pmatrix} \times \begin{pmatrix} b_1 \\ b_2 \\ \vdots \\ b_d \end{pmatrix} = \sum_{i=1}^{d} a_i b_i$$

Figure 3.2: Matrix multiplication of a row vector and column vector is the same as the inner product

The left-hand side of the schematic equation in *Figure 3.2* is a matrix multiplication, but if we calculate that matrix multiplication by hand, we get the expression on the right-hand side of the figure, which is just the inner product $\underline{a} \cdot \underline{b}$.

Matrix multiplication as a series of inner products

We can extend this connection between inner products and matrix multiplication by looking again at the right-hand side of Eq. 10. The matrix element C_{ij} looks like an inner product between a vector formed from the ith row of matrix $\underline{A}$ and the jth column of matrix $\underline{B}$. That means we can represent the matrix multiplication $\underline{A}\,\underline{B}$ schematically as follows:

$$\underline{A}_1 \longrightarrow \begin{pmatrix} A_{11} & A_{12} & \cdots & A_{1K} \\ A_{21} & A_{22} & \cdots & A_{2K} \\ \vdots & \vdots & \ddots & \vdots \\ A_{M1} & A_{M2} & \cdots & A_{MK} \end{pmatrix} \times \begin{pmatrix} B_{11} & B_{12} & \cdots & B_{1N} \\ B_{21} & B_{22} & \cdots & B_{2N} \\ \vdots & \vdots & \ddots & \vdots \\ B_{K1} & B_{K2} & \cdots & B_{KN} \end{pmatrix}$$

$$= \begin{pmatrix} \underline{A}_1^\mathsf{T}\underline{B}_1 & \underline{A}_1^\mathsf{T}\underline{B}_2 & \cdots & \underline{A}_1^\mathsf{T}\underline{B}_N \\ \underline{A}_2^\mathsf{T}\underline{B}_1 & \underline{A}_2^\mathsf{T}\underline{B}_2 & \cdots & \underline{A}_2^\mathsf{T}\underline{B}_N \\ \vdots & \vdots & \ddots & \vdots \\ \underline{A}_M^\mathsf{T}\underline{B}_1 & \underline{A}_M^\mathsf{T}\underline{B}_2 & \cdots & \underline{A}_M^\mathsf{T}\underline{B}_N \end{pmatrix}$$

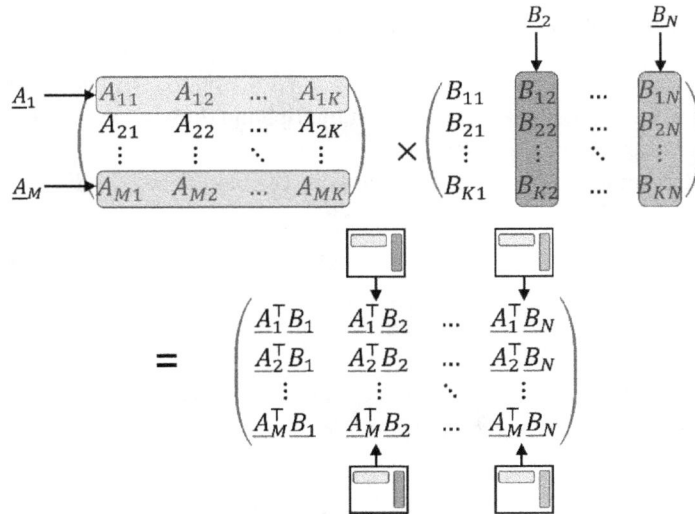

Figure 3.3: Matrix elements resulting from a matrix multiplication
can be viewed as inner product calculations

In fact, this schematic is how I remember how to do matrix multiplication, not the dry formula given in Eq. 10.

Matrix multiplication is not commutative

The matrix multiplication $\underline{A}\,\underline{B}$ is, of course, the matrix counterpart of the ordinary multiplication of real numbers that we are familiar with from school. Even the notation $\underline{A}\,\underline{B}$ gives the impression that matrix multiplication will follow the same rules and patterns as ordinary multiplication. This is not the case. There are some subtleties and nuances with matrix multiplication. One of these subtleties to be aware of is that the order of the matrices matters. In general, for two different matrices $\underline{A}$ and $\underline{B}$, we have $\underline{A}\,\underline{B} \neq \underline{B}\,\underline{A}$. We say that matrix multiplication is not commutative.

To see this more concretely, let's take an explicit example. Consider these two matrices:

$$\underline{A} = \begin{pmatrix} 1 & 4 \\ 0 & -2 \end{pmatrix} \quad \underline{B} = \begin{pmatrix} 2 & 1 \\ 1 & 1 \end{pmatrix}$$

Eq. 13

You can confirm for yourself, by doing the matrix multiplications by hand, that the following apply:

$$\underline{A}\,\underline{B} = \begin{pmatrix} 6 & 5 \\ -2 & -2 \end{pmatrix} \quad \underline{B}\,\underline{A} = \begin{pmatrix} 2 & 6 \\ 1 & 2 \end{pmatrix}$$

Eq. 14

Obviously, there are special cases where matrix multiplication does commute, for example, the trivial case when $\underline{A} = \underline{B}$, in which case $\underline{A}\,\underline{B} = \underline{B}\,\underline{A} = \underline{A}^2$. There are also cases when $\underline{A}$ and $\underline{B}$ are different and yet we have $\underline{A}\,\underline{B} = \underline{B}\,\underline{A}$. These special cases require extra conditions on the properties of the matrices $\underline{A}$ and $\underline{B}$, but for now, we can say that in general, $\underline{A}\,\underline{B} \neq \underline{B}\,\underline{A}$, so be careful you don't assume it.

The outer product as a matrix multiplication

Just as we showed that the inner product between two vectors could be written as a matrix multiplication, we can do the same for calculating the outer product between two vectors. If we have an M-component real-valued column vector $\underline{a}$, we can think of it as an $M \times 1$ matrix. Likewise, if we have an N-component real-valued row vector $\underline{b}^{\mathsf{T}}$ then we can think of it as a $1 \times N$ matrix. We can then multiply these two matrices together to get $\underline{a}\,\underline{b}^{\mathsf{T}}$. From the rules of matrix multiplication, this is an $M \times N$ matrix whose i,j matrix element is given by $a_i b_j$, which is the same as we get when we calculate the outer product, $\underline{a} \otimes \underline{b}$ between the vectors $\underline{a}$ and $\underline{b}$. Because of this, we almost always use the more succinct notation $\underline{a}\,\underline{b}^{\mathsf{T}}$ to denote the outer product $\underline{a} \otimes \underline{b}$ when $\underline{a}$ and $\underline{b}$ are real-valued. Schematically, we have the following:

$$\underline{a}\,\underline{b}^{\mathsf{T}} = \begin{pmatrix} a_1 \\ \vdots \\ a_M \end{pmatrix} \times \begin{pmatrix} b_1 & \cdots & b_N \end{pmatrix} = \begin{pmatrix} a_1 b_1 & \cdots & a_1 b_N \\ \vdots & \ddots & \vdots \\ a_M b_1 & \cdots & a_M b_N \end{pmatrix} = \underline{a} \otimes \underline{b}$$

Eq. 15

You'll recall that we could also calculate the outer product, $\underline{b} \otimes \underline{a}$, from the vectors $\underline{a}$ and $\underline{b}$. You will have guessed that we can also write this outer product as the matrix multiplication $\underline{b}\,\underline{a}^{\mathsf{T}}$ when $\underline{a}$ and $\underline{b}$ are real-valued. Again, this notation is more commonly used to represent the outer product, rather than $\underline{b} \otimes \underline{a}$.

Multiplying multiple matrices together

Once we know how to multiply two matrices together, it is a simple matter to multiply many matrices together – we simply take them two at a time. For example, if we have $N \times N$ matrices $\underline{A}$, $\underline{B}$, and $\underline{C}$, then their product can be calculated via the following:

$$\underline{A}\,\underline{B}\,\underline{C} = \underline{A} \times (\underline{B}\,\underline{C}) = (\underline{A}\,\underline{B}) \times \underline{C}$$

Eq. 16

We can either multiply $\underline{B}$ and $\underline{C}$ together first and then multiply the result by $\underline{A}$, or we can multiply $\underline{A}$ and $\underline{B}$ together first and then use the result to multiply $\underline{C}$. Either way, we get the same result. This means matrix multiplication is associative.

Transforming a vector by matrix multiplication

So far, we have learned about vectors and matrices and their basic properties. We have also learned how to multiply matrices together. We have even seen how we can consider a vector as a special kind of matrix. This immediately raises the question of what happens if we multiply a vector by a matrix – what do we get and what does that multiplication represent?

Consider an $M \times N$ matrix $\underline{A}$ and an N-component column vector $\underline{b}$. As we can think of the vector $\underline{b}$ as an $N \times 1$ matrix, we can clearly multiply $\underline{b}$ by $\underline{A}$ using the rules of matrix multiplication. In fact, we get the following:

$$\underline{A}\,\underline{b} = \begin{pmatrix} A_{11} & \cdots & A_{1N} \\ \vdots & \ddots & \vdots \\ A_{M1} & \cdots & A_{MN} \end{pmatrix} \times \begin{pmatrix} b_1 \\ \vdots \\ b_N \end{pmatrix} = \begin{pmatrix} A_{11}\,b_1 + A_{12}\,b_2 + \cdots + A_{1N}\,b_N \\ \vdots \\ A_{M1}\,b_1 + A_{M2}\,b_2 + \cdots + A_{MN}\,b_N \end{pmatrix}$$

Eq. 17

The result is an $M \times 1$ matrix, that is, an M-component column vector. So, multiplying a vector by a matrix gives us another vector. The components of this new vector are given by the expressions inside the brackets on the right-hand side of Eq. 17. The components of this new vector are (in general) different from those of vector $\underline{b}$, so the effect of multiplying a vector by a matrix is to transform that vector. From this, we can conclude that matrices represent transformations.

If we look more closely at the individual expressions in the vector on the right-hand side of Eq. 17, we can see that each component in the new vector is a linear combination of the components in the old vector $\underline{b}$. So, the matrix $\underline{A}$ represents a linear transformation. The individual matrix elements A_{11}, A_{12}, and so on tell us the weights in those linear combinations that give us the components of the new vector. In other words, the individual matrix elements encode the details of the linear transformation.

One thing we haven't spoken about yet is what effect the relative sizes of M and N has. If $M = N$, then obviously, multiplying an N-component vector $\underline{b}$ by the $N \times N$ matrix $\underline{A}$ gives us another N-component vector. Although we have transformed the vector, we have, in this case, stayed within the same N-dimensional space. However, if $M < N$, then our new vector has fewer components than the starting vector $\underline{b}$, and so we have reduced the dimensionality. Alternatively, if $M > N$, our new vector has more components than we started with, and we have increased the dimensionality.

In all the examples previously, we have been multiplying a column vector by a matrix. But we can equally multiply a matrix by a row vector. Let's stick with our vector $\underline{b}$ but we will use its row vector form $\underline{b}^{\mathsf{T}}$. Now we can think of $\underline{b}^{\mathsf{T}}$ as a $1 \times N$ matrix. So, if we have an $N \times M$ matrix $\underline{C}$, then we can perform the matrix multiplication $\underline{b}^{\mathsf{T}}\underline{C}$, and we get a $1 \times M$ matrix out of it, that is, an M-component row vector. As you might expect, this new M-component vector is just a linear transformation of our starting N-component vector $\underline{b}^{\mathsf{T}}$, with the details of the linear transformation encoded in the matrix elements C_{ij}.

Finally, we should highlight that since matrix multiplication is a linear transformation, it means that if we apply a matrix $\underline{A}$ to a combination of vectors, the result is the same as combining the results of applying $\underline{A}$ to each vector individually. In more detail, we have the following:

$$\underline{A} \times (\underline{b}_1 + \underline{b}_2 + \cdots + \underline{b}_K) = \underline{A}\,\underline{b}_1 + \underline{A}\,\underline{b}_2 + \cdots + \underline{A}\,\underline{b}_K$$

Eq. 18

We will make use of this fact shortly.

The identity matrix

Now that we have learned that matrix multiplication represents the linear transformation of vectors, let's look at some particular special cases of transformations. Consider the $N \times N$ matrix $\underline{I}_N$ given here:

$$\underline{I}_N = \begin{pmatrix} 1 & 0 & 0 & \cdots & 0 \\ 0 & 1 & 0 & \cdots & 0 \\ \vdots & \vdots & 1 & \cdots & \vdots \\ 0 & 0 & \cdots & \ddots & 0 \\ 0 & 0 & 0 & \cdots & 1 \end{pmatrix}$$

Eq. 19

The matrix $\underline{I}_N$ has 1 for each matrix element along its diagonal and 0 everywhere else. Now, what is the effect of multiplying by $\underline{I}_N$? Let's try it. Consider an N-component column vector $\underline{a}$. If we multiply $\underline{a}$ by $\underline{I}_N$, we get the result shown here:

$$\underline{I}_N\,\underline{a} = \underline{I}_N \times \begin{pmatrix} a_1 \\ \vdots \\ a_N \end{pmatrix} = \begin{pmatrix} 1 \times a_1 + 0 \times a_2 + \cdots + 0 \times a_N \\ \vdots \\ 0 \times a_1 + 0 \times a_2 + \cdots + 1 \times a_N \end{pmatrix} = \begin{pmatrix} a_1 \\ \vdots \\ a_N \end{pmatrix} = \underline{a}$$

Eq. 20

So, multiplying any vector $\underline{a}$ by $\underline{I}_N$ just gives us back $\underline{a}$ itself. We haven't done anything to the starting vector. The transformation represented by $\underline{I}_N$ is just the identity transformation, which leaves vectors untouched. Hence, $\underline{I}_N$ is called the **identity matrix**. Or, more specifically, it is the N-component identity matrix because it operates on N-component vectors.

It is a simple matter to confirm, via a similar calculation to the previous one, that if we reverse the order of the calculation, so that we multiply $\underline{I}_N$ by a row vector $\underline{a}^\top$, we leave the row vector unchanged. In terms of math notation, we have the following:

$$\underline{a}^\top \underline{I}_N = \underline{a}^\top$$

Eq. 21

Now remember that when we explained matrix multiplication as a series of inner products, we learned that we could think of a matrix as a set of column vectors, so it is not surprising that when we multiply an $N \times N$ matrix $\underline{B}$ by $\underline{I}_N$, we leave the matrix untouched. In terms of the math, we have the following:

$$\underline{I}_N \underline{B} = \underline{B}$$

Eq. 22

Again, if we multiply them in the opposite order, we also leave the matrix $\underline{B}$ unchanged, so in terms of the math, we have the following:

$$\underline{B}\,\underline{I}_N = \underline{B}$$

Eq. 23

The inverse matrix

If the identity matrix $\underline{I}_N$ leaves an $N \times N$ matrix untouched, we can think of it as the matrix analog of multiplying a number by 1. For any number a on the real number line, we have $a \times 1 = a$ and $1 \times a = a$. The number 1 here is called the **identity element**. For a number a, we also have the concept of its reciprocal, a^{-1}, which is the number we multiply a by to get the identity element, so that a^{-1} is defined by the following relationship:

$$a^{-1}a = a\,a^{-1} = 1$$

Eq. 24

For an $N \times N$ matrix $\underline{A}$, we have an analogous concept – the **inverse matrix** of $\underline{A}$, which is denoted by the symbol $\underline{A}^{-1}$. The matrix $\underline{A}^{-1}$ is an $N \times N$ matrix and, as you might have guessed, is defined as the matrix we multiply $\underline{A}$ by to get the identity element, the matrix $\underline{I}_N$ in this case. So, $\underline{A}^{-1}$ is defined by the following relationship:

$$\underline{A}^{-1}\underline{A} = \underline{A}\,\underline{A}^{-1} = \underline{I}_N$$

Eq. 25

Conceptually, we can think of $\underline{A}^{-1}$ as playing a similar role and having similar properties to the reciprocal a^{-1} in ordinary arithmetic. Just like the reciprocal in ordinary arithmetic, the inverse matrix can be extremely useful in simplifying mathematical expressions by canceling other terms out.

Note that the inverse matrix is only defined for square matrices. Non-square matrices do not have a proper inverse. However, not all square matrices necessarily have an inverse. That is, there are some square matrices, $\underline{A}$, for which there are no solutions, $\underline{A}^{-1}$, to the relation in Eq. 25. We will talk more about that later when we introduce eigen-decompositions of a square matrix.

More examples of matrices as transformations

Let's look at another specific example of a matrix and understand its effect as a transformation. Consider the 2×2 matrix here:

$$\underline{A} = \begin{pmatrix} \frac{1}{\sqrt{2}} & -\frac{1}{\sqrt{2}} \\ \frac{1}{\sqrt{2}} & \frac{1}{\sqrt{2}} \end{pmatrix}$$

Eq. 26

Clearly matrix $\underline{A}$ operates on two-component vectors that live in a two-dimensional plane. We can think of that plane as being the usual (x, y) plane. What transformation does this represent? Let's break it down. Let's look at the effect of the transformation represented by $\underline{A}$ on a specific vector. In this case, we're going to choose the vector that represents the x axis. In column vector form, this vector is as follows:

$$\begin{pmatrix} 1 \\ 0 \end{pmatrix}$$

Eq. 27

All other vectors representing points on the x axis are just multiples of the vector in Eq. 27. Now, what is the effect of $\underline{A}$ on this vector? It is easy to compute, and we find the following:

$$\underline{A} \begin{pmatrix} 1 \\ 0 \end{pmatrix} = \begin{pmatrix} \frac{1}{\sqrt{2}} & -\frac{1}{\sqrt{2}} \\ \frac{1}{\sqrt{2}} & \frac{1}{\sqrt{2}} \end{pmatrix} \times \begin{pmatrix} 1 \\ 0 \end{pmatrix} = \begin{pmatrix} \frac{1}{\sqrt{2}} \\ \frac{1}{\sqrt{2}} \end{pmatrix}$$

Eq. 28

The new vector on the right-hand side of Eq. 28 represents a point in the (x, y) plane that has identical and positive x and y components. In other words, it represents a 45° anti-clockwise rotation of our starting point, which was on the x axis.

Let's look at the effect of $\underline{A}$ on another vector. This time we're going to choose a vector that represents a point on the y axis. In column vector form, this vector is as follows:

$$\begin{pmatrix} 0 \\ 1 \end{pmatrix}$$

Eq. 29

All other vectors representing points on the y axis are just multiples of the vector in Eq. 29. The effect of $\underline{A}$ on this vector is as follows:

$$\underline{A} \begin{pmatrix} 0 \\ 1 \end{pmatrix} = \begin{pmatrix} \frac{1}{\sqrt{2}} & -\frac{1}{\sqrt{2}} \\ \frac{1}{\sqrt{2}} & \frac{1}{\sqrt{2}} \end{pmatrix} \times \begin{pmatrix} 0 \\ 1 \end{pmatrix} = \begin{pmatrix} -\frac{1}{\sqrt{2}} \\ \frac{1}{\sqrt{2}} \end{pmatrix}$$

Eq. 30

The new vector on the right-hand side of Eq. 30 represents a point in the second quadrant of the (x, y) plane, and again represents a 45° anti-clockwise rotation of our starting point on the y axis. The effect of matrix $\underline{A}$ on the vectors $\begin{pmatrix} 1 \\ 0 \end{pmatrix}$ and $\begin{pmatrix} 0 \\ 1 \end{pmatrix}$ is illustrated schematically in *Figure 3.4*:

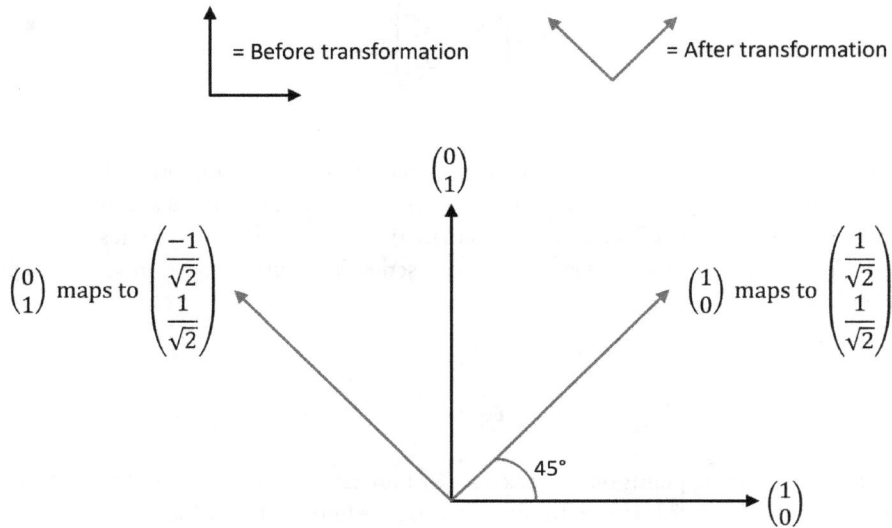

= Before transformation   = After transformation

$\begin{pmatrix} 0 \\ 1 \end{pmatrix}$

$\begin{pmatrix} 0 \\ 1 \end{pmatrix}$ maps to $\begin{pmatrix} \dfrac{-1}{\sqrt{2}} \\ \dfrac{1}{\sqrt{2}} \end{pmatrix}$ $\begin{pmatrix} 1 \\ 0 \end{pmatrix}$ maps to $\begin{pmatrix} \dfrac{1}{\sqrt{2}} \\ \dfrac{1}{\sqrt{2}} \end{pmatrix}$

45°

$\begin{pmatrix} 1 \\ 0 \end{pmatrix}$

Figure 3.4: Schematic illustration of the effect of matrix $\underline{A}$

Now, any two-dimensional vector can be written as a sum of the two vectors we have just studied. To show this, consider the following:

$$\begin{pmatrix} x \\ y \end{pmatrix} = \begin{pmatrix} x \\ 0 \end{pmatrix} + \begin{pmatrix} 0 \\ y \end{pmatrix} = x\begin{pmatrix} 1 \\ 0 \end{pmatrix} + y\begin{pmatrix} 0 \\ 1 \end{pmatrix}$$

Eq. 31

Given the effect of $\underline{A}$ on both $\begin{pmatrix} 1 \\ 0 \end{pmatrix}$ and $\begin{pmatrix} 0 \\ 1 \end{pmatrix}$ is a 45° anti-clockwise rotation, then the effect of $\underline{A}$ on any two-dimensional vector will be a 45° anti-clockwise rotation. Therefore, as a transformation, $\underline{A}$ is a matrix that represents a 45° anti-clockwise rotation.

Since any two-dimensional vector can be written as a linear combination of the vectors $\begin{pmatrix} 1 \\ 0 \end{pmatrix}$ and $\begin{pmatrix} 0 \\ 1 \end{pmatrix}$, these two vectors are called **basis vectors** – they provide a basis from which we can construct all other two-dimensional vectors. These two vectors are also **orthogonal** to each other. In geometric terms, this means they are at right-angles to each other – this is obvious in this example because one vector lies along the x axis while the other lies along the y axis. In algebraic terms, orthogonality means the inner product between the two vectors is 0. Basis vectors don't have to be orthogonal to each other. For example, the two vectors $\begin{pmatrix} 1 \\ 0 \end{pmatrix}$ and $\begin{pmatrix} 1 \\ 1 \end{pmatrix}$ can also be used to describe any point on the (x, y) plane. However, when basis vectors are orthogonal, they are easy to work with. Moving along one orthogonal basis vector does not change how far along we are on another orthogonal basis vector. For example, moving along the x axis does not affect where we are on the y axis. This means we can apply calculations along one orthogonal basis vector without having to worry about what is happening in terms of the other basis vectors. This makes orthogonal basis vectors very convenient to work with – a fact we will make use of when we move on to decompositions of matrices in the next section.

Given a set of orthogonal basis vectors $\underline{v}_1, \underline{v}_2, \ldots, \underline{v}_d$ in a d-dimensional space, we can easily work out how to represent any vector $\underline{a}$ in terms of those basis vectors. Say we have a vector $\underline{a}$ and we want to write it as follows:

$$\underline{a} = \alpha_1 \underline{v}_1 + \alpha_2 \underline{v}_2 + \cdots + \alpha_d \underline{v}_d$$

Eq. 32

Then, we can work out the values of the weights $\alpha_1, \alpha_2, \cdots, \alpha_d$ by taking the inner product of both sides of Eq. 32 with each of the basis vectors $\underline{v}_1, \underline{v}_2, \ldots, \underline{v}_d$. Doing so, we get the following:

$$\underline{a}^T \underline{v}_i = \alpha_1 \underline{v}_1^T \underline{v}_i + \alpha_2 \underline{v}_2^T \underline{v}_i + \ldots + \alpha_d \underline{v}_d^T \underline{v}_i$$

Eq. 33

Since $\underline{v}_i$ is, by definition, orthogonal to all the other basis vectors except $\underline{v}_i$ itself, then the inner products $\underline{v}_j^T \underline{v}_i = 0$, unless $j = i$. Plugging this fact into the preceding equation, we get the following:

$$\underline{a}^T \underline{v}_i = \alpha_i \underline{v}_i^T \underline{v}_i \implies \alpha_i = \frac{\underline{a}^T \underline{v}_i}{\underline{v}_i^T \underline{v}_i}$$

Eq. 34

So, we can easily work out the required weights. If the basis vectors are all of unit length, so that $\underline{v}_i^T \underline{v}_i = 1$ for every value of i, then the expression in Eq. 34 for the weights becomes even easier. It becomes $\alpha_i = \underline{a}^T \underline{v}_i$. A set of orthogonal basis vectors that are of unit length are called **orthonormal** and form an **orthonormal basis**. Using an orthonormal basis to represent our vectors is extremely convenient. In the next section of this chapter, we will show how an orthonormal basis can be extracted from any matrix and therefore can be used as an extremely convenient way of working with that matrix. But for now, let's look at how to do some of those matrix multiplications and transformations in a code example.

Matrix transformation code example

For our code example, we'll use the in-built functions in the NumPy package to do this. All the code examples that follow (and additional ones) can be found in the Code_Examples_Chap3.ipynb Jupyter notebook in the GitHub repository.

First, we'll use the numpy.matmul function to multiply two matrices together:

```
import numpy as np
# Create 3x3 matrices
A = np.array([[1.0, 2.0, 1.0], [-2.5, 1.0, 0.0], [3.0, 1.0, 1.5]])
```

```
B = np.array([[1.0, -1.0, -1.0], [5, 2.0, 3.0], [3.0, 1.0, 2.0]])

# Multiply the matrices together
np.matmul(A, B)
```

The preceding code produces the following output:

```
array([[14.,   4. ,   7. ],
       [2.5,   4.5,   5.5],
       [12.5, 0.5,   3. ]])
```

We can use the same NumPy function to multiply a vector by a matrix:

```
# Create a 4-dimensional vector
a = np.array([1.0, 2.0, 3.0, -2.0])

# Create a 3x4 matrix
A = np.array([
    [1.0, 1.0, 0.0, 1.0], [-2.0, 2.5, 1.5, 3.0],
    [0.0, 1.0, 1.0, 4.0]])

# We'll use the matrix multiplication function to calculate # A*a
np.matmul(A, a)
```

We get the following output:

```
array([ 1. ,   1.5, -3. ])
```

The NumPy package even has an in-built function for calculating the inverse of a matrix, as the following code demonstrates:

```
# Create a 4x4 square matrix
A = np.array( [[1, 2, 3, 4],
               [2, 1, 2, 1],
               [0, 1, 3, 2],
               [1, 1, 2, 2]])

# Calculate and store the inverse matrix
Ainv = np.linalg.inv(A)

# Multiply the matrix by its inverse.
# We should get the identity matrix
#[[1,0,0,0], [0,1,0,0], [0,0,1,0], [0,0,0,1]]
# up to numerical precision
np.matmul(Ainv, A)
```

These simple code examples of matrix transformations bring this section neatly to a close, so let's recap what we have learned in this section.

What we learned

In this section, we have learned the following:

- How to multiply matrices together
- How to multiply a vector by a matrix and vice versa
- What the identity matrix is and its effect on any other matrix
- What the inverse of a matrix is and why it is useful
- How a matrix represents a linear transformation
- How sets of orthonormal vectors provide a convenient basis on which we can express any other vector

Having learned the basics of matrix multiplication and how matrices represent transformations, we'll now learn some standard ways of representing or decomposing matrices. These decompositions help us to understand in more detail the effect of a matrix and provide convenient ways to work with and manipulate matrices.

Matrix decompositions

The word decomposition means breaking something down into smaller parts. In this case, a matrix decomposition means breaking down a matrix into a sum of simpler matrices. By simpler matrices, we mean matrices whose properties are more convenient or efficient to work with. So, while a decomposition of a matrix still just gives us the same matrix, working with the component parts of the decomposition allows us to prove things more easily mathematically, such as derive a new algorithm, or to implement a calculation more efficiently in code.

We shall learn about two of the most important matrix decompositions in data science: the eigen-decomposition and the SVD. We won't try to prove the decompositions – that is beyond the scope of this book. Instead, we shall state the decompositions and then show you their resulting properties and how they are useful.

Eigen-decompositions

We start with the eigen-decomposition of a square matrix. As this suggests, eigen-decompositions are limited to square matrices, that is, matrices that have the same number of rows as columns. Despite this limitation, eigen-decompositions are extremely useful because they allow us to understand the effect of a square matrix in terms of a set of simpler transformations that are independent of each other.

To explain the eigen-decomposition of a square matrix, we must first explain what the **eigenvectors** and **eigenvalues** of a matrix are.

Eigenvector and eigenvalues

If we have a square $N \times N$ matrix $\underline{M}$, then an eigenvector of $\underline{M}$ is a vector $\underline{v}$, which satisfies the following equation:

$$\underline{M}\,\underline{v} = \lambda \underline{v}$$

Eq. 35

The number λ is called an eigenvalue and is the eigenvalue associated with the eigenvector $\underline{v}$. An $N \times N$ matrix will have N eigenvectors, that is, N solutions to Eq. 35, although it is possible that multiple of these eigenvectors will have the same eigenvalue, so there are N or less distinct eigenvalues. By convention, eigenvectors are of unit length, that is, $\underline{v}^\top \underline{v} = 1$, since the length of any eigenvector that isn't of unit length can just be absorbed into the eigenvalue.

So, what does Eq. 35 tell us? It tells us that the effect of $\underline{M}$ on $\underline{v}$ is to leave it untouched, except for the scaling by the constant λ. For many matrices $\underline{M}$, the eigenvectors form an orthonormal basis. We will assume that is the case from now on – see point 1 in the *Notes and further reading* section at the end of this chapter. So, now we can express any vector $\underline{a}$ in terms of the eigenvectors of $\underline{M}$, and since we understand the effect of $\underline{M}$ on each of these eigenvectors, understanding the effect of $\underline{M}$ on $\underline{a}$ is easy. Let's make that explicit. We can represent any vector $\underline{a}$ in terms of the eigenvectors $\underline{v}_1, \underline{v}_2, \ldots, \underline{v}_N$ as follows:

$$\underline{a} = \alpha_1 \underline{v}_1 + \alpha_2 \underline{v}_2 + \ldots + \alpha_N \underline{v}_N$$

Eq. 36

Since the eigenvectors $\underline{v}_1, \underline{v}_2, \ldots, \underline{v}_N$ are orthogonal to each other and of unit length, the coefficients α_i in Eq. 36 are easy to calculate and are given by the formula $\alpha_i = \underline{a}^\top \underline{v}_i$, as we showed in Eq. 34. The effect of $\underline{M}$ on $\underline{a}$ is then easily calculated as follows:

$$\underline{M}\,\underline{a} = \underline{M}\sum_{i=1}^{N}\alpha_i \underline{v}_i = \sum_{i=1}^{N}\lambda_i \alpha_i \underline{v}_i$$

Eq. 37

So, the effect of $\underline{M}$ is simply to stretch or shrink along each of the eigenvectors. Whether we stretch or shrink along a particular eigenvector direction depends upon whether the magnitude of the corresponding eigenvalue λ_i is bigger or smaller than 1. We can think of the eigenvectors of $\underline{M}$ as giving us a set of directions for an orthogonal coordinate system – the N-dimensional equivalent of the (x, y) plane. The values $(\alpha_1, \alpha_2, \ldots, \alpha_N)$ are just the coordinates of the point $\underline{a}$ in this coordinate system, and the right-hand side of Eq. 37 tells us what happens to those coordinates when we apply the transformation $\underline{M}$ to the point $\underline{a}$. Some of the coordinates increase in magnitude, some decrease in magnitude, and some get flipped if the corresponding eigenvalue is negative.

Eigen-decomposition of a square matrix

If we have an $N \times N$ matrix $\underline{M}$ and it has N linearly independent eigenvectors, meaning we can't write any eigenvector as a linear combination of the other eigenvectors, then we can write $\underline{M}$ as follows:

$$\underline{M} = \underline{P}\underline{\Delta}\underline{P}^{-1}$$

Eq. 38

This is the **eigen-decomposition** of $\underline{M}$. Here, $\underline{P}$ is an $N \times N$ matrix whose columns are the eigenvectors of $\underline{M}$. The matrix $\underline{\Delta}$ is an $N \times N$ diagonal matrix; the off-diagonal elements are all 0 and the ith diagonal element of $\underline{\Delta}$ is the eigenvalue λ_i.

Within data science, a lot of the matrices we deal with are derived from data, and so will be real-valued. This can make things simpler, particularly if our square matrix is also **symmetric**. A symmetric square matrix $\underline{M}$ is one whose matrix elements satisfy $M_{ij} = M_{ji}$, and so $\underline{M}$ is equal to its own transpose, that is, $\underline{M} = \underline{M}^{\mathsf{T}}$.

For a real symmetric matrix $\underline{M}$, the inverse matrix $\underline{P}^{-1}$ is given by the transpose of $\underline{P}$, that is, $\underline{P}^{-1} = \underline{P}^{\mathsf{T}}$. Firstly, this means that $\underline{P}\underline{P}^{\mathsf{T}} = \underline{P}^{\mathsf{T}}\underline{P} = \underline{I}_N$ and $\underline{P}$ is termed a **unitary matrix** because of this property. Secondly, it means we can write the eigen-decomposition as follows:

$$\underline{M} = \underline{P}\underline{\Delta}\underline{P}^{\mathsf{T}}$$

Eq. 39

If we write this out longhand in terms of the eigenvectors $\underline{v}_1, \underline{v}_2, \ldots, \underline{v}_N$, we have the following:

$$\underline{M} = \sum_{i=1}^{N} \lambda_i \underline{v}_i \underline{v}_i^{\mathsf{T}}$$

Eq. 40

You'll recall that the outer product $\underline{v}_i\underline{v}_i^{\mathsf{T}}$ gives a matrix, so the eigen-decomposition in Eq. 40 tells us we can write $\underline{M}$ as a sum of simpler matrices, as follows:

$$\underline{M} = \sum_{i=1}^{N} \lambda_i \underline{M}_i \qquad \underline{M}_i = \underline{v}_i \underline{v}_i^{\mathsf{T}}$$

Eq. 41

Since each of those matrices $\underline{M}_i$ contains just the transformation along a single eigenvector, it represents a very simple transformation. Secondly, since all the eigenvectors are orthogonal to each other, the effect of $\underline{M}_i$ is independent of the effect of $\underline{M}_j$ when $j \neq i$. When applying $\underline{M}_i$ to a vector $\underline{a}$, it only affects the bit of $\underline{a}$ that lies along eigenvector $\underline{v}_i$. This is the main benefit of the eigen-decomposition of the matrix. We have used the eigen-decomposition of $\underline{M}$ to understand its effect by breaking down $\underline{M}$ into a set of much simpler transformations that are easier to understand.

The sum in Eq. 40 also hints at another use of the eigen-decomposition. We can assume, without loss of generality, that the eigenvalues λ_i are ordered in magnitude, so that $|\lambda_1| \geq |\lambda_2| \geq ... \geq |\lambda_N|$. An eigenvalue with a small magnitude will contribute very little to the sum in Eq. 40. This suggests we can approximate $\underline{M}$ by truncating the summation in Eq. 40 to include only the largest-magnitude eigenvalues. Let's say, for example, the first three eigenvalues were much bigger in magnitude than the others, then we could approximate the following:

$$\underline{M} \simeq \lambda_1 \underline{v}_1 \underline{v}_1^\mathsf{T} + \lambda_2 \underline{v}_2 \underline{v}_2^\mathsf{T} + \lambda_3 \underline{v}_3 \underline{v}_3^\mathsf{T}$$

Eq. 42

We shall see an example of this use of eigen-decompositions when we introduce PCA later in this chapter.

Our last point on the eigen-decomposition of a square matrix concerns the inverse matrix. Consider Eq. 38. Now let's define another matrix, $\underline{K}$, constructed as $\underline{K} = \underline{P} \underline{\Delta}^{-1} \underline{P}^{-1}$. If we multiply $\underline{M}$ by $\underline{K}$, we get the following:

$$\underline{K}\underline{M} = \underline{P}\underline{\Delta}^{-1}\underline{P}^{-1}\underline{P}\underline{\Delta}\underline{P}^{-1} = \underline{P}\underline{\Delta}^{-1}\underline{\Delta}\underline{P}^{-1} = \underline{P}\underline{P}^{-1} = \underline{I}_N$$

Eq. 43

A similar calculation shows that $\underline{M}\underline{K} = \underline{I}_N$. These two results tell us that $\underline{K}$ is in fact the inverse matrix of $\underline{M}$ (and vice versa). Now we already have the matrices $\underline{P}, \underline{\Delta}$, and $\underline{P}^{-1}$ from the eigen-decomposition of $\underline{M}$. So, to calculate the inverse of $\underline{M}$, all we need to do is calculate the inverse of the diagonal matrix $\underline{\Delta}$. Calculating the inverse of a diagonal matrix is easy – we just take the ordinary reciprocal of the diagonal matrix elements. So, $\underline{\Delta}^{-1}$ is also diagonal and its ith diagonal element is $1/\lambda_i$. So, the inverse of $\underline{M}$ can be written as follows:

$$\underline{M}^{-1} = \underline{P} \begin{pmatrix} \lambda_1^{-1} & \cdots & 0 \\ \vdots & \ddots & \vdots \\ 0 & \cdots & \lambda_N^{-1} \end{pmatrix} \underline{P}^{-1}$$

Eq. 44

The first thing to notice is that we can only do the calculation in Eq. 44 if we can take the reciprocal of each of the eigenvalues λ_i. If any of the eigenvalues are 0, then we can't perform this calculation and the inverse of $\underline{M}$ does not exist. This is what we meant when we introduced the definition of the inverse matrix and said that some square matrices do not have an inverse.

The second thing to notice is when we have a real symmetric matrix. If $\underline{M}$ is real and symmetric, then, as we have said, its eigen-decomposition takes the simpler form $\underline{M} = \underline{P}\underline{\Delta}\underline{P}^\mathsf{T}$. As you might expect, its inverse matrix is then $\underline{M}^{-1} = \underline{P}\underline{\Delta}^{-1}\underline{P}^\mathsf{T}$, provided, of course, that none of the eigenvalues λ_i are 0. This means that we can write $\underline{M}^{-1}$ out in longhand, as a sum of simpler matrices, just as we did in Eq. 40, and we get the following:

$$\underline{M}^{-1} = \sum_{i=1}^{N} \lambda_i^{-1} \underline{v}_i \underline{v}_i^\mathsf{T}$$

Eq. 45

So, just as Eq. 40 helped us understand what the transformation $\underline{M}$ represented, Eq. 45 helps us understand what $\underline{M}^{-1}$ represents.

Let's look at how to calculate an eigen-decomposition of a square matrix with a code example.

Eigen-decomposition code example

Once again, we'll use the in-built functions in the NumPy package to do this. All the following code examples (and additional ones) can be found in the Code_Examples_Chap3.ipynb Jupyter notebook in the GitHub repository.

Eigen-decomposition of a square array can be done using the numpy.linalg.eig function. However, since in this code example we'll focus on calculating the eigen-decomposition of a real symmetric matrix, we can use the more specific numpy.linalg.eigh function:

```
import numpy as np
# Create a 4x4 symmetric square matrix
A = np.array( [[1, 2, 3, 4],
               [2, 1, 2, 1],
               [3, 2, 3, 2],
               [4, 1, 2, 2]])


# Calculate the eigen-decomposition
eigvals, eigvecs = np.linalg.eigh(A)
```

eigvals is a NumPy array that holds the eigenvalues. The eigenvalues are shown here:

```
array([-2.84382794, -0.22727708, 1.02635, 9.04475502])
```

We can also check that the eigenvectors are orthogonal to each other:

```
# Check that the eigenvectors are orthogonal to each other
# and are of unit length. We can do this by calculating the
# inner product for each pair of eigenvectors
for i in range(A.shape[0]):
    for j in range(i,A.shape[0]):
        print("i=",i, "j=",j, "Inner product = ",
              np.inner(eigvecs[:, i], eigvecs[:, j]))
```

This gives the following output:

```
i= 0 j= 0 Inner product =  1.0
i= 0 j= 1 Inner product =  -2.7755575615628914e-17
i= 0 j= 2 Inner product =  -5.551115123125783e-17
i= 0 j= 3 Inner product =  -1.6653345369377348e-16
i= 1 j= 1 Inner product =  1.0
```

```
i= 1 j= 2 Inner product =  5.551115123125783e-17
i= 1 j= 3 Inner product =  0.0
i= 2 j= 2 Inner product =  0.9999999999999993
i= 2 j= 3 Inner product =  -8.326672684688674e-17
i= 3 j= 3 Inner product =  0.9999999999999999
```

That code example concludes our explanation of the eigen-decomposition of a square matrix, so we'll now move on to another important matrix decomposition technique.

Singular value decomposition

The eigen-decomposition works for square matrices. What happens if our matrix is not square? Are there any matrix decompositions for non-square matrices that break the matrix down into orthonormal vectors? The answer is yes. Any matrix, whether square or otherwise, has an **singular value decomposition (SVD)**.

We will initially restrict ourselves to real matrices. Again, in a data science setting, this is not too much of a restriction as most of the matrices you will encounter will be data-related and hence real. We will return to the more general case of complex matrices later. The SVD of a real $N \times M$ matrix $\underline{\underline{A}}$ that has $N > M$ (more rows than columns) is given as follows:

$$\underline{\underline{A}} = \underline{\underline{U}}\,\underline{\underline{D}}\,\underline{\underline{V}}^{\mathsf{T}}$$

Eq. 46

This decomposition is shown schematically in *Figure 3.5*.

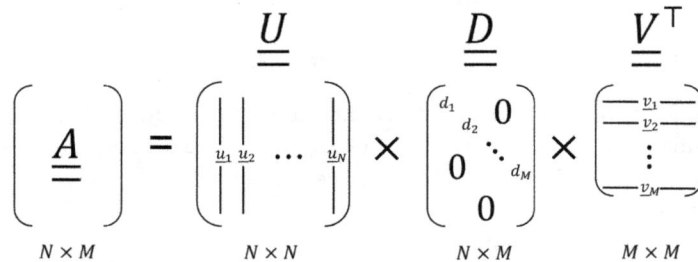

Figure 3.5: The SVD of a real-valued matrix

Here, $\underline{\underline{U}}$ is an $N \times N$ matrix whose columns are a set of N orthonormal vectors, $\underline{u}_1, \underline{u}_2, \dots, \underline{u}_N$, so that $\underline{u}_i^{\mathsf{T}}\underline{u}_i = 1$ for all i and $\underline{u}_i^{\mathsf{T}}\underline{u}_j = 0$ for $i \neq j$. In fact, the matrix $\underline{\underline{U}}$ is unitary, so $\underline{\underline{U}}\,\underline{\underline{U}}^{\mathsf{T}} = \underline{\underline{U}}^{\mathsf{T}}\underline{\underline{U}} = \underline{\underline{I}}_N$. The vectors $\underline{u}_1, \underline{u}_2, \dots, \underline{u}_N$ are known as the **left singular vectors** of $\underline{\underline{A}}$.

Likewise, the matrix $\underline{\underline{V}}$ is an $M \times M$ matrix whose columns are a set of M orthonormal vectors $\underline{v}_1, \underline{v}_2, \dots, \underline{v}_M$, so that $\underline{v}_i^{\mathsf{T}}\underline{v}_i = 1$ for all i, and $\underline{v}_i^{\mathsf{T}}\underline{v}_j = 0$ if $i \neq j$. $\underline{\underline{V}}$ is also a unitary matrix, so $\underline{\underline{V}}\,\underline{\underline{V}}^{\mathsf{T}} = \underline{\underline{V}}^{\mathsf{T}}\underline{\underline{V}} = \underline{\underline{I}}_M$. The vectors $\underline{v}_1, \underline{v}_2, \dots, \underline{v}_M$ are known as the **right singular vectors** of $\underline{\underline{A}}$.

The matrix $\underline{D}$ is a diagonal matrix – all the off-diagonal elements are 0. Matrix $\underline{D}$ has M values on its leading diagonal. The ith diagonal element of $\underline{D}$ is d_i and is called the ith **singular value**.

Okay, the preceding explanation is the textbook definition. Let's unpack the SVD a bit further to understand what the singular vectors $\underline{u}_i, \underline{v}_i$ and singular value d_i represent. We'll look at the $N \times N$ symmetric matrix $\underline{A}\underline{A}^\mathsf{T}$. Using the SVD of $\underline{A}$, we can write this as follows:

$$\underline{A}\underline{A}^\mathsf{T} = \underline{U}\underline{D}\underline{V}^\mathsf{T}(\underline{U}\underline{D}\underline{V}^\mathsf{T})^\mathsf{T} = \underline{U}\underline{D}\underline{V}^\mathsf{T}\underline{V}\underline{D}^\mathsf{T}\underline{U}^\mathsf{T}$$

Eq. 47

Here, we have used the fact that the transpose of the product of matrices is the same as taking the product of the transposes, but with the order reversed, that is, $(\underline{X}_1\underline{X}_2)^\mathsf{T} = \underline{X}_2^\mathsf{T}\underline{X}_1^\mathsf{T}$. Now, if you also recall that since the columns of $\underline{V}$ are a set of orthonormal vectors, then $\underline{V}^\mathsf{T}\underline{V} = \underline{I}_M$, and so we have the following:

$$\underline{A}\underline{A}^\mathsf{T} = \underline{U}\underline{D}\underline{D}^\mathsf{T}\underline{U}^\mathsf{T}$$

Eq. 48

Since $\underline{D}$ is a diagonal matrix, the matrix $\underline{D}\underline{D}^\mathsf{T}$ is an $N \times N$ diagonal matrix, and so by comparing it to Eq. 39, we can recognize Eq. 48 as the eigen-decomposition of the square matrix $\underline{A}\underline{A}^\mathsf{T}$. So, the left singular vectors, $\underline{u}_1, \underline{u}_2, \ldots, \underline{u}_M$, are also eigenvectors of the matrix $\underline{A}\underline{A}^\mathsf{T}$.

A similar calculation, where we consider the $M \times M$ symmetric matrix $\underline{A}^\mathsf{T}\underline{A}$, shows the following:

$$\underline{A}^\mathsf{T}\underline{A} = \underline{V}\underline{D}^\mathsf{T}\underline{D}\underline{V}^\mathsf{T}$$

Eq. 49

With $\underline{D}^\mathsf{T}\underline{D}$ being an $M \times M$ diagonal matrix, we recognize Eq. 49 as an eigen-decomposition and so the right singular vectors, $\underline{v}_1, \underline{v}_2, \ldots, \underline{v}_M$, are also eigenvectors of $\underline{A}^\mathsf{T}\underline{A}$. So, in a similar way to what we did with the eigen-decomposition of a matrix, it helps to write out the SVD in a longhand form:

$$\underline{A} = \sum_{i=1}^{M} d_i \underline{u}_i \underline{v}_i^\mathsf{T}$$

Eq. 50

The first thing to notice is that because we only have M singular values, the sum in Eq. 50 is restricted to just M terms and only the first M left singular vectors contribute. This means that sometimes you will see the SVD of a matrix defined in a more compact form, with $\underline{U}$ being an $N \times M$ matrix whose columns are $\underline{u}_1, \underline{u}_2, \ldots \underline{u}_M$, and $\underline{D}$ being an $M \times M$ diagonal matrix whose diagonal elements are $d_1, d_2, \ldots, d_M$. This slightly more compact form of the SVD is shown schematically in *Figure 3.6*.

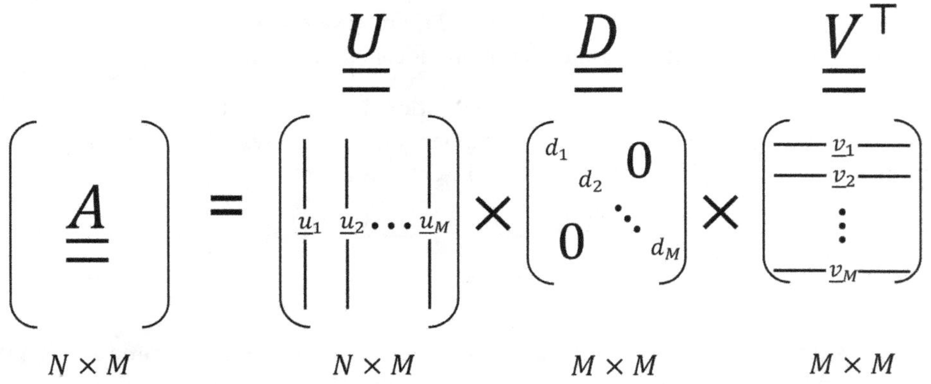

$$
\left(\underline{\underline{A}} \right) = \left(\underline{u}_1\ \underline{u}_2 \cdots \underline{u}_M \right) \times \begin{pmatrix} d_1 & & & 0 \\ & d_2 & & \\ & & \ddots & \\ 0 & & & d_M \end{pmatrix} \times \begin{pmatrix} - v_1 - \\ - v_2 - \\ \vdots \\ - \underline{v}_M - \end{pmatrix}
$$

$$
N \times M \qquad\qquad N \times M \qquad\qquad M \times M \qquad\qquad M \times M
$$

Figure 3.6: The compact form of the SVD of a real-valued matrix

Secondly, we can see that Eq. 50 allows us to write $\underline{\underline{A}}$ as a sum of independently acting matrices,

$$
\underline{\underline{A}} = \sum_{i=1}^{M} d_i \underline{\underline{A}}_i \quad \underline{\underline{A}}_i = \underline{u}_i \underline{v}_i^{\mathsf{T}}
$$

Eq. 51

The individual matrix $\underline{\underline{A}}_i$ contains just the transformation from the ith left and right singular vector pair $\underline{u}_i, \underline{v}_i$. Applying $\underline{\underline{A}}_i$ to a vector $\underline{a}$ is easily calculated and is given by the following:

$$
\underline{\underline{A}}_i \underline{a} = \underline{u}_i \underline{v}_i^{\mathsf{T}} \underline{a} = \underline{u}_i \times \underline{v}_i^{\mathsf{T}} \underline{a}
$$

Eq. 52

Since the vectors $\underline{u}_i$ and $\underline{v}_i$ are unit length, Eq. 52 tells us that the effect of $\underline{\underline{A}}_i$ on $\underline{a}$ is to project $\underline{a}$ onto $\underline{v}_i$ and use the result of that projection to output a scaled version of the left singular vector $\underline{u}_i$. Also, since each left singular vector is orthogonal to every other left singular vector, and each right singular vector is orthogonal to every other right singular vector, this means each matrix $\underline{\underline{A}}_i$ acts in an orthogonal fashion, that is, independently of the others. Just as the eigen-decomposition helps us to understand a square matrix, the SVD of $\underline{\underline{A}}$ allows us to understand more easily the effect of the transformation that $\underline{\underline{A}}$ represents.

As with the eigen-decomposition, the larger the magnitude of the singular value d_i, the more important the corresponding pair of left and right singular vectors $\underline{u}_i$, $\underline{v}_i$ in the sum in Eq. 50. The SVD of $\underline{\underline{A}}$ shows us how we can approximate $\underline{\underline{A}}$ by truncating the sum in Eq. 50 to just the largest-magnitude singular values.

The SVD of a complex matrix

So far, we have restricted our discussion of the SVD to real matrices, since those are most of the matrices we encounter as data scientists. Let's now return to a general matrix, that is, one that potentially contains matrix elements that are complex numbers. For such matrices, we still have an SVD. For a general matrix $\underline{A}$, its SVD is as follows:

$$\underline{A} = \underline{U}\underline{D}\underline{V}^\dagger$$

Eq. 53

In this more general form, $\underline{U}$ and $\underline{V}$ are complex unitary matrices. Also, the other new aspect here is that we have $\underline{V}^\dagger$ instead of $\underline{V}^\mathsf{T}$. What is this new symbol? What does $\underline{V}^\dagger$ represent? It is called the **Hermitian conjugate** of the matrix $\underline{V}$. The Hermitian conjugate is also the conjugate transpose of $\underline{V}$, obtained by taking the transpose of $\underline{V}$ and then taking the complex conjugate of each matrix element. The order of those operations can be reversed, that is, take the complex conjugate of each element of $\underline{V}$ and then the transpose – the combined result is the same.

We'll finish our explanation of the SVD with a code example. We'll look at how to calculate the SVD of a real-valued matrix.

SVD code example

For our code example, we'll again make use of the in-built functions in the NumPy package to do this. The following code example (and additional ones) can be found in the `Code_Examples_Chap3.ipynb` Jupyter notebook in the GitHub repository.

We'll use the `numpy.linalg.svd` function to perform the SVD. This function has a `full_matrices` Boolean argument, which, when set to `False`, computes the compact form of the SVD shown in *Figure 3.6*, with the smaller form of the matrix $\underline{U}$. When `full_matrices` is set to `True`, we get the full form of the SVD shown in *Figure 3.5*. We'll calculate both forms of the SVD for a 5 × 3 real matrix $\underline{A}$. We start by creating the matrix $\underline{A}$:

```
import numpy as np
# Create a 5x3 matrix
A = np.array([[1.0, 2.0, 3.0],
              [-1.0, 0.0, 1.0],
              [3.0, 3.0, 2.0],
              [2.0, 4.0, 7.0],
              [1.0, -0.5, -2.0]])
```

Now, we can calculate the full SVD of the matrix. Calling the `numpy.linalg.svd` function returns a triple of arrays corresponding to the $\underline{U}$, $\underline{D}$ and $\underline{V}^\mathsf{T}$ matrices. This is shown here:

```
# Calculate SVD. In this instance the matrix
# u will be a square matrix with the same number of rows
# as the input matrix A has
u, d, v_transpose = np.linalg.svd(A, full_matrices=True)
```

We can also calculate the compact form of the SVD by simply changing the value of the `full_matrices` argument. This is shown here:

```
# Calculate the compact form of the SVD. In this instance
# the matrix u_compact will have the same shape as the input
# matrix A
u_compact, d_compact, v_transpose_compact = np.linalg.svd(
    A, full_matrices=False)
```

Let's check that the two different SVD calculations do indeed have different shapes for the matrix $\underline{U}$. We do this as follows:

```
# Let's check that the matrix U has different shapes in
# compact and non-compact forms of the SVD
print(u.shape, u_compact.shape)
```

This gives the following output:

```
(5, 5) (5, 3)
```

With that short code example, we'll end our explanation of the SVD. It is time to recap what we have learned about matrix decompositions.

What we learned

In this section, we have learned the following:

- About the eigenvectors and eigenvalues of a square matrix

- How to calculate the eigen-decomposition of a square matrix

- How the eigen-decomposition of a square matrix provides us with a simpler way to understand the transformation represented by real symmetric square matrices

- How to calculate the SVD of a matrix

- How the SVD of a matrix gives us a simpler way to understand the transformation represented by any matrix

Having learned about the two most important decomposition methods that can be applied to matrices, we will learn about some key metrics or properties that are used to summarize square matrices and how those metrics are calculated in terms of the eigenvalues of the matrix.

Matrix properties

Having defined the eigenvectors and eigenvalues of a matrix, we can now introduce some additional matrix properties that can be useful later. Specifically, we will introduce the **trace** and **determinant** of a square matrix. These two quantities quantify some useful aspects of a matrix. Since we are simply giving their definitions here, this will be a relatively short section.

Trace

The trace of a square $N \times N$ matrix $\underline{A}$ is simply the sum of its diagonal elements. If $\underline{A}$ has matrix elements A_{ij}, $i,j = 1,\ldots,N$, then the trace of $\underline{A}$ is calculated as follows:

$$\text{trace}(\underline{A}) = \text{tr}(\underline{A}) = \sum_{i=1}^{N} a_{ii}$$

Eq. 54

Note the abbreviated notation, tr, that is commonly used when denoting the trace of a matrix. Now it turns out that the trace of a square matrix is also equal to the sum of its eigenvalues (we state this without proof), so that for a square matrix $\underline{A}$, which has eigenvalues $\lambda_1, \lambda_2, \ldots, \lambda_N$, we can calculate its trace using the following formula:

$$\text{tr}(\underline{A}) = \sum_{i=1}^{N} \lambda_i$$

Eq. 55

Often, when working with data science algorithms involving square matrices, we need to calculate the sum of the eigenvalues, and so we use $\text{tr}(\underline{A})$ as a convenient shorthand for this quantity. So, now you will know what $\text{tr}(\underline{A})$ means when you see it in these algorithms.

Determinant

The determinant of an $N \times N$ square matrix $\underline{A}$ can also be expressed in terms of its eigenvalues. While the trace is defined as the sum of the eigenvalues, the determinant is defined as the product of the eigenvalues. This means we calculate the determinant of $\underline{A}$ as follows:

$$determinant(\underline{A}) = det(\underline{A}) = \prod_{i=1}^{N} \lambda_i$$

Eq. 56

Again, note the abbreviated notation, det, which is commonly used when denoting the determinant of a matrix.

There are alternative ways of calculating the determinant of a square matrix in terms of the individual matrix elements, but the formula in Eq. 56 is probably the most succinct.

That's it for this section. We said it would be short. Let's briefly recap what we have learned before moving on to the next section.

What we learned

In this section, we have learned the following:

- How to calculate the trace of a square matrix in terms of its eigenvalues
- How to calculate the determinant of a square in terms of its eigenvalues

Having learned about the eigen-decomposition and the SVD, we will learn how to apply them to **dimensionality reduction** by using them to identify low-dimensional structures within datasets.

Matrix factorization and dimensionality reduction

In this section, we will learn techniques for handling data with many features or variables. We will make extensive use of the matrix decomposition techniques we learned about in the previous sections. An example of such data is shown in *Table 3.1*.

Data point	X1	X2	X3	X4
1	3.486	3.977	0.242	0.626
2	2.024	2.807	1.651	1.566
3	-0.653	0.250	2.912	2.009
4	3.036	3.993	1.982	1.010
5	0.001	0.662	0.955	-0.323
6	1.518	2.468	3.042	1.971
7	1.104	1.509	4.669	-0.377

Table 3.1: Example rows from a dataset to illustrate dimensionality reduction

Each row of the table represents a four-dimensional datapoint or observation, consisting of four feature values. There are N datapoints in total, so the table has N rows. The columns of the table are the features. The table looks very similar to a matrix, so much so that we refer to it as a data matrix and use the matrix symbol $\underline{X}$ to represent the data in the table. In general, the matrix $\underline{X}$ is of size $N \times d$ if we have N data points with d features. We use x_{kj} to denote its (k, j) matrix element, and this corresponds to the jth feature value of the kth datapoint.

Typically, the first thing we want to do with any dataset, such as the one in the previous table, is visualize it and simplify it. When we have many features, this can be difficult. One way to do this would be to just drop some features, but which ones? Is there a better way in which we can reduce the number of features we are working with? The answer is yes. This is the goal of dimensionality reduction.

Dimensionality reduction

Dimensionality reduction does what it says. It takes data that is d-dimensional and converts it into data that is of lower dimensionality. Why would we want to do that? Well, many datasets are already effectively of lower dimensionality than the number of features or variables they contain. Take the example dataset shown in *Figure 3.7*.

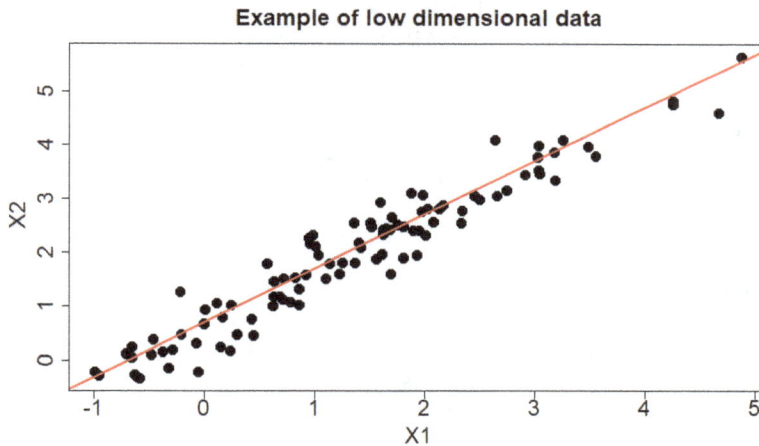

Figure 3.7: An example dataset with low-dimensional structure

The datapoints clearly tend to be spread out along the red line. There is some scatter in the data perpendicular to the red line, but it is considerably less than the variation along the red line. To all intents and purposes, this data is one-dimensional. Clearly, it would make a lot more sense to construct a new feature or variable that could measure how far along the red line each datapoint would be. This new feature would have two benefits:

- Visualizing the data would be easier because we would only have to plot data on a single axis (corresponding to the red line)

- We would only have to work with one feature instead of two, and analyzing data, extracting insight, and building predictive models is always a lot easier the fewer features you have to deal with

You may believe that by having only one feature we have lost some information that was in the original features, X1 and X2. Yes, we have lost information, but very little. The value of X1 effectively predicts what the value of X2 is. The value of X2 provides us with very little new information. Ultimately, this is because X1 and X2 are strongly correlated. There is structure in the data.

How do we go about finding this structure? When we have a small number of original features, it can be straightforward, as the visual example in *Figure 3.7* demonstrates. We can just plot the data. When we have many features, this approach does not work so well. We need to use algorithms that can automatically identify this correlated structure.

In general, we want to take our original d-dimensional data and construct $q < d$ new features, with $q \geq 1$. In some cases, those q new features or dimensions will represent a genuine low-dimensional structure present the data – a q-dimensional generalization of the one-dimensional red line in *Figure 3.7*. In some other cases, there may not be a clear-cut lower-dimensional structure present in the original data, but we may still want to perform dimensionality reduction to simply reduce the number of features we use in our predictive models, or, more typically, to extract a small enough number of new features that we are able to plot. This last use case of dimensionality reduction, that of enabling visualization of the original high-dimensional data, is one of the most common data science use cases of dimensionality reduction.

We'll start with one of the most frequently used dimensionality reduction algorithms – **principal component analysis (PCA)**.

Principal component analysis

If we are going to reduce the dimension of our data, we are going to lose some information. We must decide what the information is in the data that we want to keep. What do we consider the most important aspect of the data?

One potential way forward is to say we want to choose our new coordinates so that we retain the variation in the original data as much as possible. This assumes that the directions of the greatest variation we see in the original data correspond to some sort of signal we are interested in studying or uncovering. This is an assumption. It may not be true. With that proviso and warning, let's proceed.

Our goal, then, is to find new directions, such that when we project our original data onto these new directions, we retain as much of the total variance in the original data as possible. The first question is how do we measure the total variation in the original data? Using the sample variance calculation we introduced in *Chapter 2*, of course! Since the sample variance is calculated as the variation around the sample mean, we need to calculate the mean of the data. Let's do that.

The sample mean of the ith feature is as follows:

$$m_i = \frac{1}{N}\sum_{k=1}^{N}x_{ki}$$

Eq. 57

We have used the symbol m_i to represent this sample mean. Calculating the sample mean for each feature gives us a vector, $\underline{m} = (m_1, m_2, \ldots, m_d)$, which we call the mean vector. We can center each feature x_i by subtracting its sample mean. This gives us a new feature, Δx_i, whose values in our dataset are given by $\Delta x_{ki} = x_{ki} - m_i$. Because of this centering, the sample mean of Δx_i is 0. To see this, we'll calculate it as follows:

$$\text{Sample mean of } \Delta x_i = \frac{1}{N}\sum_{k=1}^{N}\Delta x_{ki} = \frac{1}{N}\sum_{k=1}^{N}(x_{ki} - m_i) = \frac{1}{N}\sum_{k=1}^{N}x_{ki} - \frac{1}{N}\sum_{k=1}^{N}m_i = m_i - \frac{N}{N}m_i = 0$$

Eq. 58

Just as we formed a matrix, $\underline{X}$, from all our feature values, x_{ki}, we can form a new matrix from these centered feature values. We'll denote the new matrix $\underline{\Delta X}$, and its matrix elements are just the centered feature values Δx_{ki}.

Now, the variation in the original data is just the variance of these centered features. We'll take a slight digression to introduce the concept of **covariance**. As the name suggests, it measures the amount of covariation between two different variables, while variance measures the amount of variation between a variable and itself. The math definition of covariance between features Δx_i and Δx_j is intuitive and a natural extension of the formula for sample variance, namely the following:

$$\text{Cov}(\Delta x_i, \Delta x_j) = \widehat{C}_{ij} = \frac{1}{N-1}\sum_{k=1}^{N}\Delta x_{ki}\Delta x_{kj}$$

Eq. 59

The variance of the ith-centered feature Δx_i is given by $\widehat{C}_{ii}$. You may also recognize the right-hand side of Eq. 59 as being in the form of a matrix multiplication. Because of this, we define a new matrix, $\widehat{\underline{C}}$, called the **sample covariance matrix**, whose matrix elements are $\widehat{C}_{ij}$. In matrix notation, $\widehat{\underline{C}}$ is calculated as follows:

$$\widehat{\underline{C}} = \frac{1}{N-1}\underline{\Delta X}^{\mathsf{T}}\underline{\Delta X}$$

Eq. 60

The matrix $\widehat{\underline{C}}$ is a square symmetric $d \times d$ matrix. The variance of the ith-centered feature is just the ith diagonal element of $\widehat{\underline{C}}$, and so the total variance in the dataset is given by the following:

$$\sum_{i=1}^{N}\widehat{C}_{ii} = \text{tr}(\widehat{\underline{C}})$$

Eq. 61

Since $\widehat{\underline{C}}$ is a square matrix, its trace is given by the sum of its eigenvalues. This starts to give us a hint that determining optimal new coordinate directions may be related to the eigen-decomposition of $\widehat{\underline{C}}$, and indeed it is. The optimal new directions turn out to be the eigenvectors of $\widehat{\underline{C}}$.

To see this and understand it in more detail, let's consider the first new direction, which we denote by the vector $\underline{a}$. Since we only need to determine the direction of this new coordinate, we restrict $\underline{a}$ to being of unit length, so that $\underline{a}^\top \underline{a} = 1$. The projection of the kth-centered observation onto this new direction $\underline{a}$ is given by the following:

$$\sum_{i=1}^{d} \Delta x_{ki} a_i$$

Eq. 62

The projection in Eq. 62 represents the value of the kth-centered observation in the new coordinate direction $\underline{a}$. The mean, across all the observations, of these new coordinate values is easily calculated as follows:

$$\frac{1}{N} \sum_{k=1}^{N} \sum_{i=1}^{d} \Delta x_{ki} a_i = \frac{1}{N} \sum_{i=1}^{d} a_i \sum_{k=1}^{N} \Delta x_{ki} = \frac{1}{N} \sum_{i=1}^{d} a_i \times 0 = 0$$

Eq. 63

The result on the right-hand side in Eq. 63 is because the features Δx_{ki} have a sample mean of 0. Consequently, the variance of these new coordinate values is just the average of their squares, that is, it is calculated as follows:

$$\frac{1}{N-1} \sum_{k=1}^{N} \left[\left(\sum_{i=1}^{d} \Delta x_{ki} a_i \right) \times \left(\sum_{j=1}^{d} \Delta x_{kj} a_j \right) \right]$$

Eq. 64

Re-arranging the order of the summations, we can write Eq. 64 as follows:

$$\frac{1}{N-1} \sum_{i,j} a_i a_j \sum_{k=1}^{N} \Delta x_{ki} \Delta x_{kj}$$

Eq. 65

In matrix notation, we write Eq. 65 as follows:

$$\underline{a}^\top \hat{\underline{C}} \, \underline{a}$$

Eq. 66

Now, in Eq. 66, we have calculated the variance of the original data along our new direction, $\underline{a}$, but how should we choose $\underline{a}$? Since we want our new direction, $\underline{a}$, to retain as much of the original variance as possible, we want to choose $\underline{a}$ so that it maximizes the expression in Eq. 66. Remember, however, that we have the constraint $\underline{a}^\top \underline{a} = 1$. How do we impose the constraint whilst doing the maximization? We use a **Lagrange multiplier**. We won't explain the theory behind Lagrange multipliers here, but simply explain how they are used.

To impose the constraint, we add a term of the form $\phi \times$ Constraint, to the thing we are optimizing. So, in our case, we want to maximize:

$$\underline{a}^\top \widehat{\underline{C}} \, \underline{a} + \phi \underline{a}^\top \underline{a}$$

Eq. 67

The parameter ϕ is called the Lagrange multiplier because it multiplies the constraint term. We solve for the value of $\underline{a}$ that maximizes the preceding expression. The solution, the optimal value of $\underline{a}$, will be a function of ϕ. We then adjust the value of ϕ so that the optimal value of $\underline{a}$ satisfies the desired constraint.

Using basic calculus, we can find the value of $\underline{a}$ that maximizes the expression in Eq. 67 by taking the derivative with respect to $\underline{a}$ and setting that derivative to 0 (see *Chapter 1* for a recap of this). This means we choose $\underline{a}$, which solves the following equation:

$$\frac{\partial}{\partial \underline{a}}\left[\underline{a}^\top \widehat{\underline{C}} \, \underline{a} + \phi \, \underline{a}^\top \underline{a}\right] = 2\widehat{\underline{C}} \, \underline{a} + 2\phi \underline{a} = \underline{0}$$

Eq. 68

This gives the following equation:

$$\widehat{\underline{C}} \, \underline{a} = -\phi \underline{a}$$

Eq. 69

You'll recognize Eq. 69 as an eigenvector equation (compare it to Eq. 35). So, the optimal choice of our new direction $\underline{a}$ is one of the eigenvectors of the sample covariance matrix $\widehat{\underline{C}}$. In this case, since the eigenvectors of $\widehat{\underline{C}}$ are, by convention, of unit length, we would only need to adjust ϕ so that $-\phi$ matches the corresponding eigenvalue.

You'll naturally be asking, okay so our new coordinate direction $\underline{a}$ should correspond to an eigenvector of $\widehat{\underline{C}}$, but which eigenvector? There are d eigenvectors of $\widehat{\underline{C}}$, after all! We'll tackle this in the obvious way. We'll calculate how much variance in the data there is along each of the eigenvectors and choose the largest. We've already worked out the formula for how much variance is captured by a vector $\underline{a}$. It is given by Eq. 66. We'll now use Eq. 66 but where $\underline{a}$ is explicitly equal to an eigenvector of $\widehat{\underline{C}}$. Doing so, we have the following:

$$\underline{a}^\top \widehat{\underline{C}} \, \underline{a} = \underline{a}^\top \lambda \underline{a} = \lambda \underline{a}^\top \underline{a} = \lambda$$

Eq. 70

Here, λ is the eigenvalue corresponding to the eigenvector $\underline{a}$. In deriving Eq. 70, we have made use of the fact that $\underline{a}$ is an eigenvector of $\widehat{\underline{C}}$ and so $\widehat{\underline{C}} \, \underline{a} = \lambda \underline{a}$, and also the fact that $\underline{a}$ is of length 1, so $\underline{a}^\top \underline{a} = 1$. From Eq. 70, we can see that the variance in the original data captured by eigenvector $\underline{a}$ is simply the corresponding eigenvalue λ. So, to capture as much of the variation in the original data as possible, using this single new coordinate direction $\underline{a}$, we choose $\underline{a}$ to be the eigenvector of $\widehat{\underline{C}}$ that has the largest eigenvalue. This eigenvalue will be positive – see note 2 in the *Notes and further reading* section at the end of the chapter.

Excellent, we are making good progress. But wait a minute! What happens if we want more than one new coordinate direction? How do we work out the extra new coordinate directions?

The second new coordinate direction is chosen to be orthogonal to the first new direction. It is also chosen to, as much as possible, retain the variation in the original data that hasn't already been captured by the first new direction. As you might have guessed, the second new direction turns out to be given by the eigenvector of $\widehat{\underline{C}}$ with the second largest eigenvalue. We could prove this by continuing the preceding derivation method, but for brevity, we will not do so. What about if we want more than two new coordinate directions? We shall merely state the result, which should be apparent by now – the q new coordinate directions that retain the maximum amount of variation in the original data are the eigenvectors of $\widehat{\underline{C}}$ that correspond to the q largest eigenvalues.

The scores and loadings

The new coordinate directions are called the **principal components** because they are "main" or "principal" directions in the dataset. You'll often see this abbreviated to PC, so that PC1 means the first principal component, PC2 means the second principal component, and so on. As we have said, the principal components are given by the leading eigenvectors of the sample covariance matrix $\widehat{\underline{C}}$. But how do we plot the original data in these new coordinates? In other words, how do we reduce the dimensionality? We do this by projecting each observation onto each of the new coordinate vectors. So, if $\underline{a}_1$ is the first eigenvector of $\widehat{\underline{C}}$ and $\underline{o}_k^\top = (\Delta x_{k1}, \Delta x_{k2}, \ldots, \Delta x_{kd})$ is a row vector of the kth-centered observation, then the coordinate of this k^{th} observation along this new coordinate direction is $\underline{o}_k^\top \underline{a}_1$. Likewise, if we want the coordinate value of $\underline{o}_k$ along PC2, that is, along the second leading eigenvector, $\underline{a}_2$, then we calculate $\underline{o}_k^\top \underline{a}_2$. These projections are called the **scores** for the kth observation since we are scoring it against each of the principal components. If we are reducing the dimensionality by using $q < d$ principal components, then for each d-dimensional observation $\underline{o}_k$, we will have q scores, or in other words, a point in the new q-dimensional coordinate system.

Let's look more closely at how a principal component contributes to the score for an observation. We'll denote the ith score, $i = 1, 2, \ldots, q$, of the kth observation by the symbol s_{ki}. As we have already said, the scores are calculated as the projection of the original centered data, $\underline{o}_k$, onto the new coordinate directions, $\underline{a}_i$, so the score, s_{ki}, is calculated as follows:

$$s_{ki} = \underline{o}_k^\top \underline{a}_i = \sum_{j=1}^{d} \Delta x_{kj} a_{ij}$$

Eq. 71

Here, a_{ij} is the jth component value of the ith principal component. So, if a_{ij} is very small or 0, it means that the jth feature will not contribute much, or anything at all, to the new i^{th} coordinate values. The magnitude of a_{ij} gives us a measure of how important the jth old coordinate or feature is in determining the new i^{th} coordinate or feature. Because of this, the a_{ij} values are called the **loadings**.

The PCA score plot

Once we have run a PCA algorithm, we typically look at the score plot. For the two-dimensional data shown in *Figure 3.7*, we can have two principal components. The scores of the original data along these principal components are shown on the right-hand side of *Figure 3.8*, along with the original plot on the left-hand side. We have added the directions of the principal components to the plot of the original data on the left-hand side of *Figure 3.7*, so you can see how the score plot is related to the original data. You may be wondering why the data no longer looks so one-dimensional in the right-hand-side PCA plot. That is because the scale of the *y*-axis in the right-hand plot is much smaller than the scale of the *x*-axis. The right-hand plot doesn't look one-dimensional because we have made a square plot area for demonstration purposes only.

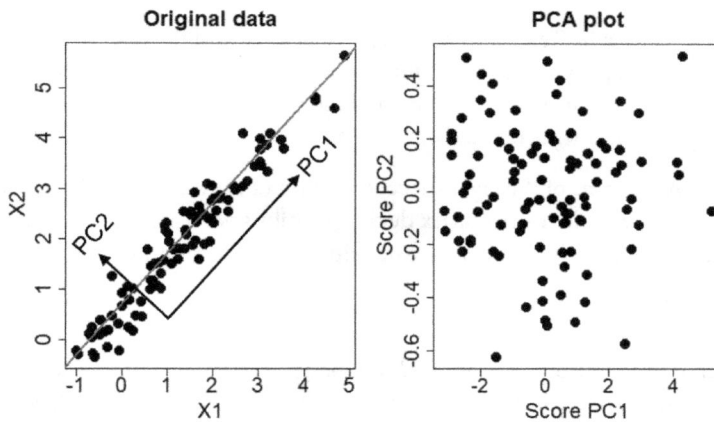

Figure 3.8: PCA plot and original data

The plot on the right-hand side also highlights the relationship between dimensionality reduction and PCA. The right-hand plot of *Figure 3.8* is displaying all the principal components that we can calculate for this data. No reduction in dimensionality has occurred. We started with the two-dimensional data on the left and we have two-dimensional data on the right. When we use all the principal components, we are just transforming (rotating and scaling) the original data, but we still effectively have the original data. It is not until we make the choice to keep only a subset of the principal components that we are performing dimensionality reduction. So, not until we choose a value $q < d$ are we performing dimensionality reduction. What value for q should we choose?

Choosing a value $q < d$ involves the loss of some information that is in the original data, but how much? Since the variance captured by each principal component is given by the corresponding eigenvalue, then the total variance captured by the $q < d$ principal components is just as follows:

$$\sum_{i=1}^{q} \lambda_i$$

Eq. 72

It is more useful to express this as a fraction of the total variance in the original data, that is, follows:

$$\text{Proportion total of variance captured} = \frac{\sum_{i=1}^{q}\lambda_i}{\sum_{i=1}^{d}\lambda_i} = \frac{\sum_{i=1}^{q}\lambda_i}{\text{tr}(\widehat{C})}$$

Eq. 73

Often, you will also see a percentage of variance captured or explained by each individual principal component. For example, the proportion of total variance explained by PC1 is as follows:

$$\text{Proportion of variance explained by PC1} = \frac{\lambda_1}{\text{tr}(\widehat{C})}$$

Eq. 74

In the two-dimensional dataset in *Figure 3.7*, PC1 accounts for 98.3% of the total variance in the original data. By only using PC1 to approximate our original data and dropping PC2, we would be losing only a very small amount of the variation in the original data. It is common to see these percentages for the individual principal components printed on the axes of the PCA score plot itself.

We have explained PCA in terms of the eigen-decomposition of the sample covariance matrix $\widehat{C}$. But eigen-decompositions weren't the only matrix decompositions we learned about. We also learned about the SVD. So, next we'll see how the SVD can help us to understand PCA.

PCA via the SVD

Our centered matrix of observations, $\underline{\Delta X}$, is an $N \times d$ matrix. It is not square, so we cannot calculate an eigen-decomposition for it. But we can calculate an SVD for it. We'll write this SVD in compact form as follows:

$$\underline{\Delta X} = \underline{U}\,\underline{W}\,\underline{V}^{\top}$$

Eq. 75

We've chosen the symbol $\underline{W}$ to represent the $d \times d$ matrix of singular values to avoid clashing with other symbols we've already used. Individual singular values we'll represent by $w_1, w_2, \ldots, w_d$. Now the sample covariance matrix that we use to do PCA is given by the following:

$$\widehat{C} = \frac{1}{N-1}\underline{\Delta X}^{\top}\underline{\Delta X}$$

Eq. 76

If we express $\widehat{C}$ in terms of the SVD of $\underline{\Delta X}$, we get the following:

$$\widehat{C} = \frac{1}{N-1}\underline{V}\,\underline{W}^{2}\,\underline{V}^{\top}$$

Eq. 77

Since $\widehat{C}$ does have an eigen-decomposition and the right-hand side of Eq. 77 is of the correct form for an eigen-decomposition, it means we can read off the quantities we need for PCA directly from the SVD of the data matrix $\underline{\underline{\Delta X}}$. Specifically, if λ_i is the ith eigenvalue of $\widehat{C}$, then $\lambda_i = w_i^2/(N-1)$. The loadings for the PCA are given by the columns of $\underline{V}$ and the score vector for the kth observation is given by the kth row of $\underline{U}\,\underline{W}$.

The link between the SVD of the data matrix $\underline{\underline{\Delta X}}$ and PCA helps highlight two important points:

- Since we can get all the information we need about the sample covariance matrix $\widehat{C}$ from the SVD of $\underline{\underline{\Delta X}}$, it raises the question of whether there are other matrices related to $\underline{\underline{\Delta X}}$ that can give us the information to perform a PCA of $\underline{\underline{\Delta X}}$. The answer is yes. Consider the matrix $\underline{G}$, defined as follows:

$$\underline{G} = \underline{\underline{\Delta X}}\,\underline{\underline{\Delta X}}^{\mathsf{T}}$$

Eq. 78

$\underline{G}$ is a real symmetric $N \times N$ matrix and is sometimes called the **Gram matrix**. From the SVD of $\underline{\underline{\Delta X}}$, we find the following:

$$\underline{G} = \underline{U}\,\underline{W}^2\,\underline{U}^{\mathsf{T}}$$

Eq. 79

This is of the form of an eigen-decomposition. This means that from the eigen-decomposition of matrix $\underline{G}$, we can extract the matrices $\underline{U}$ and $\underline{W}$. If we re-arrange the SVD in Eq. 75, we can get the following:

$$\underline{W}^{-1}\,\underline{U}^{\mathsf{T}}\underline{\underline{\Delta X}} = \underline{V}^{\mathsf{T}}$$

Eq. 80

This means we can also construct the matrix $\underline{V}$ once we have $\underline{U}$ and $\underline{W}$. The consequences of this are that from the matrix $\underline{G}$, we can effectively perform PCA. Why is this important? Well, let's look at the matrix elements of $\underline{G}$. These are as follows:

$$G_{kl} = \sum_i \Delta x_{ki}\,\Delta x_{li} = \underline{o}_k^{\mathsf{T}}\,\underline{o}_l$$

Eq. 81

So, the matrix element G_{kl} is the inner product between the kth-centered datapoint and the lth-centered datapoint. Remember that the inner product between two vectors is effectively a measure of similarity between those two vectors. From this, we conclude that we can effectively do PCA entirely in terms of similarities between the datapoints in our dataset. This important idea is what we will exploit when we introduce a non-linear form of PCA, called **kernel PCA**, in *Chapter 12*.

- If we return to the SVD of the data matrix $\underline{\underline{\Delta X}}$ and absorb the matrix of singular values $\underline{\underline{W}}$ into the matrix of scores $\underline{\underline{U}}$, by defining a new matrix $\underline{\underline{\tilde{U}}} = \underline{\underline{U}}\,\underline{\underline{W}}$, then the data matrix can be written as follows:

$$\underline{\underline{\Delta X}} = \underline{\underline{\tilde{U}}}\,\underline{\underline{V}}^{\top}$$

Eq. 82

Schematically, this decomposition of $\underline{\underline{\Delta X}}$ is shown in *Figure 3.9*. We have factorized $\underline{\underline{\Delta X}}$ as a product of two matrices.

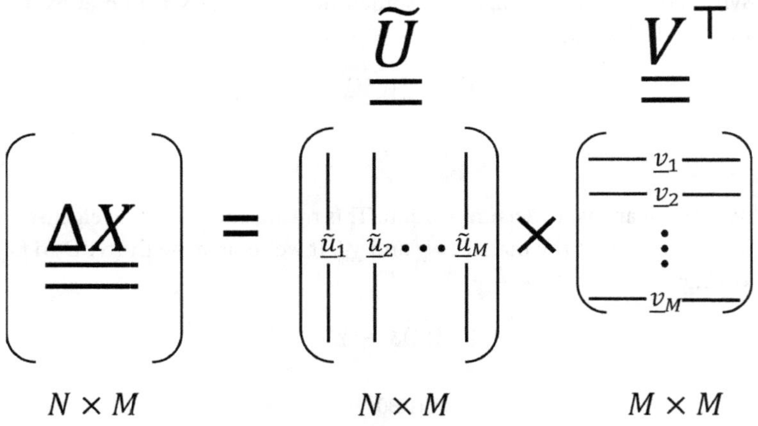

Figure 3.9: The SVD as a product of two matrices

If we expand the expression in Eq. 82 for the kth row, we have the following:

$$(\Delta x_{k1}, \Delta x_{k2}, \ldots, \Delta x_{kd}) = \sum_{j=1}^{d} \tilde{U}_{kj}\underline{v}_{j}^{\top}$$

Eq. 83

This tells us that the kth (centered) observation can be written as a sum of special vectors, $\underline{v}_1, \underline{v}_2, \ldots, \underline{v}_d$, multiplied by weights whose values are specific to that observation. The special vectors are general to the entire dataset. When we perform PCA, we are truncating this sum of vectors to a smaller number $q < d$, and so we are approximating the original data as follows:

$$(\Delta x_{k1}, \Delta x_{k2}, \ldots, \Delta x_{kd}) \simeq \tilde{U}_{k1} \underline{v}_1^\top + \tilde{U}_{k2} \underline{v}_2^\top + \ldots + \tilde{U}_{kq} \underline{v}_q^\top$$

Eq. 84

This idea of representing or approximating a dataset by a weighted sum of a small number of vectors that are generic to the entire dataset, but where we have observation-specific weights, is a common one in data science. In fact, we will use this idea in the next section when we encounter **Non-negative Matrix Factorization** (NMF).

It is time to look at how to do PCA on some real data with a code example.

PCA code example

For our code example, we'll use the in-built functions in the `scikit-learn` package to do this. The following code example can be found in the `Code_Examples_Chap3.ipynb` Jupyter notebook in the GitHub repository. We'll use the `sklearn.decomposition.PCA` function to perform the PCA.

We will use stock market prices from the S&P500 index between 2010 and 2013. The data can be found in the `SP500_log_returns.csv` file in the `Data` folder of the GitHub repository. The data consists of daily log closing returns for each stock in the index for which we have a full history between January 1st 2010 and December 31st 2013. That gives us 453 stocks and 1,258 observation points. The stocks are the columns of the dataset, and the rows are the timepoints. The stocks are labeled S1, S2, ..., S453. Let's read in the data and fit a PCA object to the data:

```
import pandas as pd
import numpy as np
import matplotlib.pyplot as plt
from sklearn.decomposition import PCA

## Read in the raw data into a pandas dataframe.
sp500_returns = pd.read_csv('../Data/SP500_log_returns.csv')
## Instantiate a PCA object and fit it to the SP500 returns matrix.
## Note that we convert the pandas dataframe to a numpy array first
pca = PCA()
pca.fit(sp500_returns.to_numpy())
```

The `scikit-learn` PCA uses an SVD of the data matrix to do the PCA. The singular values tell us the standard deviations, not the variances, of the projections of the datapoints onto each principal component. We can see that the first principal component accounts for much more variation than any other PC. We can look at the first five singular values using the following code:

```
pca.singular_values_[0:5]
```

This gives the following output:

```
array([6.81525175, 1.91949441, 1.63241764, 1.50134004, 1.28434791])
```

Let's look at the cumulative percentage of variance:

```
# Convert standard deviations into variances
pc_variance = np.power(pca.singular_values_, 2.0)

# Calculate total variance
total_variance = np.sum(pc_variance)

# Calculate cumulative total variance
cumulative_variance = 100.0*np.cumsum(pc_variance)/total_variance

# Plot
plt.plot(np.arange(1, 21), cumulative_variance[0:20],
        color='b', marker='o')
plt.title('Cumulative variance explained by first 20 PCs')
plt.xlabel('PC')
plt.ylabel('Cumulative % of total variance')
plt.show()
```

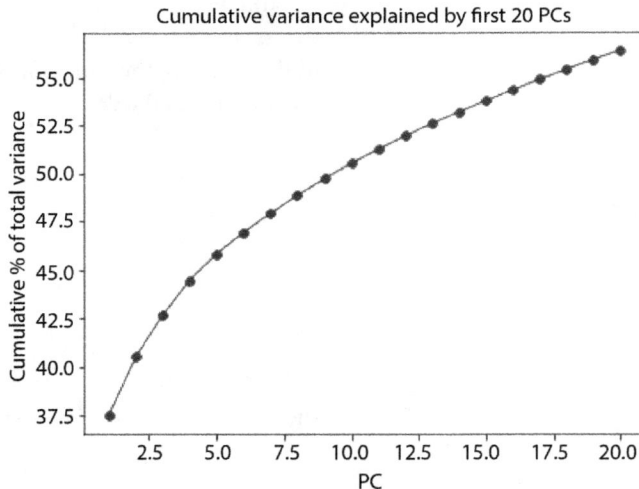

Figure 3.10: Cumulative variance explained

The first PC accounts for nearly 38% of the total variance in the dataset. We'll plot the loading values of the first principal component. In the plot, we'll ignore the sign of the loading values, which we can do without loss of generality because if a is a valid principal component, then so is $-a$:

```
# Plot
plt.rcParams["figure.figsize"] = (20,6)
```

```
plt.stem(np.arange(1, pca.components_.shape[1] + 1),
        np.abs(pca.components_[0, :]), markerfmt=' ')
plt.title('Loading values for PC1', fontsize=24)
plt.xlabel('Stock', fontsize=20)
plt.ylabel('Loading', fontsize=20)
plt.show()
```

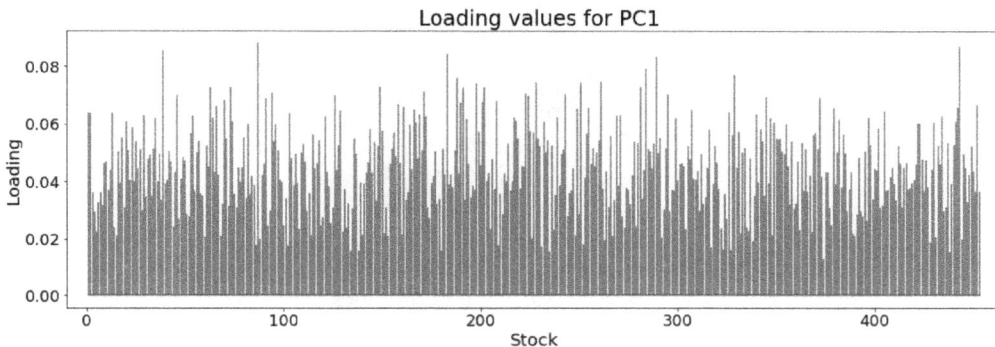

Figure 3.11: Loadings for PC1

The loading plot in *Figure 3.11* shows that every loading value for the first principal component is of the same sign and similar magnitude. This means that the first PC represents variation in the data (the stock prices) where every stock contributes to the variation and to a similar degree. This means that the first PC represents a pattern of variation whereby if one stock increases, all stocks do. This is called "a market mode" and simply reflects the fact that in a rising stock market, all the stocks tend to go up together. In hindsight, this is obvious and so less interesting than it might seem, so let's look at the other principal components.

We'll look at the PCA plot of the non-market-mode principal components, that is, PC3 versus PC2. First, we'll need to calculate the scores of the data. We do this using the transform method of the scikit-learn PCA object. The code is shown here:

```
#Plot
plt.scatter( x=scores[:,1], y=scores[:,2])
plt.title('PC3 score vs PC2 score', fontsize=24)
plt.xlabel('PC2 :' + str(round(100.0*(
    pc_variance[1]/total_variance),2)) + '%', fontsize=20)
plt.ylabel('PC3 :' + str(round(100.0*(
    pc_variance[2]/total_variance),2)) + '%', fontsize=20)
plt.xticks(fontsize=18, rotation=45)
plt.yticks(fontsize=18)
plt.show()
```

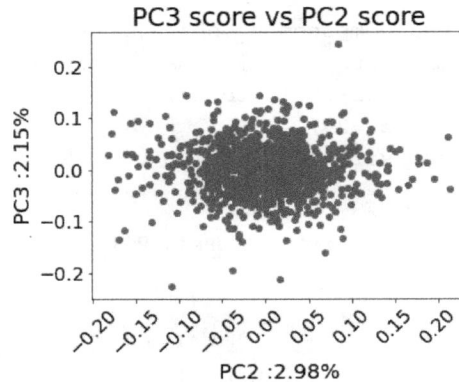

Figure 3.12: PC3 versus PC2 plot

We can see from *Figure 3.12* that the percentage of variance explained by PC2 and PC3 is a lot less than the market mode. This means most of the variation in the stock prices is caused by factors that affect all stocks, that is, global factors. In contrast, PC2 and PC3 are likely to represent sector-specific variation, for example, variation in the oil and gas or IT and technology sectors. Looking at the loadings of PC2 and PC3 would help us identify which sectors of stocks they represent.

With that demonstration of PCA on real data completed, we wrap up this sub-section and move on to our next dimensionality reduction technique, NMF.

Non-negative matrix factorization

When we used PCA, we did not have any restrictions on the values in the data matrix $\underline{X}$, other than that they should be real. However, in many real-world examples, the data we're analyzing is positive. For example, the data matrix may be recording, with a simple 0/1 variable, whether a customer has bought a particular product or not from an online retail site. In this example, the rows of the data matrix would represent customers and the columns would represent products. The data matrix elements essentially represent "likeliness to buy." An example data matrix is shown in *Table 3.2*.

Customer number	Product 1 bought	Product 2 bought	Product 3 bought	Product 4 bought
#1	1	1	1	0
#2	1	1	1	1
#3	0	1	0	0
#4	1	1	0	0
#5	1	0	1	1
#6	1	1	0	1
#7	1	1	1	1

Table 3.2: An example customer-product purchase data matrix

Now we're going to approximate the data matrix using a factorization of the form illustrated in *Figure 3.13*, which is of the form we introduced in Eq. 84 when we showed how to use the SVD to do PCA.

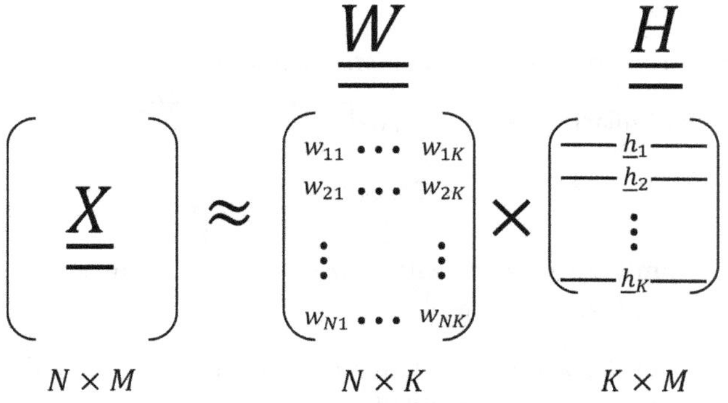

Figure 3.13: Schematic showing NMF

In matrix notation, we're approximating $\underline{\underline{X}} \simeq \underline{\underline{W}}\,\underline{\underline{H}}$. Writing this out in longhand gives us the ith observation:

$$(x_{i1}, x_{i2}, \ldots, x_{iM}) \simeq w_{i1}\underline{h}_1 + w_{i2}\underline{h}_2 + \ldots + w_{iK}\underline{h}_K$$

Eq. 85

In other words, we're approximating the data for our i^{th} customer as a sum of generic vectors, $\underline{h}_1, \underline{h}_2, \ldots, \underline{h}_K$, with customer-specific weights $w_{i1}, w_{i2}, \ldots, w_{iK}$. The dimensionality reduction occurs because we will set $K < M$, and in doing so this forces the vectors $\underline{h}_1, \underline{h}_2, \ldots, \underline{h}_K$ to represent general patterns in the data matrix, and so the vectors capture behavior that is generic to all customers.

If we want to approximate the data matrix as weighted sums of the form in Eq. 85, then we probably want those generic vectors $\underline{h}_1, \underline{h}_2, \ldots, \underline{h}_K$ to have the same properties as the original data. For our product purchase data matrix, we would want the components of those vectors to also be non-negative. In that way, we can still interpret each vector $\underline{h}_j$ as representing a "likeliness-to-buy" pattern. Similarly, we also want the weights $w_{i1}, w_{i2}, \ldots, w_{iK}$ to be positive. After all, how would we interpret a negative weight applied to a "likeliness-to-buy" pattern? This means we have the restriction that we want the matrix elements of the matrices $\underline{\underline{W}}$ and $\underline{\underline{H}}$ to be positive. In other words, we want to factorize a positive matrix as the product of two other positive matrices – hence the name **non-negative matrix factorization (NMF)**.

Now, how do we work out the best values for the matrix elements of $\underline{\underline{W}}$ and $\underline{\underline{H}}$? As we have said, we want the product $\underline{\underline{W}}\,\underline{\underline{H}}$ to approximate the data matrix $\underline{\underline{X}}$ as much as possible. Just as we did with PCA, we maximize the similarity between $\underline{\underline{X}}$ and the approximation $\underline{\underline{W}}\,\underline{\underline{H}}$. Equivalently, we minimize the difference between the two. How should we measure this difference? We use the **Root Mean Squared Error (RMSE)** of the individual matrix elements. In math notation, this is as follows:

$$\text{Dissimilarity measure} = \sqrt{\sum_{i=1}^{N}\sum_{j=1}^{M}\left[X_{ij} - (\underline{W}\underline{H})_{ij}\right]^2}$$

Eq. 86

We can write this in a more compact form using matrix notation:

$$\text{Dissimilarity measure} = \sqrt{\text{tr}([\underline{X} - \underline{W}\underline{H}]^\top[\underline{X} - \underline{W}\underline{H}])}$$

Eq. 87

We then minimize this dissimilarity measure with respect to $\underline{W}$ and $\underline{H}$ subject to the constraints that the matrix elements of $\underline{W}$ and $\underline{H}$ are positive.

We won't go into how this constrained minimization is done. Fortunately, we don't have to as there are Python packages that will do the NMF calculation for us.

Once we have done the NMF calculation, we can construct the approximation to the original data matrix. This is where we reap the potential benefits of NMF for our customer purchase data. Since the approximation $\underline{W}\underline{H}$ has forced the vectors $\underline{h}_1, \underline{h}_2, \ldots, \underline{h}_K$ to learn general patterns in the data matrix $\underline{X}$, it means that if the approximation gives a large matrix element value, then the NMF approximation thinks there is a high likeliness to buy for that customer-product combination. We can see whether the customer has already bought the product, and if not, we have a potential recommendation for the customer. This is why NMF forms the basis of many recommender system algorithms, although we have simplified the explanation a lot here. Let's look at NMF in action with a code example.

NMF code example

We'll use the in-built functions in the `scikit-learn` package to do this. The following code example (and additional ones) can be found in the `Code_Examples_Chap3.ipynb` Jupyter notebook in the GitHub repository.

We'll use the `sklearn.decomposition.NMF` function to perform NMF. We'll start with the small example data matrix in *Table 3.2*:

```
import numpy as np
from sklearn.decomposition import NMF
# Create a customer-purchase matrix. We'll use the one
# in the main text.
X = np.array([[1, 1, 1, 0],
              [1, 1, 1, 1],
              [0, 1, 0, 0],
              [1, 1, 0, 0],
              [1, 0, 1, 1],
              [1, 1, 0, 1],
              [1, 1, 1, 1]])
```

```
# Instantiate the NMF model
model = NMF(n_components=2, init='random', random_state=0)

# Fit the model to the data matrix
W = model.fit_transform(X)

# Extract the component vectors
H = model.components_

# Compute the difference between NMF approximation and the #data
np.matmul(W, H) - X
```

This gives the following output:

```
array([[-0.09412173,  0.04403871, -0.46340574,  0.53659426],
       [ 0.16151778, -0.07560645, -0.06281629, -0.06281629],
       [ 0.38384462, -0.17957389,  0.          0.        ],
       [-0.34976125,  0.16368387,  0.13600481,  0.13600481],
       [-0.14594226,  0.06843165,  0.05675037,  0.05675037],
       [-0.09412173,  0.04403871,  0.53659426, -0.46340574],
       [ 0.16151778, -0.07560645, -0.06281629, -0.06281629]])
```

The highlighted cell indicates a customer-product combination where the NMF model predicts a high value relative to the data. Since the actual data is just 0 or 1 values, this means the data is 0, that is, the first customer has not bought the fourth product yet, but the NMF model is predicting that the customer should – so we have a product recommendation for this customer.

That completes our demonstration of NMF and with it this section on matrix factorization and dimensionality reduction techniques. Let's recap what we've learned and summarize the chapter overall.

What we learned

In this section, we have learned the following:

- How to perform PCA on a dataset
- How to use PCA as a dimensionality reduction technique
- How to interpret the PCA scores and loadings
- How to perform NMF to approximately factorize a data matrix with positive entries
- How to use NMF as a dimensionality reduction technique

Summary

You've made it this far; well done! The effort will be worth it. Along with random variables and probability distributions, linear algebra is one of the core math building blocks for all data science algorithms.

Vectors are a natural way to represent data, and matrices are a natural way to encode transformations that act on that data. And it is those transformations that are a core part of what a data scientist does – shaping, aggregating, and manipulating data. Explanations of matrix algebra are often dry, hiding what the matrices are doing. We have tried to correct that in this chapter. Along the way, we have learned the following:

- How to calculate inner and outer products of pairs of vectors
- How to do matrix multiplication
- How a matrix represents a transformation
- The inverse and identity matrices
- The two core matrix decomposition methods: the eigen-decomposition and the SVD
- How to calculate the trace and determinant of a square matrix from its eigenvalues
- How to perform PCA and use it as a tool for performing dimensionality reduction of a dataset
- How to perform NMF of a positive matrix and use it as a tool for dimensionality reduction of a dataset

The last section of this chapter has shown how linear algebra techniques can be explicitly applied to data, to help us understand that data. PCA and NMF confirm that the matrix algebra formalities of the preceding sections are genuinely useful. They will continue to be useful when we begin to build predictive models in *Chapter 5*, when we use matrices to transform between input features and output predictions. To build models, we must first learn how to recognize when we have a good model and when we have a poor model. We do that in the next chapter, where we introduce loss functions and methods to optimize those loss functions.

Exercises

The following is a series of short exercises. Answers to all the exercises are given in the `Answers_ to_Exercises_Chap3.ipynb` Jupyter notebook in the GitHub repository:

1. We have some sales data taken at several different timepoints. For each timepoint, we have the number of units sold for five different products. An example of the data is given in *Table 3.3*:

Timepoint	Units sold product 1	Units sold product 2	Units sold product 3	Units sold product 4	Units sold product 5
1	10	1	1	2	25
2	12	0	2	5	23
3	7	1	4	9	18
4	3	2	7	2	10
5	5	1	15	4	3

Table 3.3: Product sales timepoints

Write code to store the preceding data as a two-dimensional NumPy array. We'll call this data matrix $\underline{X}$. Construct a vector, called $\underline{a}$, that when multiplied by data matrix $\underline{X}$ returns a vector consisting of the total revenue value at the various timepoints. The prices of the different products are shown in *Table 3.4*:

Product	Price
1	$1.25
2	$2.40
3	$1.75
4	$4.70
5	$0.50

Table 3.4: Product prices

2. Read in the data in the `Data_Q2_Chap3.csv` file in the `Data` folder of the GitHub repository. The data is for a 10 × 10 matrix. The matrix is real and symmetric. We'll call the matrix $\underline{X}$. Calculate its eigen-decomposition. Check that the first two eigenvalues are much greater than the remainder.

    Construct an approximation to $\underline{X}$ using Eq. 40 and just the first two eigenvalues. We'll call the approximation $\underline{X}^{(approx)}$. For this 10 × 10 matrix, the RMSE is calculated using the following formula:

    $$\text{RMSE} = \frac{1}{10} \sqrt{\sum_{i,j} (X_{ij} - X_{ij}^{(approx)})^2}$$

    Eq. 88

    See whether you can calculate the RMSE using just matrix operations. You may find the NumPy function `numpy.trace` useful.

3. The data in the `Data/Data_Cortex_Nuclear.csv` file in the GitHub repository contains measurements of protein expression levels in the cerebral cortex of mice. Each row of the data corresponds to a different mouse. The mice are divided into groups according to their genotype (genome), which treatment the mouse received, and which behavioral group they belong to. The data is a subset of the publicly available dataset held in the UCI Machine Learning Repository (`https://archive.ics.uci.edu/datasets.php`). The original data can be found in the Mice Protein Expression dataset (`https://archive.ics.uci.edu/dataset/342/mice+protein+expression`).

    Read in the data from the `Data/Data_Cortex_Nuclear.csv` file in the GitHub repository and perform PCA on the data.

    Confirm that there are two singular values of the data matrix that are significantly larger than the others.

    Plot the PC2 score (*y*-axis) against the PC1 score (*x*-axis) for the data points. Label the points according to which behavioral class the datapoint belongs to. What do you notice?

Notes and further reading

1. Outlining the situations when the eigenvectors of $\underline{M}$ can or cannot be used to construct an N-dimensional orthonormal basis is beyond the scope of this chapter.

2. The eigenvalues of the sample covariance matrix are guaranteed to be non-negative, that is, 0 or strictly positive. We won't go into the proof in this book, but it does mean that the largest-magnitude eigenvalue is also the largest eigenvalue.

4

Loss Functions and Optimization

In *Chapter 2* and *Chapter 3*, we focused on the two most important and core math concepts that are at the heart of virtually all of data science. In this chapter, we are going to move on to math concepts behind specific, but still very important, data science activities. Specifically, we are going to lay some of the groundwork for building predictive models.

At the end of the last chapter, we hinted that one of the key concepts when building models is knowing or measuring how good a model is. When we train or fit a **machine learning** (ML) model, we adjust the parameter values of the model so that it gives a "better" fit or explanation of the data. But this raises the question: What do we mean by "better"? Without an exact quantitative definition of what we mean when we say that one set of parameter values gives a better fit to the data than another, we cannot construct an objective and quantitative training process. This is where loss functions come in. They measure how well a model fits the training data. This chapter goes into the details behind loss functions and their use in the training of models. We do this by covering the following topics:

- *Loss functions – what are they?*: In this section, we learn the basics of loss functions and risk functions

- *Least squares (LS)*: In this section, we learn at a high level about least squares minimization as a general technique for estimating model parameters

- *Linear models*: In this section, we learn how to use least squares minimization for fitting linear models via **ordinary least squares** (OLS) regression

- *Gradient descent*: In this section, we learn a powerful and general technique for minimizing risk functions and objective functions

Technical requirements

All code examples given in this chapter (and additional examples) can be found at the GitHub repository, `https://github.com/PacktPublishing/15-Math-Concepts-Every-Data-Scientist-Should-Know/tree/main/Chapter04`. To run the Jupyter notebooks, you will need a full Python installation including the following packages:

- `pandas` (>=2.0.3)
- `numpy` (>=1.24.3)
- `scipy` (>=1.11.1)
- `scikit-learn` (>=1.3.0)
- `matplotlib` (>=3.7.2)
- `statsmodels` (>=0.14.0)

Loss functions – what are they?

A **loss function** takes two inputs; for example, a model prediction and the corresponding ground-truth value. It then compares the two inputs and summarizes this comparison into a single number.

Let's take that example further. We'll denote the ground-truth value by y and the model prediction by $\hat{y}$. A loss function in this example would then be a function of both y and $\hat{y}$, which returns a single real number. Let's call that loss function $L(y, \hat{y})$. We'll meet a concrete example of a loss function in the next section. But for now, it suffices to say that a loss function $L(y, \hat{y})$ attempts to measure how similar $\hat{y}$ is to y, with a loss function value of zero indicating that $\hat{y}$ is identical to y.

In general, a lower value of the function $L(y, \hat{y})$ means that $\hat{y}$ is closer or more similar to y, while a higher value of the loss function means that $\hat{y}$ is further from or less similar to y.

When training a model, our training data will consist of lots of ground-truth values, so we will also have lots of predictions. If our training set consists of N datapoints, then we will have N ground-truth values, $y_1, y_2, \ldots, y_N$. We can represent these by the vector $\underline{y} = (y_1, y_2, \ldots, y_N)$. Similarly, we can represent the corresponding set of predictions by the vector $\underline{\hat{y}} = (\hat{y}_1, \hat{y}_2, \ldots, \hat{y}_N)$. Once we have chosen a particular form for the loss function $L(y, \hat{y})$ we can use it to measure how close the whole set of predictions $\underline{\hat{y}}$ are to their corresponding ground-truth values $\underline{y}$ by simply calculating the average loss over the entire set. In other words, we calculate the following quantity:

$$\frac{1}{N} \sum_{i=1}^{N} L(y_i, \hat{y}_i)$$

Eq. 1

The value of the quantity in *Eq. 1* tells us how close our fitted model values are to the ground-truth values.

When training our model, we adjust the model parameters so that the quantity in *Eq. 1* is minimized. However, no model is perfect. Even a suitably trained model that has not been overfitted to the training data will not have $\hat{y}$ identical to y. Using $\hat{y}$ to represent y is an approximation. It will be an imperfect approximation in the sense that there will be some loss of the information that was present in y, hence the name "loss function." A loss function enables us to measure how much loss we suffer when representing y by $\hat{y}$. Or rather, it attempts to quantify how much loss we suffer when we represent the true process that generates the real data y by our model, which produces the predictions $\hat{y}$. In this way, the loss function measures how well our model represents the true data generation process – it is a measure of how good our model is.

Risk functions

The quantity in *Eq. 1* is an example of a **risk function**. A risk function is the expectation value of a loss function; that is, its average value.

Why would we want to calculate the expectation of a loss function? Let's take a closer look at what a loss function is. For our example, our loss function $L(y, \hat{y})$ is a function of data. The model prediction value $\hat{y}$ is a function of the feature vector x that we input into the model, so $\hat{y}$ is a function of data. Similarly, the ground-truth value y is also data. This makes the loss function value $L(y, \hat{y}(x))$ a random variable – recall from *Chapter 2* that all data contains a random component, and anything derived from data contains a random component.

As we learned in *Chapter 2*, a random variable can take a range of values, and by taking the expectation value, we get a single, deterministic, quantity. When constructing a measure of how good a model is or constructing a measure to minimize as part of a training process, a scalar deterministic quantity is always easier to work with than a random quantity. The risk function associated with $L(y, \hat{y}(x))$ is defined as the expectation of $L(y, \hat{y}(x))$ over the values of the feature vector x and the ground-truth value y. So, the risk is calculated using the following formula:

$$\text{Risk} = \mathbb{E}_{x,y}\left(L(y, \hat{y}(x))\right) = \int L(y, \hat{y}(x))\, p(x, y) dx dy$$

Eq. 2

Since we have integrated over all possible values of x and y, the risk function defined by *Eq. 2* is now a function of the model parameters only. It is now a function we can use to work out optimal model parameter values because it is deterministic – whenever we minimize the risk function in *Eq. 2* with respect to the model parameters we will always get the same answer.

In practice, calculating the risk function in *Eq. 2* can be tricky, so instead, if we have a training dataset, we can approximate the expectation value in *Eq. 2* by the sample average of the loss function. That is, we approximate the following:

$$\text{Risk} \simeq \frac{1}{N}\sum_{i=1}^{N} L(y_i, \hat{y}_i)$$

Eq. 3

You'll recognize the quantity in *Eq. 3* as being the same as that in *Eq. 1*, which is why we said the quantity in *Eq. 1* was a risk function. As we learned in *Chapter 2*, we can think of a sample average as an expectation value calculated using the empirical density function. In this case, the empirical density function is calculated as the following:

$$\text{Empirical density function} = \frac{1}{N} \sum_{i=1}^{N} \delta(x - x_i)\delta(y - y_i)$$

Eq. 4

We can use the empirical density function in *Eq. 4* to approximate the density function $p(x, y)$ in *Eq. 2*. Plugging it into *Eq. 2*, we then get *Eq. 3*. Because of this, we call the risk function defined by *Eq. 3* (and *Eq. 1*) the **empirical risk function**. Training a model by minimizing the quantity in *Eq. 3* with respect to the model parameters is called **empirical risk minimization**.

Note that the empirical risk function still has a dependence on data, because we have approximated the population distribution $p(x, y)$ using a finite sample of data. The empirical risk function is a function of the entire training dataset, and so its value will be a random variable. If the dataset is large, then the variance of the empirical risk function may be small enough that we can confidently ignore this variation, but clearly, in general, the model parameter values that result from minimizing the empirical risk function will be sensitive to (depend on) the precise details of the training set used.

Finally, we should point out that because an empirical risk function is just a sum of loss function values, the terminology "risk function" and "loss function" are often used interchangeably in a loose fashion. So, sometimes, you may hear someone speak of a loss function when they are referring to a risk function or vice versa. However, the intent is usually clear – they are referring to a function that is to be minimized in order to find good values for the model parameters.

There are many loss functions

So far, we have been vague about the details of our loss function. We have referred to it simply as $L(y, \hat{y})$, but we haven't given an actual formula for our loss function. This is deliberate because there are many potential choices of loss function formula, and so we wanted to keep the explanation of how loss functions are used very general to encompass all these potentially different loss function formulas.

One of the simplest loss function formula for $L(y, \hat{y})$ is to take the difference between y and $\hat{y}$ and square it, to ensure a positive quantity. That is, we use the following formula:

$$L(y, \hat{y}) = (y - \hat{y})^2$$

Eq. 5

The loss function formula in *Eq. 5* is called the **squared loss**. It is a very simple formula. Because of that, it is an extremely widely used loss function, with some convenient properties. Consequently, in the next section, we go into more detail about this loss function.

Different loss functions = different end results

The squared loss is just one way of combining y and $\hat{y}$ into a positive number. We could just as well have chosen to use the formula $L(y, \hat{y}) = |y - \hat{y}|$, which is called the **absolute loss**. Different choices of loss function will lead to different results. If we used an absolute-loss loss function in the empirical risk function in *Eq. 3* when doing our model training, we would end up with different model parameter estimates compared to if we had used a squared-loss function. How the parameter estimates would differ depends on the different properties of the two loss functions. This highlights the following:

- The properties of the parameter estimates depend upon the properties of the loss function. Some of these properties are advantageous, and so often, we will choose a particular loss function precisely because we want the parameter estimates to have certain properties.

- The end results of training a model depend not only on the choice of model and choice of training data but also the whole training process, so also on such things as the choice of loss function; that is, how we choose to measure how good a model is. The end results can also depend on the algorithm we choose to minimize the empirical risk once we make our choice of loss function.

Loss functions for anything

Up till now, we have been talking about loss functions $L(y, \hat{y})$ that measure the difference between a ground-truth value y and the corresponding model prediction $\hat{y}$. But at the very start of this section, we just said that a loss function took in two inputs and compared them. From this, you're probably beginning to realize that we can use loss functions to compare more than just predictions and ground-truth values. For example, we might want to measure the difference between our estimates of some model parameters and what should be the true model parameter values. If we denote a true model parameter value by β and our estimate of it by $\hat{\beta}$, then we could measure how close $\hat{\beta}$ is to β using the squared loss, as follows:

$$\text{Similarity between } \beta \text{ and } \hat{\beta} = \left(\beta - \hat{\beta}\right)^2$$

Eq. 6

In the preceding example, we have effectively defined a loss function $L(\beta, \hat{\beta})$ that takes model parameter values as input. But it is still just comparing two real numbers. Again, we highlight the fact that at the start of this section, we said that a loss function just takes two inputs. There is no reason why those inputs must always be two real numbers. There are, in fact, many situations where we want to compare other mathematical objects. For example, we may want to compare two continuous probability distributions, particularly if the model we are building is not a model of the ground-truth values y, but is a model of a probability distribution.

One of the most common functions for comparing two probability distributions is the **Kullback-Leibler (KL)** divergence. If the two continuous probability distributions we wish to compare have probability density functions $p(x)$ and $q(x)$, and they are distributions of a thing denoted by x, then the KL divergence is defined as follows:

$$KL(p \parallel q) = \int p(x) \log\left(\frac{p(x)}{q(x)}\right) dx$$

Eq. 7

As the KL divergence is the average of the logarithmic difference between $p(x)$ and $q(x)$, we can think of the KL divergence as a risk function.

We haven't been specific about what the thing x represents, so you'll realize that the KL divergence can be used for measuring the similarity of distributions of many exotic mathematical objects – vectors, networks, matrices, and so on. We won't go into any more details. We'll meet the KL divergence again in *Chapter 13*, but for now, we'll leave it by just saying that the KL divergence can be used to measure the expected loss that occurs when we use $q(x)$ to approximate $p(x)$.

Hopefully, by now, you'll have realized that the concept of a loss function is a very general one. A loss function measures the similarity between two mathematical objects. We can do this for many different types of mathematical objects. Even for a given type of mathematical object, there are many different potential choices of formula for the loss function we use, with each different formula leading to subtle differences and nuances in the final results when we use the loss function for, say, model training.

A loss function by any other name

Since the concept of a loss function is a very general one and they just compare two mathematical objects, it is not surprising that such functions occur in many different branches of science and mathematics and consequently have different names for the same or related concepts. Therefore, you may also see the following concepts and terminology used:

- **Cost function**: Since our model predictions $\hat{y}$ are an imperfect representation of y, there may be some consequences or costs to that imprecision. For example, in a business setting, using $\hat{y}$ as an imperfect prediction of y could result in lost revenue or overstocking of inventory, and so has a real cost to the business. Consequently, a loss function is also sometimes called a *cost function*. Clearly, we would want to choose the model parameter values to minimize this cost as much as possible, so we also talk of minimizing a cost function.

- **Objective function**: When minimizing a risk function with respect to our model's parameters, minimizing the risk function is the aim or *objective* of the whole exercise. Consequently, the risk function is also referred to as the *objective function*. This terminology is particularly common in the field of mathematical optimization, which studies methods and general algorithms for optimizing functions.

That note on the widespread use of loss functions across different mathematical fields is a good place to stop for now and recap what we have learned in this section.

What we learned

In this section, we have learned the following:

- How a loss function measures the loss incurred when we approximate one mathematical object by another

- How a risk function is constructed as the expectation value of a loss function

- How to calculate the empirical risk function from a loss function and a dataset

- How the optimal parameter values of a model can be estimated by minimizing the empirical risk function

- How the choice of loss function changes the properties of the model parameter estimates obtained through the empirical risk minimization process

Having learned about loss functions in general, in the next section, we are going to focus on one loss function in particular – the squared-loss function. This is because the squared-loss function is so ubiquitous. As we will see in further sections, it is one of the common data science methods for estimating model parameters.

Least Squares

Least squares or least squares regression is probably a term you've heard before. Why is that so? It is because it is an extremely versatile but simple technique. These characteristics of least squares stem from the properties of the squared-loss function. So to start we'll delve into the squared-loss function in a bit more detail.

The squared-loss function

The squared-loss function in *Eq. 5* is a function of the difference $y - \hat{y}$, and so we can write the squared loss in a slightly simpler form:

$$\mathrm{L}_{squared-loss}(y, \hat{y}) = f_{sq}(y - \hat{y}) \quad \text{with} \quad f_{sq}(x) = x^2$$

Eq. 8

The form of the function $f_{sq}(x)$ is shown in *Figure 4.1*:

Squared-loss function

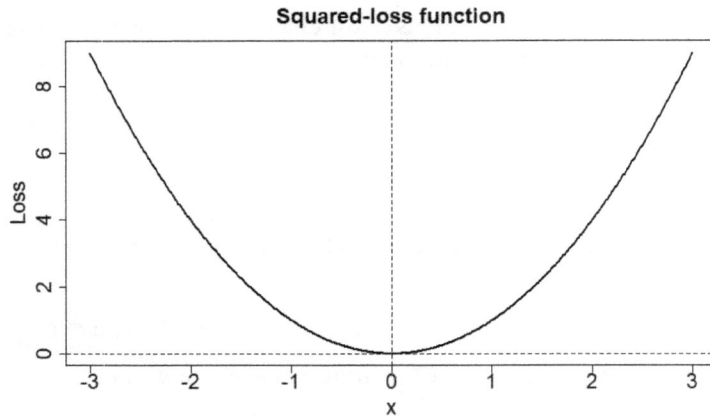

Figure 4.1: The shape of the squared-loss function

For the squared loss, the empirical risk function can be written as follows:

$$\frac{1}{N}\sum_{i=1}^{N}f_{sq}(y_i - \hat{y}_i) = \frac{1}{N}\sum_{i=1}^{N}(y_i - \hat{y}_i)^2$$

Eq. 9

The model prediction, $\hat{y}_i$, obviously depends upon the model parameters, which we'll denote by the vector $\underline{\beta}$, and the vector of feature values, $\underline{x}$, for which we are making the prediction. So, we denote our model as $\hat{y}(\underline{x} \mid \underline{\beta})$. The vertical bar in that mathematical expression means "given," so we can read this mathematical expression as the value of $\hat{y}$ evaluated at $\underline{x}$ and given the model parameter values $\underline{\beta}$. The specific model prediction $\hat{y}_i$ is obtained by plugging the feature vector $\underline{x}_i$ into this expression for our model, so $\hat{y}_i = \hat{y}(\underline{x}_i \mid \underline{\beta})$. If we determine $\underline{\beta}$ by minimizing the empirical risk function with respect to $\underline{\beta}$, this is equivalent to minimizing the following:

$$\sum_{i=1}^{N}\left(y_i - \hat{y}(\underline{x}_i \mid \underline{\beta})\right)^2$$

Eq. 10

We have dropped the pre-factor of $\frac{1}{N}$ in *Eq. 10* because it is a constant and therefore makes no difference to the value of $\underline{\beta}$ thatminimizes *Eq. 10*. The difference $y_i - \hat{y}_i$ is the ith **residual** of the model, and so the quantity in *Eq. 10* is the **sum-of-squared-residuals**, which is often shortened to **sum-of-squares**. When we determine the model parameters $\underline{\beta}$ by minimizing the sum-of-squares in *Eq. 10*, we are adjusting $\underline{\beta}$ until the sum-of-squares reaches its least possible value. Hence this technique for determining a model's parameters is known as **least squares** or **least squares minimization**.

We have said very little about the mathematical form of our model $\hat{y}(\underline{x} \mid \underline{\beta})$. This is because we haven't needed to. This makes least squares a very general technique for estimating the parameters of a model. We can apply the idea to very many different types of models and very many different situations. We already encountered least squares minimization in disguise when we minimized the dissimilarity between two matrices in *Chapter 3* when we introduced **Non-negative Matrix Factorization (NMF)**.

The idea of least squares minimization is a very intuitive one – simply construct a mathematical expression for your model predictions, which depends on the model parameters, use it to calculate the residuals $y_i - \hat{y}_i$, then minimize the sum of the squared residuals. However, least squares minimization is a heuristic idea, meaning that, at the moment, we have not provided a formal or rigorous justification of why we should minimize the sum-of-squared residuals to determine the model parameters. Why, for example, do we square the residuals and not raise them to the fourth power instead? We have given no proof that squaring the residuals is the best choice we can make to turn the residual into a positive quantity. We will provide a formal justification of least squares minimization when we introduce probabilistic models in *Chapter 5*, but for now, we will stick with the heuristic viewpoint – it is a very general, powerful, and extremely useful technique. So, let's dive into least squares minimization in a bit more detail.

OLS regression

The scatter plot in the left-hand plot of *Figure 4.2* shows some example data that we would like to build a model of. The scatter plot suggests a linear relationship, so we'll use a linear model to capture the relationship between the x and y values. For this 1D data (we have only one feature, x), a linear model is of the following form:

$$\hat{y} = \beta_0 + \beta_1 x$$

Eq. 11

The intercept β_0 and gradient β_1 are the parameters of our model, so our parameter vector $\underline{\beta} = (\beta_0, \beta_1)$. An example model is shown on the left-hand side of *Figure 4.2* by the solid red line. In fact, this line is the optimal or least squares choice for $\underline{\beta}$:

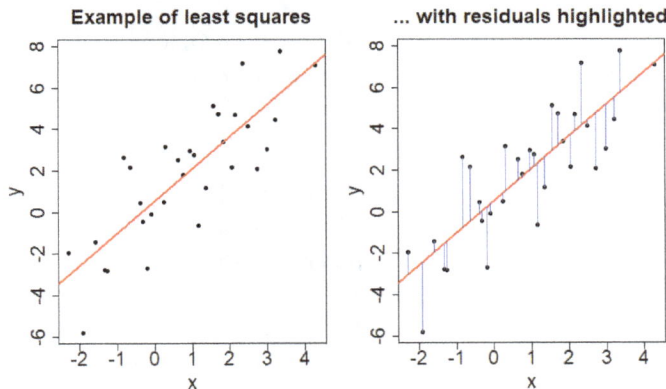

Figure 4.2: least squares model optimization for a simple linear model

On the right-hand side of *Figure 4.2*, we have reproduced the scatter plot and least squares optimal model line, but we have also highlighted, with vertical blue line segments, the residuals of each of the datapoints. A residual line segment above the red line indicates a positive residual, while a residual line segment below the red line indicates a negative residual. We can see that there is a mix of positive and negative residuals. At any given value of x the red line is passing approximately through the middle of the y values that are located at that value of x. In other words, the red line, or the least squares model, is attempting the estimate the mean value of y given the value of x. Because we are estimating parameter values β that make our linear model predict the mean of y given x, the overall estimation process is called regression. And because we are using least squares to estimate the optimal values of β, we call the overall process **least squares regression**. Furthermore, because we are using the vertical residuals we call this **ordinary least squares (OLS) regression**.

You may be wondering what is ordinary about OLS regression. This refers to the fact that we are using least squares regression in its most common setting – fitting a linear model and using the vertical residuals. There are other types of least squares regression that you may encounter. For example, if we have a non-linear model but still use the vertical residuals, this is unsurprisingly called **non-linear least squares regression** (**NLS**). Alternatively, if we still want to model a linear relationship but want the relationship to capture how the x and y values vary together, then the squares of the orthogonal distances from the model line to the datapoints, rather than the vertical distances, are a better way to measure the loss. This is called **total least squares** (**TLS**) regression. We already encountered TLS regression in disguise when we learned about **principal component analysis** (**PCA**) in *Chapter 3*, when we chose our principal components to minimize the variance lost by approximating a full dataset through dimensionality reduction.

For now, we're going to focus only on OLS regression. It sounds like a great data science technique to have in our toolkit, right? It is, but that doesn't mean it doesn't have its weaknesses. We'll explore one of its main weaknesses next.

OLS, outliers, and robust regression

The scatter plot in *Figure 4.3* shows the influence of outliers on OLS regression. The dataset clearly has two outlier values (with high y values) toward the right-hand end of the scatter plot. The solid red line shows the OLS model when we use all 31 datapoints, while the red dashed line shows the OLS model when we exclude the two outlier points.

Including the outlier points in the OLS regression has clearly pulled the regression line upward despite the number of outliers being small. One look at the shape of the loss function in *Figure 4.1* explains why this is so. The quadratic shape of the loss function means that outliers – that is, points with large residuals – contribute significantly more to the sum-of-squared residuals value. A residual, r, of size 1.0 contributes a value of 1.0 when we plug it into the squared-loss function $f_{sq}(r)$. However, a residual of size $r = 2.0$ contributes a value of 4.0 when we plug it into $f_{sq}(r)$. Since OLS regression works by minimizing the sum-of-squared residuals, the OLS algorithm is going to pay disproportionately more attention to the outlier contributions when adjusting the model parameters β.

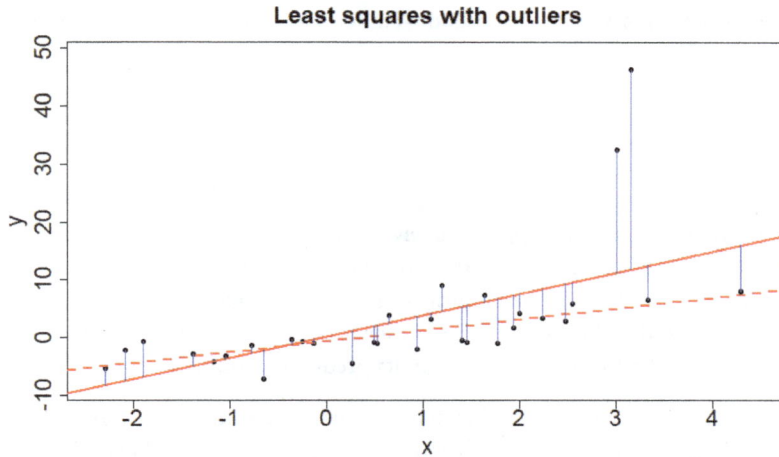

Figure 4.3: The effect of outliers on OLS regression

This tells us that OLS is sensitive to the effect of outliers. Including the outliers in the OLS regression in *Figure 4.3* has led to a model that is a poor fit for most of the data in the scatterplot, and importantly, it has led to a model that will predict poorly for new datapoints. Can we rectify this? Yes, we can. One way to do so would be to modify the shape of our loss function so that it wasn't quadratic at large residual values. Such a loss function is shown by the solid black line in *Figure 4.4*. For comparison, we have also plotted in *Figure 4.4* the squared-loss function of *Eq. 8*, as the dashed red line:

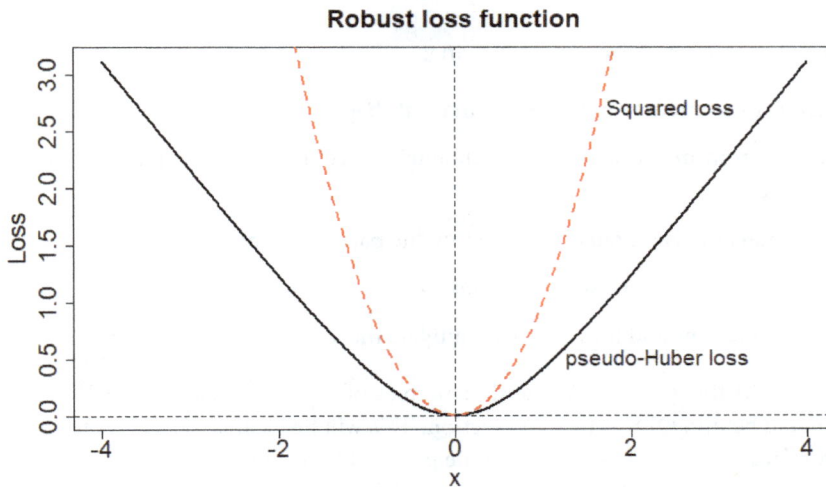

Figure 4.4: The shape of the pseudo-Huber robust loss function

This particular loss function is known as a pseudo-Huber loss function, and its mathematical form is the following:

$$f(x) = \sqrt{1 + x^2} - 1$$

Eq. 12

At large values of x, this loss function is approximately linear. The contrast to the squared-loss function at large values of x is marked. Using the pseudo-Huber loss function in *Figure 4.4* would mean that outlier values would still contribute to the empirical risk function, but not disproportionately so. The regression algorithm would then be robust to the presence of outliers in the dataset, and importantly, we would produce a model that is more accurate in its predictions on new values of x.

Not surprisingly, loss functions of the shape shown in *Figure 4.4* are studied as a part of statistics known as **robust statistics**. A detailed explanation of robust statistics techniques is beyond the scope of this book. However, the very fact that robust regression techniques are available to us may make you ask why we still use and study OLS regression. The answer lies in something we haven't yet spoken about – for OLS, how do we actually do the minimization of the empirical risk function? What are the details of the algorithm we use? For OLS regression, the combination of a linear model with a squared-loss function leads to an extremely efficient solution to the empirical risk minimization problem. It is this solution we will cover in the next section, but for now, let's summarize what we have learned about the squared-loss function and least squares.

What we learned

In this section, we have learned about the following:

- The squared-loss function and its mathematical shape
- Least squares minimization as a general heuristic technique for estimating optimal model parameter values
- OLS regression as a technique for estimating the parameters of linear models
- The sensitivity of OLS regression to outliers
- Robust loss functions and how they can mitigate the sensitivity of OLS regression to outliers

Having learned about the general ideas and principles of OLS regression, we will delve into the mathematical detail behind OLS in the next section. This will be useful because i) OLS is one of the workhorse algorithms of data science, ii) it will help to highlight some ideas about the optimization of objective functions in general that we will want to make use of later on.

Linear models

We've already introduced, at a high level, the idea of OLS regression for a linear model. But this particular combination of squared loss for measuring the risk and a linear model for $\hat{y}$ has some very convenient and simple-to-use properties. This simplicity means that OLS regression is one of the most widely used and studied data science modeling techniques. That is why we are going to look in detail at fitting linear models to data using OLS regression.

To start with, we'll revisit the squared-loss empirical risk function in *Eq. 10* and look at what happens to it when we have a linear model $\hat{y}$. To recap, the squared-loss empirical risk is given by the following:

$$\text{Risk} = \frac{1}{N} \sum_{i=1}^{N} (y_i - \hat{y}_i)^2$$

Eq. 13

Now, for a linear model with d features, $x_1, x_2, ..., x_d$, we can write the model as follows:

$$\hat{y} = \beta_0 + \beta_1 x_1 + \beta_2 x_2 + \cdots + \beta_d x_d$$

Eq. 14

The vector of model parameters is $\underline{\beta}^T = (\beta_0, \beta_1, ..., \beta_d)$. We can write the features in vector form as well. We'll write it as a row-vector, $\underline{x} = (1, x_1, x_2, ..., x_d)$. Doing so allows us to write *Eq. 14* in the following form:

$$\hat{y} = \underline{x} \, \underline{\beta}$$

Eq. 15

We can think of the extra 1 in the feature vector $\underline{x} = (1, x_1, x_2, ..., x_d)$ as being a feature value x_0 that multiplies the intercept β_0 in the linear model in Eq. 14. For the ith datapoint the feature values can be written in vector form, $\underline{x}_i = (x_{i0}, x_{i1}, x_{i2}, ..., x_{id})$, with obviously $x_{i0} = 1$ for all i. We can combine all the feature vectors $\underline{x}_i$ from all the datapoints into a data matrix:

$$\underline{X} = \begin{pmatrix} x_{10} & x_{11} & x_{12} & \cdots & x_{1d} \\ x_{20} & x_{21} & x_{22} & \cdots & x_{2d} \\ \vdots & \vdots & \vdots & \ddots & \vdots \\ x_{N0} & x_{N1} & x_{N2} & \cdots & x_{Nd} \end{pmatrix} = \begin{pmatrix} 1 & x_{11} & x_{12} & \cdots & x_{1d} \\ 1 & x_{21} & x_{22} & \cdots & x_{2d} \\ \vdots & \vdots & \vdots & \ddots & \vdots \\ 1 & x_{N1} & x_{N2} & \cdots & x_{Nd} \end{pmatrix}$$

Eq. 16

If we also put all the observed values, y_i, into a vector $\underline{y}^T = (y_1, y_2, ..., y_N)$, then we can write the risk function in *Eq. 13* in a very succinct form as follows:

$$\text{Risk} = \frac{1}{N} (\underline{y} - \underline{X}\,\underline{\beta})^T (\underline{y} - \underline{X}\,\underline{\beta})$$

Eq. 17

The data matrix $\underline{X}$ is also called the **design matrix**. This terminology originates from statistics, where often the datapoints and, hence, feature values were part of a scientific experiment to quantify the various influences on the response variable y. Being part of a scientific experiment, the feature values x_{ij} were planned in advance; that is, *designed*.

The optimal values of the model parameters β are obtained by minimizing the right-hand side of *Eq. 17* with respect to β. We'll denote the optimal values of β by the symbol $\hat{\beta}$. We can use the differential calculus we recapped in *Chapter 1* to do the minimization. Differentiating the right-hand side of *Eq. 17* with respect to β and setting the derivatives to zero gives us the following:

$$\left.\frac{\partial \text{Risk}}{\partial \underline{\beta}}\right|_{\beta=\hat{\beta}} = \frac{2}{N}\underline{X}^{\mathsf{T}}\left(\underline{y} - \underline{X}\,\hat{\underline{\beta}}\right) = \underline{0}$$

Eq. 18

Re-arranging *Eq. 18*, we get the following:

$$\underline{X}^{\mathsf{T}}\underline{X}\,\hat{\underline{\beta}} = \underline{X}^{\mathsf{T}}\underline{y}$$

Eq. 19

We can solve *Eq. 19* by applying $(\underline{X}^{\mathsf{T}}\underline{X})^{-1}$ to both the left- and right-hand sides of *Eq. 19* to get the following:

$$\hat{\underline{\beta}} = (\underline{X}^{\mathsf{T}}\underline{X})^{-1}\underline{X}^{\mathsf{T}}\underline{y}$$

Eq. 20

This solution is very efficient. It is in a closed-form, meaning we have an equation with the thing we want, $\hat{\underline{\beta}}$, on its own on the left-hand side, and a mathematical expression that doesn't involve $\hat{\underline{\beta}}$ on the right-hand side. There is no iterative algorithm required. We just perform a couple of matrix calculations, and we have our optimal parameter estimates $\hat{\underline{\beta}}$. That we can obtain a closed-form expression for the parameter estimates is one of the most attractive aspects of OLS regression and part of the reason it is so widely used. We'll walk through some code examples in a moment to illustrate how easy it is to perform OLS regression.

Practical issues

This doesn't mean the closed-form expression in *Eq. 20* doesn't cause problems. Firstly, you'll recall from *Chapter 3* on linear algebra that we can have square matrices that do not have an inverse. It is very possible that the matrix $(\underline{X}^{\mathsf{T}}\underline{X})^{-1}$ does not exist. This happens when there are linear dependencies between the columns of the design matrix $\underline{X}$; for example, if one feature is simply a scaled version of another feature, or where combining several features together gives the same numerical value as another feature. In these circumstances, one or more of the features are redundant since they add no new information.

Secondly, in a modern-day data science setting where we might have many thousands of features in a model, the $d \times d$ matrix $\underline{\underline{X}}^\top\underline{\underline{X}}$ can be unwieldy to work with if d is of the order of several thousand.

How to deal with these computational issues is beyond the scope of the book, but they are something you should be aware of in case they crop up in a problem you are trying to solve.

The model residuals

Once we have obtained an estimate $\hat{\underline{\beta}}$ for the model parameters, using *Eq. 20*, we can calculate the residuals. If we denote the ith residual by r_i, then obviously we have the following:

$$r_i = y_i - \hat{y}_i = y_i - \underline{x}_i\hat{\underline{\beta}} = y_i - \left(\underline{\underline{X}}\,\hat{\underline{\beta}}\right)_i$$

Eq. 21

What happens if we sum up all the residuals? To answer this question, we make use of *Eq. 19* and recall that the first row of $\underline{\underline{X}}^\top$ is all ones if our model has an intercept. So, *Eq. 19* tells us the following:

$$\sum_{i=1}^{N}\left(\underline{\underline{X}}\,\hat{\underline{\beta}}\right)_i = \sum_{i=1}^{N}y_i \;\Rightarrow\; \sum_{i=1}^{N}\left[y_i - \left(\underline{\underline{X}}\,\hat{\underline{\beta}}\right)_i\right] = 0$$

Eq. 22

So, the sum of all the residuals is zero if our model has an intercept.

Let's see these ideas in action with a code example.

OLS regression code example

The data in the `Data/power_plant_output.csv` file in the GitHub repository contains measurements of the power output from electricity generation plants. The power (PE) is generated from a combination of gas turbines, steam turbines, and heat recovery steam generators, and so is affected by environmental factors in which the turbines operate, such as the ambient temperature (AT) and the steam turbine exhaust vacuum level (V). The dataset consists of 9,568 observations of the PE, AT, and V values. The data is a subset of the publicly available dataset held in the *UCI Machine Learning Repository* (https://archive.ics.uci.edu/datasets). The original data can be found at https://archive.ics.uci.edu/ml/datasets/Combined+Cycle+Power+Plant.

We'll use the data to build a linear model of the power output PE as a function of the AT and V values. We will build the linear model in two ways – i) using the Python `statsmodels` package, ii) using *Eq. 20* via an explicit calculation. The following code example can be found in the `Code_Examples_Chap4.ipynb` notebook in the GitHub repository. To begin, we need to read in the data:

```
import pandas as pd
import numpy as np
import matplotlib.pyplot as plt
```

```
import statsmodels.formula.api as smf

# Read in the raw data
df = pd.read_csv("../Data/power_plant_output.csv")
```

We'll do a quick inspection of the data. First, we'll compute some summary statistics of the data:

```
# Use pd.describe() to get the summary statistics of the data
df.describe()
```

	AT	V	PE
Count	9568.000000	9568.000000	9568.000000
Mean	19.651231	54.305804	454.365009
Std	7.452473	12.707893	17.066995
Min	1.810000	25.360000	420.260000
25%	13.510000	41.740000	439.750000
50%	20.345000	52.080000	451.550000
75%	25.720000	66.540000	468.430000
Max	37.110000	81.560000	495.760000

Table 4.1: Summary statistics for the power-plant dataset

Next, we'll visualize the relationship between the response variable (the target variable) and the features. We'll start with the relationship between power output (PE) and ambient temperature (AT):

```
# Scatterplot between the response variable PE and the AT feature.
# The linear relationship is clear.
plt.scatter(df.AT, df.PE)
plt.title('PE vs AT', fontsize=24)
plt.xlabel('AT', fontsize=20)
plt.ylabel('PE', fontsize=20)
plt.xticks(fontsize=18)
plt.yticks(fontsize=18)
plt.show()
```

PE vs AT

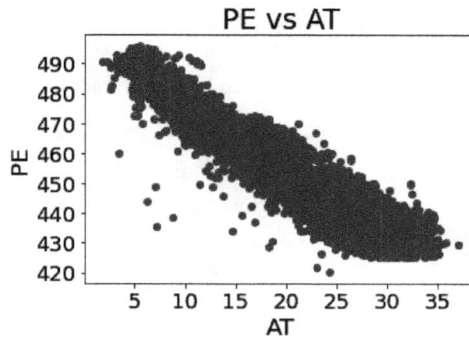

Figure 4.5: Plot of power output (PE) versus ambient temperature (AT)

Now, let's look at the relationship between power and vacuum (V):

```
# Scatterplot between the response variable PE and the V feature.
# The linear relationship is clear, but not as strong as the
# relationship with the AT feature.
plt.scatter(df.V, df.PE)
plt.title('PE vs V', fontsize=24)
plt.xlabel('V', fontsize=20)
plt.ylabel('PE', fontsize=20)
plt.xticks(fontsize=18)
plt.yticks(fontsize=18)
plt.show()
```

PE vs V

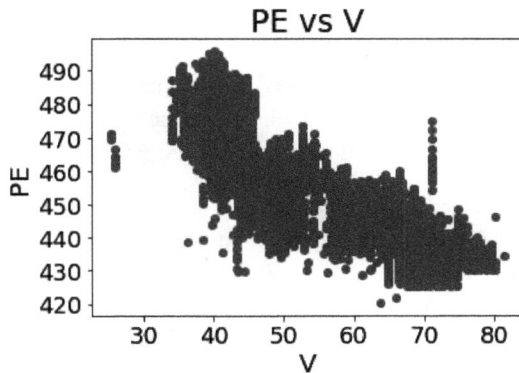

Figure 4.6: Plot of power output (PE) versus vacuum level (V)

Now, we'll fit a linear model using the `statsmodels` package. The linear model formula is specified in statistical notation as PE $\sim$ AT + V. You can think of it as the statistical formula equivalent of the mathematical formula PE $= \beta_0 + \beta_{AT} x_{AT} + \beta_V x_V$. We do this fitting using the following code:

```
# First we specify the model using statsmodels.formula.api.ols
model = smf.ols(formula='PE ~ AT + V', data=df)

# Now we fit the model to the data, i.e. we minimize the sum-of-
# squared residuals with respect to the model parameters
model_result = model.fit()

# Now we'll look at a summary of the fitted OLS model
model_result.summary()
```

This gives the following parameter estimates for our linear model:

OLS Regression Results

	coef	std err	T	P>|t|	[0.025	0.975]
Intercept	505.4774	0.240	2101.855	0.000	505.006	505.949
AT	-1.7043	0.013	-134.429	0.000	-1.729	-1.679
V	-0.3245	0.007	-43.644	0.000	-0.339	-0.310

Table 4.2: OLS regression parameter estimates for our power-plant linear model

We can see from *Table 4.2* that, as expected, we get negative estimates for the parameters corresponding to the AT and V features. Now, we'll repeat the calculation explicitly using the formula in *Eq. 20*. We'll use the linear algebra functions available to us in numpy. First, we need to extract the data from the pandas DataFrame to appropriate numpy arrays:

```
# We extract the design matrix as a 2D numpy array.
# This initially corresponds to the feature columns of the dataframe.
# In this case it is all but the last column
X = df.iloc[:, 0:(df.shape[1]-1)].to_numpy()

# Now we'll add a column of ones to the design matrix.
# This is the feature that corresponds to the intercept parameter
# in the moddel
X = np.c_[np.ones(X.shape[0]), X]
```

```
# For convenience, we'll create and store the transpose of the
# design matrix
xT = np.transpose(X)

# Now we'll extract the response vector to a numpy array
y = df.iloc[:, df.shape[1]-1].to_numpy()
```

Now, we can calculate the OLS parameter estimates using the formula $\hat{\beta} = (\underline{X}^T\underline{X})^{-1}\underline{X}^Ty$::

```
# Calculate the inverse of xTx using the numpy linear algebra
# functions
xTx_inv = np.linalg.inv(np.matmul(xT, X))

# Finally calculate the OLS model parameter estimates using the
# formula (xTx_inv)*(xT*y).
# Again, we use the numpy linear algebra functions to do this
ols_params = np.matmul(xTx_inv, np.matmul(xT, y))
```

We can compare the OLS parameter estimates obtained from statsmodels with those obtained from the explicit calculation:

```
# Now compare the parameter estimates from the explicit calculation
# with those obtained from the statsmodels fit
df_compare = pd.DataFrame({'statsmodels': model_result.params,
                           'explicit_ols':ols_params})
df_compare
```

	statsmodels	explicit_ols
Intercept	505.477434	505.477434
AT	-1.704266	-1.704266
V	-0.324487	-0.324487

Table 4.3: A comparison of the parameter estimates from the statsmodels
packages and explicit calculation using the OLS formula

The parameter estimates from the two different OLS regression codes are identical to more than 6 decimal places.

This walkthrough of a real example highlights the power of the closed-form OLS regression formula in *Eq. 20*. This closed-form arises from the linear (in β) nature of the optimality criterion in *Eq. 18*, which itself arises from the quadratic nature of the risk function in *Eq. 17*, which ultimately is a consequence of the quadratic form of the squared-loss function in *Eq. 8*.

But what if we don't want to use a linear model or a squared-loss function? Firstly, we can't use OLS regression! Secondly, a different choice of loss function, such as the absolute loss or the pseudo-Huber robust loss function in *Eq. 12*, will not lead to a closed-form solution for $\hat{\beta}$ if we minimize the empirical risk in *Eq. 3*. So, how do we minimize the empirical risk to obtain optimal model parameter estimates in these situations? We'll learn how to address this question in the next section, but for now, let's review what we have learned in this section.

What we learned

In this section, we have learned the following:

- How to write the empirical risk for OLS regression in matrix notation
- How to derive a closed-form expression for OLS model parameter estimates
- Some of the properties and practical limitations of OLS regression
- How to perform OLS regression using available Python packages such as `statsmodels`
- How to perform OLS regression by explicitly calculating the closed-form formula for OLS model parameter estimates

Having learned how to perform OLS regression, we'll now learn how to perform least squares regression in more general settings by using gradient descent techniques to minimize the empirical risk.

Gradient descent

As we just hinted at the end of the last section, we aren't always in a position where we can use the closed-form OLS solution of *Eq. 20*. What are our options? To construct a more general approach to empirical risk minimization, we'll have to revisit the shape of the empirical risk function so that we can understand how to locate its minima.

Locating the minimum of a simple risk function

To understand the shape of the empirical risk function, let's take a simple example with a model that has a single parameter. We'll use the risk function for a linear model and a squared-loss function. We'll use a linear model with a single feature, and so it is of the following form:

$$\hat{y} = \beta x$$

Eq. 23

The model has a single parameter, β, which multiplies the single feature x. In *Figure 4.7* we have plotted the shape of the empirical risk function against the value of β, and where we have calculated the empirical risk on a dataset of 100 datapoints. The dataset has been generated via simulation and using a model of the mathematical form in *Eq. 23*. The model we are going to fit to this data by minimizing the empirical risk is of the correct mathematical form (by construction); it is just that we don't know the true value of β – well, I do, but I'm not going to tell you just yet. To estimate the true value of β that underlies the data, we'll have to use the model form in *Eq. 23* and minimize the empirical risk:

Empirical Risk vs β

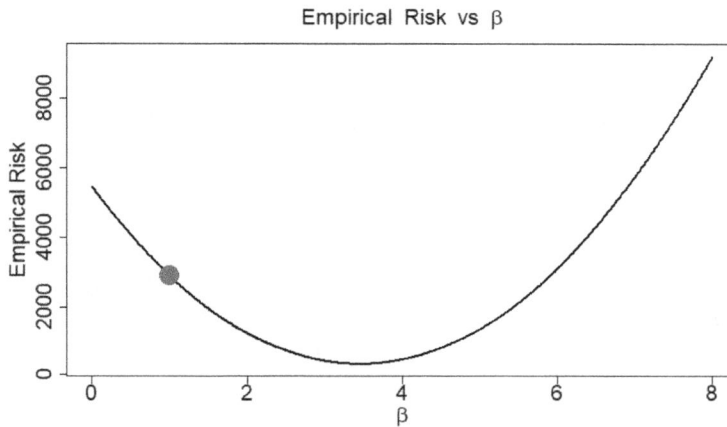

Figure 4.7: Empirical risk function and starting parameter estimate for our simulated dataset

From the shape of the risk function in *Figure 4.7*, you can see that there is a single minimum close to $\beta = 3.5$. As you may have guessed, $\beta = 3.5$ is the value of β that I used to generate the simulated data. Let's say we start with a guess for the true value of β that is 1. That is, we're going to initially set our estimate $\hat{\beta} = 1$. How good is that initial estimate? We have also plotted the position of this estimate as a red dot in *Figure 4.7* so that you can see how good an estimate it is by seeing how close it is to the minimum. If it were the optimal (best) estimate, we would be at the minimum of the empirical risk; that is, we would have the following:

$$\left. \frac{\partial \text{Risk}}{\partial \beta} \right|_{\beta = \hat{\beta}} = 0$$

Eq. 24

We can easily derive a formula for the derivative on the left-hand side of *Eq. 24*. It is the following:

$$\frac{\partial \text{Risk}}{\partial \beta} = \frac{1}{100} \sum_{i=1}^{100} \frac{\partial}{\partial \beta} (y_i - \beta x_i)^2 = -\frac{2}{100} \sum_{i=1}^{100} (y_i - \beta x_i) x_i$$

Eq. 25

So, we can just plug our current estimate $\hat{\beta} = 1$ into *Eq. 25* to see how close we are to the optimality criterion in *Eq. 24*. If we are within a specified tolerance of the criterion in *Eq. 24*, our estimate $\hat{\beta}$ is good.

What happens if we are not within the specified tolerance? How should we adjust our estimate $\hat{\beta}$? Clearly, we want to adjust our estimate $\hat{\beta}$ by moving it downhill toward the minimum. How do we work out which direction is downward – that is, which direction reduces the empirical risk? Well, this is what the derivative in *Eq. 25* tells us. Recall from *Chapter 1* that the derivative of a function tells us the gradient or slope of a function. So, the numerical value of the calculation in *Eq. 25* evaluated at $\hat{\beta}$ tells us in which direction we should adjust $\hat{\beta}$. If the derivative is positive then increasing $\hat{\beta}$ will increase the empirical risk, so we want to decrease $\hat{\beta}$. Conversely, if the derivative is negative then increasing $\hat{\beta}$ will decrease the empirical risk as desired.

The size of the derivative also gives us a guide by how much we should adjust $\hat{\beta}$. If the absolute value of the derivative is large, it tells us we are high on the steep slopes of the risk function, as depicted by the red dot starting point in *Figure 4.7*, and so we are a long way from the minimum. Overall, the bigger the absolute value of the derivative, the more we should adjust $\hat{\beta}$.

Putting together the arguments from the previous two paragraphs, we should adjust our estimate of $\hat{\beta}$ by descending according to the direction and size of the gradient of our objective function. This approach is called **gradient descent** and tells us we should adjust our estimate of $\hat{\beta}$ according to an update rule:

$$\hat{\beta} \leftarrow \hat{\beta} - \eta \left. \frac{\partial \text{Risk}}{\partial \beta} \right|_{\beta = \hat{\beta}}$$

Eq. 26

The quantity η controls how quickly we adjust $\hat{\beta}$, and so is called the learning rate, since it controls how quickly we move toward or learn the true optimal value of $\hat{\beta}$. A very small value of η will mean we make relatively small adjustments to $\hat{\beta}$ even when the derivative $\frac{\partial \text{Risk}}{\partial \beta}$ is relatively large. Conversely, a larger value of η will mean we take larger steps toward the minimum of the empirical risk.

Let's see what happens when we use the update rule in *Eq. 26* for a few iterations. The results are shown in *Figure 4.8*:

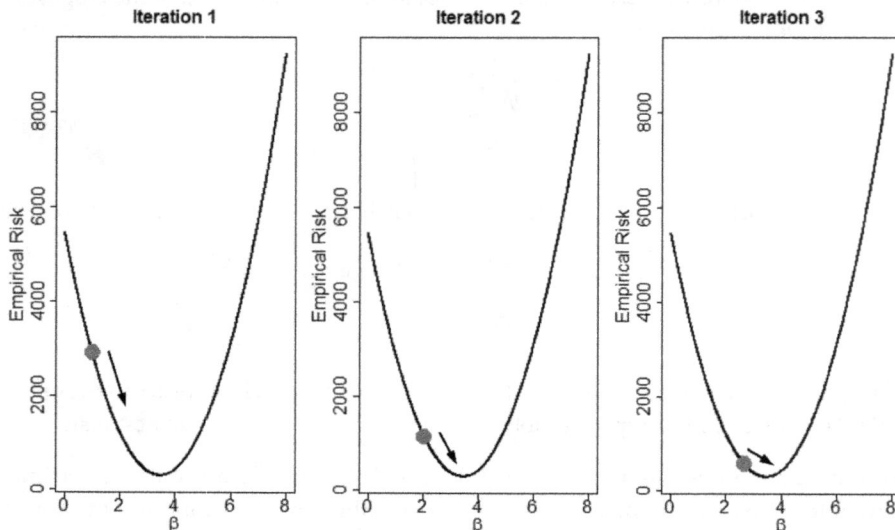

Figure 4.8: Evolution of the model parameter estimate as we iterate the gradient descent update rule

In the left-hand panel of *Figure 4.8*, we see our starting estimate of $\hat{\beta}=1$, shown by the position of the red dot. After one iteration, we have updated our estimate of $\hat{\beta}$ according to the update rule in *Eq. 26* (and using a learning rate of $\eta = 0.05$). This gives us a new value of $\hat{\beta} = 2.062$ for our estimate $\hat{\beta}$. This is shown in the middle panel of *Figure 4.8*, with again the red dot showing the position of $\hat{\beta}$ and its corresponding empirical risk value. In the right-hand panel, the red dot shows the updated value of $\hat{\beta}$ at $\hat{\beta} = 2.664$ (and corresponding empirical risk) after iteration 2. For each of the iterations in *Figure 4.8*, the direction and size of the update to the current estimate of $\hat{\beta}$ is shown schematically by the arrow. We can see that the updates all move the estimate $\hat{\beta}$ closer to the minimum, but the magnitude of the updates decrease as we get closer to the minimum. What happens if we carry on iterating? Let's look at how we would do that in a code example.

Gradient descent code example

The simulated data is in the `Data/gradient_descent_example.csv` file in the GitHub repository. You'll also find the following code example (and more) in the `Code_Examples_Chap4.ipynb` Jupyter notebook in the repository.

To begin, we'll read in the data:

```
import pandas as pd
import numpy as np

# Read in the raw data
df_risk = pd.read_csv("../Data/gradient_descent_example.csv")

# Extract the feature and response values to
# numpy arrays
x=df_risk['x'].to_numpy()
y=df_risk['y'].to_numpy()
```

Then, we'll define some functions to calculate the empirical risk and its derivative:

```
# Define functions for performing gradient descent

def risk(x, y, beta):
    '''
    Function to compute the empirical risk.
    x is a 1D numpy array of the feature values,
    y is a 1D numpy array of the response values.
    beta is the model parameter value.
    '''
    # Initialize the risk value
    risk = 0.0
```

```
    # Loop over the data an increment the risk with
    # a squared-loss
    for i in range(x.shape[0]):
        risk += np.power(y[i]-(beta*x[i]), 2.0)

    risk /= x.shape[0]

    return risk


def derivative_risk(x, y, beta):
    '''
    Function to compute the derivative of the empirical risk
    with respect to the model parameter.
    x is a 1D numpy array of the feature values,
    y is a 1D numpy array of the response values.
    beta is the model parameter value.
    '''
    derivative_risk = 0.0

    for i in range(x.shape[0]):
        derivative_risk += - (2.0*x[i]*(y[i]-(beta*x[i])))

    derivative_risk /= x.shape[0]

    return derivative_risk
```

Now, we'll perform 20 iterations of the gradient descent update rule:

```
# Set the learning rate and the number of iterations we want to
# perform
eta=0.05
n_iter=20

# Initialize arrays to hold the sequence of
# parameter estimates and empirical risk values
beta_learn=np.full(1+n_iter, np.nan)
risk_learn=np.full(1+n_iter, np.nan)

# Set the starting estimate for the
# model parameter
beta_learn[0]=1.0
```

```
# Iterate using the gradient descent update rule
for iter in range(n_iter):
    risk_learn[iter] = risk(x,y,beta_learn[iter])
    beta_learn[iter+1] = beta_learn[iter]
    beta_learn[iter+1] -= (eta*derivative_risk(x,y,beta_learn[iter]))
```

Finally, we can plot the trajectory of the parameter estimates:

```
# Plot parameter estimates at each iteration
plt.plot(beta_learn, marker="o")
plt.title(r'$\hat{\beta}$ vs Iteration', fontsize=24)
plt.xlabel('Iteration', fontsize=20)
plt.ylabel(r'$\hat{\beta}$', fontsize=20)
plt.xticks(fontsize=14)
plt.yticks(fontsize=14)
plt.show()
```

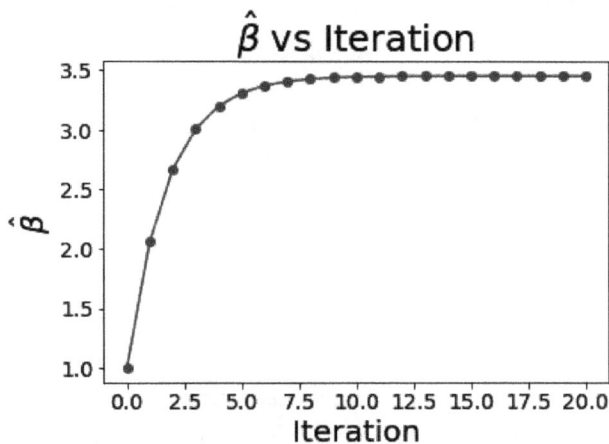

Figure 4.9: Plot of gradient descent trajectory of the model parameter estimate

We can see from *Figure 4.9* that as we perform more gradient descent iterations, the parameter estimate $\hat{\beta}$ appears to converge to a value close to 3.5. What is the value that $\hat{\beta}$ converges to? If you run the code in the notebook, you'll find that $\hat{\beta}$ converges to 3.453. This is not the true value of $\beta = 3.5$ that was used to generate the data. Why is this so? What has happened? The answer is that the empirical risk is a function of the dataset, and the data contains a random component. Because of this, the value of β that minimizes the empirical risk will be close to but not precisely the same as the true value that was used to generate the data.

Gradient descent is a general technique

The example we just walked through, including the code, is very simplistic. What it does highlight, though, is how general the update rule in *Eq. 26* is. We can extend the update rule to when we have multiple model parameters, which we'll denote by the vector β. The new update rule is the following:

$$\hat{\beta} \leftarrow \hat{\beta} - \eta \left. \frac{\partial \text{Risk}}{\partial \beta} \right|_{\beta = \hat{\beta}}$$

Eq. 27

The gradient descent update rule in *Eq. 27* can be applied to any empirical risk function, meaning we have a method of constructing optimal parameter estimates $\hat{\beta}$ for any model and any choice of loss function.

To illustrate this, imagine we wanted to use the pseudo-Huber loss function of *Eq. 12*. Our empirical risk function is given by the following:

$$\text{Empirical Risk} = \frac{1}{N} \sum_{i=1}^{N} \left[\sqrt{1 + \left(y_i - \hat{y}_i(x_i | \beta) \right)^2} - 1 \right]$$

Eq. 28

So, we have a gradient descent update rule of the following form:

$$\beta \leftarrow \beta + 2 \frac{\eta}{N} \sum_{i=1}^{N} \left[\frac{\left(y_i - y_i(x_i | \beta) \right)}{\sqrt{1 + \left(y_i - y_i(x_i | \beta) \right)^2}} \frac{\partial y_i(x_i | \beta)}{\partial \beta} \right]_{\beta = \hat{\beta}}$$

Eq. 29

Beyond simple gradient descent

Although being an iterative technique rather than a closed-form solution, gradient descent is a very general technique. This makes it a very powerful technique. Consequently, it has been widely studied, modified, and adapted in different ways. One of the main drivers of this is the fact that variants of gradient descent are used in the training of **neural networks** (**NNs**), including **deep learning** (**DL**) NNs. We don't have space here to go into the full mathematical detail of the various adaptations of gradient descent. Instead, we will briefly describe some of the most important modifications and concepts:

- **Local minima**: The risk function shown in *Figure 4.7* has a single minimum. For real-world datasets, it is typical for a risk function to have multiple minima. Since the process of gradient descent is akin to the red dot shown in *Figure 4.7* rolling down the slope of the risk function, it means gradient descent will roll downhill to the minimum closest to its starting point. Consequently, the result of a simple gradient descent can vary according to where we start. One pragmatic solution to this is to run the gradient descent multiple times and keep the results from the run with the lowest final value of the risk function.

- **Stochastic gradient descent (SGD)**: If we have a large training dataset, calculating even just a single iteration of the update rule given by *Eq. 27* may be computationally costly. An alternative approach is to evaluate the empirical risk (and hence the update rule) for just a random subset of the training data. At one extreme, we can just select a single training datapoint to update our model parameter estimate $\hat{\beta}$. This leads to quicker updating of $\hat{\beta}$ but we must perform multiple updates (that is, random selections of single training datapoints) to obtain a reliable overall estimate $\hat{\beta}$. We may cycle through the entire training data but in a random order. Because of this random order in which we use the training data to calculate parameter updates, this approach is called SGD. An alternative version of SGD is to use bigger random subsets of the training data to calculate the empirical risk in the update rule in *Eq. 27*. This approach is known as **mini-batch** SGD since we are using the training data in small batches, rather than just its entirety as we did in the simple version of gradient descent. In fact, the simple version of gradient descent is also known as batch gradient descent because we are using the entire batch of training data in one go.

- **Adaptive gradient descent algorithms**: The simple version of gradient descent we demonstrated had a fixed learning rate. We explained that this learning rate had to be chosen well, but we didn't explain how to do this. Tuning the learning rate can be an art. However, adaptive gradient descent algorithms attempt to automatically tune parameters such as the learning rate, so we have an adaptive learning rate whose value depends on where we are on the empirical risk function surface. There are a number of these adaptive gradient descent algorithms. Some of the more well-known (and well-used) ones include the AdaGrad optimizer and the Adam optimizer. The Adam optimizer also makes use of the concept of **momentum**, whereby the updates to the parameter estimate $\hat{\beta}$ are based not just on the current risk gradient $\frac{\partial \text{Risk}}{\partial \beta}\big|_{\beta=\hat{\beta}}$ but also on the gradient at the preceding iterations.

That brief summary of some of the improvements on simple gradient descent is a good place to stop and recap what we have learned in this section and summarize the chapter overall.

What we learned

In this section, we have learned the following:

- How to use the derivative of the empirical risk to estimate model parameters with any choice of loss function

- How the simple idea of gradient descent has been extended to create sophisticated gradient-based general optimization algorithms

Summary

This chapter has focused on a single, but important, concept – loss functions. Loss functions are important because they help us measure how good our predictive models are and, more generally, how well one mathematical object approximates another. They are also important because we can minimize them with respect to our model parameters, and so we can use loss functions, or more specifically risk functions, to fit our models to training data. In this chapter, we have learned about the different aspects of risk functions and how to minimize them. Specifically, we have learned about the following:

- What a loss function is and what it measures
- That a risk function is the expectation value of a loss function
- What the empirical risk function is and how it is calculated from training data
- How least squares minimization is a form of empirical risk minimization and can be used to estimate optimal parameter values for a model
- How OLS regression performs least squares minimization for linear models
- How to derive the closed-form formula for the OLS parameter estimates of a linear model
- How to perform OLS regression using specialized regression packages such as `statsmodels` and via explicit calculation using the closed-form formula
- How empirical risk minimization can also be performed very generally via gradient descent
- Variants of simple gradient descent, such as SGD, and also adaptive gradient descent algorithms such as the Adam optimizer

Now we have learned how to measure how good a model is, we'll move on to the task of learning the math behind the building of those models. Since those models will be built on data and data always contains a random component, it is natural for the building of models to use the probability concepts we introduced in *Chapter 2*. That is why, in the next chapter, we'll learn about probabilistic models.

Exercises

Next is a series of exercises. Answers to all the exercises are given in the `Answers_to_Exercises_Chap4.ipynb` Jupyter notebook in the GitHub repository:

1. Look at the documentation for the `scikit-learn` class named `sklearn.linear_model.LinearRegression`, which can fit a linear model using OLS regression. See if you can use it to fit a linear model to the power-plant output data that we analyzed in the code example in the *Linear models* section of this chapter. Do you get the same parameter estimates as when we used the `statsmodels` package?

2. The data plotted in *Figure 4.3* is stored in the `Data/outliers_example.csv` file of the GitHub repository. Using the pseudo-Huber loss function in *Eq. 12* and a learning rate of $\eta = 0.05$, see if you can use the simple gradient descent algorithm to construct robust estimates for both the intercept and the slope for a linear model of the data.

3. The data in the `Data/nls_example.csv` file of the GitHub repository contains data that has been generated according to the following relationship:

$$y = A + Be^{-Cx}$$

Eq. 30

The y values in the dataset have been corrupted by noise; that is, they also contain an additive random component. Use least squares minimization to estimate suitable values for the parameters A, B, and C. That is, minimize the empirical risk using a squared-loss function and a model of the form in the preceding equation. You can assume that the parameter C is strictly positive; that is, $C > 0$.

5

Probabilistic Modeling

At the heart of probabilistic modeling is the idea that because data is random, and so follows a probability distribution, our models of that data must also follow a probability distribution and be probabilistic models from the outset. To understand how to build those models, we must first understand the probability distribution that the data follows. From this, we can calculate the distribution that our model parameters follow by using one of the most famous theorems in probability theory. To do all of this, we will cover the following topics:

- *Likelihood*: In this section, we will learn about the probability distribution of the data given a model

- *Bayes' theorem*: In this section, we will learn how to work with conditional probabilities and calculate the probability of a model given the data

- *Bayesian modeling*: In this section, we will learn how to use the probability of the model given the data to make useful inferences

- *Bayesian modeling in practice*: In this section, we will learn practical computational techniques and mathematical approximations for doing Bayesian modeling

Technical requirements

All code examples given in this chapter (and additional examples) can be found at the GitHub repository: https://github.com/PacktPublishing/15-Math-Concepts-Every-Data-Scientist-Should-Know/tree/main/Chapter05. To run the Jupyter notebooks, you will need a full Python installation, including the following packages:

- `pandas` (>=2.0.3)

- `numpy` (>=1.3)

- `scipy` (>=1.1)

- `matplotlib` (>=3.7.2)

Likelihood

To understand the probability distribution that the data follows, we'll look at an explicit example of how a random component is incorporated into data.

A simple probabilistic model

We'll start with the simplest way in which we can introduce a random component into our observations of the response (target) variable y, namely by adding noise to a deterministic quantity. In fact, we'll just consider the observations y_i in our dataset to be noise-corrupted versions of a model output $\hat{y}(\underline{x}_i \mid \underline{\beta})$. So, we have this relationship:

$$y_i = \hat{y}(\underline{x}_i \mid \underline{\beta}) + \varepsilon_i$$

Eq. 1

Here, ε_i is the noise value that has been added to the model output $\hat{y}_i = \hat{y}(\underline{x}_i \mid \underline{\beta})$ to get the observation y_i for the ith datapoint. The value ε_i is a random variable. Without loss of generality, we can assume its expectation value is zero, so we have $\mathbb{E}(\varepsilon_i) = 0$. We can make this assumption because if the expectation of ε_i was non-zero, it would mean we have a non-zero deterministic average contribution from ε_i that we could just absorb into the definition of $\hat{y}_i$. This is just the same as saying that we can write:

$$\varepsilon_i = a + \tilde{\varepsilon}_i$$

Eq. 2

$a = \mathbb{E}(\varepsilon_i)$, and $\tilde{\varepsilon}_i$ is just a new random variable that does have zero expectation (by construction). We then just add a into the definition of our model $\hat{y}$.

The simple scenario we shall consider is where the noise values ε_i for the different data points are uncorrelated with each other. Mathematically, we write this assumption as $\mathbb{E}(\varepsilon_i \varepsilon_{i'}) = 0$, for all $i' \neq i$.

Now, if each noise value ε_i, $i = 1, \ldots, N$ is a random variable; what probability distributions should those random noise values come from? Again, we consider the simplest situation, where each noise value ε_i is drawn from the same distribution. Note that this doesn't mean that all the noise values are identical; it means that they are identically distributed.

What should we use for this common noise distribution? Let's use one of the most common distributions, the Gaussian (or normal) distribution. So, we can write the probability density of ε_i as follows:

$$p(\varepsilon_i) = \frac{1}{\sqrt{2\pi\sigma^2}} \exp\left(-\frac{1}{2\sigma^2}\varepsilon_i^2\right)$$

Eq. 3

The quantity σ is the standard deviation of the noise value ε_i, and this is the same for all data points i. The bigger the value of σ, the bigger the typical values of the noise will be, i.e., the more the observations y_i will differ from the deterministic output from our model.

Since we are using a Gaussian distribution for each of the noise values, the fact that different noise values are uncorrelated means they are statistically independent. So, our noise values are **independently and identically distributed**, or **i.i.d.**

In *Chapter 2*, we said that anything derived from a random variable was itself a random variable. Well, if ε_i is a random variable, then Eq. 1 means that y_i is a random variable – as we would expect, it is data after all. But what is the probability distribution for y_i? The combination of Eq. 1 and Eq. 3 tells us. If we re-arrange Eq. 1 we get this:

$$\varepsilon_i = y_i - \hat{y}\left(\underline{x}_i \mid \underline{\beta}\right) = y_i - \hat{y}_i \; .$$

Eq. 4

If we plug the expression for ε_i in Eq. 4 into Eq. 3, we get this:

$$p\left(y_i \mid \hat{y}_i\right) = \frac{1}{\sqrt{2\pi\sigma^2}} \exp\left(-\frac{1}{2\sigma^2}(y_i - \hat{y}_i)^2\right)$$

Eq. 5

This means that y_i is a Gaussian random variable distributed around a mean of $\hat{y}_i$ with a standard deviation of σ. Since $\hat{y}_i$ is a deterministic function of the input features $\underline{x}_i$ and the model parameters $\underline{\beta}$, then $p\left(y_i \mid \hat{y}_i\right)$ means $p\left(y_i \mid \underline{x}_i, \underline{\beta}\right)$, so we can re-write the probability density for y_i in Eq. 5 as follows:

$$p\left(y_i \mid \underline{x}_i, \underline{\beta}\right) = \frac{1}{\sqrt{2\pi\sigma^2}} \exp\left(-\frac{1}{2\sigma^2}\left(y_i - \hat{y}\left(\underline{x}_i \mid \underline{\beta}\right)\right)^2\right)$$

Eq. 6

This is the probability density for the ith observation y_i. Since the observations are independent, then the probability density for the whole set of observations is just the product of the individual densities, and so we have this:

$$p\left(\underline{y} \mid \underline{X}, \underline{\beta}\right) = \prod_{i=1}^{N} p\left(y_i \mid \underline{x}_i, \underline{\beta}\right)$$

Eq. 7

This probability density function gives the probability density of the observed data y. It is called the **likelihood of the data**, or simply the **likelihood**, and is usually denoted by the symbol L. If we were dealing with discrete observations instead of continuous ones then we would compute the probability of the data, rather than the probability density of the data, but in either case, we refer to the resulting function as the likelihood.

The expression in Eq. 7 is the general form for the likelihood when we have independent observations. If we plug in the expression in Eq. 6 into Eq. 7, we obtain an explicit expression for the likelihood L for our specific model. In this case, the likelihood is as follows:

$$L = \prod_{i=1}^{N} \frac{1}{\sqrt{2\pi\sigma^2}} \exp\left(-\frac{1}{2\sigma^2}\left(y_i - \hat{y}\left(x_i \mid \beta\right)\right)^2\right)$$

Eq. 8

In shorthand, whether we are dealing with a probability or a probability density, we often refer to the likelihood using the notation $P(\text{Data} \mid \text{Model})$, $p(\text{Data} \mid \text{Model})$, or $p\left(\text{Data} \mid \beta\right)$. This is sometimes shortened to $P(\text{D} \mid \text{Model})$, $p(\text{D} \mid \text{Model})$, or $p\left(\text{D} \mid \beta\right)$, with D standing for data.

The likelihood is an extremely useful quantity. In fact, it is the central quantity when developing probabilistic models, and as we have learned, all models should be probabilistic. Given the importance of the likelihood, we're going to look at some of its properties in detail.

Log likelihood

The expression in Eq. 8 looks complicated. It's a horrible, complicated product of expressions. We can simplify it a bit by converting it to a sum. We do this by taking the logarithm. From our rules of logarithms recapped in *Chapter 1*, we have this:

$$\log \text{Likelihood} = \log L = \sum_{i=1}^{N} \log p\left(y_i \mid x_i, \beta\right)$$

Eq. 9

And for our example in Eq. 8, we have this:

$$\log \text{Likelihood} = \log L = \sum_{i=1}^{N} \left[-\frac{1}{2}\log(2\pi\sigma^2) - \frac{1}{2\sigma^2}\left(y_i - \hat{y}\left(x_i \mid \beta\right)\right)^2\right]$$

$$= -\frac{N}{2}\log(2\pi\sigma^2) - \frac{1}{2\sigma^2}\sum_{i=1}^{N}\left(y_i - \hat{y}\left(x_i \mid \beta\right)\right)^2$$

Eq. 10

The first thing to point out is that the typical value of the log-likelihood scales linearly with the number of data points N. For independent observations, each data point will make a contribution

$\log p(y_i \mid \hat{y}_i)$, and these contributions build up with each observation. Even when the observations are not independent, but are correlated, we would still expect the $\mathbb{E}$(log Likelihood) to scale linearly with N, as $N \to \infty$, in most cases. The correlations between observations will reduce the amount of information that each observation adds to the log-likelihood, but each observation will still add some information. Overall, this means that the larger the N, the larger the magnitude of the log-likelihood.

The second thing we can point out about the log-likelihood is that it is not only a function of the model parameters β, but also of the parameters that control the random component in our data; in this case the parameter σ.

Maximum likelihood estimation

The likelihood is $p(\text{Data} \mid \beta)$, where we mean either probability or probability density depending on whether we are talking about discrete data or continuous data. How can we use the likelihood to estimate model parameter values that are a good fit for the data?

If we have a good model, we expect the probability of the observed data to be high. After all, we have observed the data. If we had parameter values for which $p(\text{Data} \mid \beta)$ is low, that would say the probability of the data we have actually observed is low. This would seem strange and perhaps just the result of the random sampling variation – it is possible but unlikely. In contrast, parameter values for which $p(\text{Data} \mid \beta)$ is high appear to be more appropriate and a more credible explanation of the data. This gives us a means of estimating the model parameters β. We estimate β by maximizing $p(\text{Data} \mid \beta)$ with respect to β. This is called **maximum likelihood estimation**.

> **Pro tip**
>
> As a technique, maximum likelihood is widely used in data science. However, maximum likelihood was used long before the term data science was coined. Maximum likelihood is taught as part of traditional courses in statistics. The abbreviation **ML** for **maximum likelihood** is still used in some research areas. In data science, this has the potential to confuse, since most practitioners of data science would interpret **ML** to mean **machine learning**. So, be aware there may be data science circumstances where **ML** means **maximum likelihood**.

Least squares as an example of maximum likelihood

We'll look at what happens if we estimate the parameters of our model $\hat{y}(x \mid \beta)$ via maximum likelihood.

You'll recall from the recap in *Chapter 1* that the location of the maximum of a function is the same as the location of the maximum of the logarithm of the function. This is because the logarithm is a monotonically increasing transformation. Maximizing the likelihood L with respect to the model parameters β is the same as maximizing the log-likelihood, $\log L$, with respect to the model parameters β. The latter is easier to do. For our model, the log-likelihood is given by Eq. 10. So, let's maximize Eq. 10.

Technically, we should maximize $\log L$ with respect to all the parameters of the log-likelihood, so with respect to β and σ. We will do this, but for now, let's take σ as given, i.e., fixed. We can, in theory, maximize with respect to β at fixed σ to find the maximum likelihood values $\hat{\beta}(\sigma)$ as a function of σ and plug $\hat{\beta}(\sigma)$ back into Eq. 10, and then finally maximize the resulting expression with respect to σ. However, for now σ is fixed. From the relatively simple form of Eq. 10, we can see that maximizing Eq. 10 with respect to β at fixed σ is the same as maximizing the following:

$$-\sum_{i=1}^{N}\left(y_i - \hat{y}\left(x_i \mid \underline{\beta}\right)\right)^2$$

Eq. 11

This is clearly the same as minimizing (on dropping the minus sign):

$$\sum_{i=1}^{N}\left(y_i - \hat{y}\left(x_i \mid \underline{\beta}\right)\right)^2$$

Eq. 12

So, to maximize the likelihood, we minimize the expression in Eq. 12. Wait a minute! Haven't we seen the expression in Eq. 12 somewhere before, and even minimized it? The answer is yes. Estimating the model parameters β by minimizing the expression in Eq. 12 is least squares estimation!

We covered least squares in *Chapter 4*, so we don't need to go into further detail in this chapter. However, it does show that we can begin to put least squares onto a more rigorous footing.

So far, we have kept the noise standard deviation σ fixed. Do we still have this equivalence between maximum likelihood estimation and least squares estimation if we maximize the likelihood with respect to all parameters, not just β? Let's see. To locate the maximum of $\log L$, we use the differential calculus we recapped in *Chapter 1*. Specifically, we solve the following equations:

$$\left.\frac{\partial \log L}{\partial \beta}\right|_{\beta=\hat{\beta},\ \sigma=\hat{\sigma}} = 0 \qquad \left.\frac{\partial \log L}{\partial \sigma}\right|_{\beta=\hat{\beta},\ \sigma=\hat{\sigma}} = 0$$

Eq. 13

Solving the equations in Eq. 13 will give us the maximum likelihood estimates $\hat{\beta}$ and $\hat{\sigma}$. If we take the left-hand equation in Eq. 13 for our particular log-likelihood expression in Eq. 10, then differentiating with respect to β gives the following equation:

$$-\frac{1}{2\hat{\sigma}^2}\frac{\partial}{\partial\beta}\sum_{i=1}^{N}\left(y_i - \hat{y}\left(x_i \mid \beta\right)\right)^2\Bigg|_{\beta=\hat{\beta}} = 0$$

Eq. 14

We can re-arrange this to get the following:

$$\frac{\partial}{\partial \beta} \sum_{i=1}^{N} \left(y_i - \hat{y}\left(x_i \mid \beta \right) \right)^2 \Bigg|_{\beta = \hat{\beta}} = 0$$

Eq. 15

This is the criterion we would get if we were minimizing Eq. 12. So, in this instance, solving for the maximum likelihood value of β is the same as least squares minimization, irrespective of the particular value of σ.

If we take the right-hand equation in Eq. 13 for our log-likelihood expression in Eq. 10, then differentiating with respect to σ gives the following equation:

$$-\frac{N}{\hat{\sigma}} + \frac{1}{\hat{\sigma}^3} \sum_{i=1}^{N} \left(y_i - \hat{y}\left(x_i \mid \hat{\beta} \right) \right)^2 = 0$$

Eq. 16

We can re-arrange this to get the following:

$$\hat{\sigma}^2 = \frac{1}{N} \sum_{i=1}^{N} \left(y_i - \hat{y}\left(x_i \mid \hat{\beta} \right) \right)^2$$

Eq. 17

Eq. 17 tells us that our estimate of the noise variance is just the mean squared residual once we have solved Eq. 15 to get the value of $\hat{\beta}$. So, we can determine the maximum likelihood estimate $\hat{\sigma}$ after and separately from calculating the maximum likelihood estimate $\hat{\beta}$.

This convenient separation of the maximum likelihood estimation into the estimation of the model parameters β distinctly from the estimation of the parameters controlling the random component is a consequence of the noise being Gaussian additive. For many modeling problems, the random component is not Gaussian additive. So, what does maximum likelihood estimation look like in these situations?

Maximum likelihood estimation for non-additive randomness

We often start with the form of the distribution that we think the data follows. For example, for an observation y_i that is a positive and continuous quantity, we might have good reason to model the observation using an exponential distribution. The exponential distribution has a single parameter, λ. In terms of λ we would write this:

$$y_i \sim \text{Exponential}(\lambda)$$

Eq. 18

The mean of an exponential distribution is easily calculated in terms of λ, and so we have the following:

$$\mu = \mathbb{E}(y_i) = \frac{1}{\lambda}$$

Eq. 19

Obviously, the mean μ is the typical value we would see for y_i, and so μ specifies the systematic component in the data. Now, our model $\hat{y}_i$ is deterministic, so we should relate it to the systematic component of y_i in some form. We often construct our model $\hat{y}_i$ as being a model of the mean of y_i, so we can write $\hat{y}_i = \mathbb{E}(y_i)$. Given the relation in Eq. 19 and equating $\hat{y}_i = \mathbb{E}(y_i)$, we can re-arrange to get $\lambda = 1/\hat{y}_i$. Plugging this into Eq. 18 gives us this:

$$y_i \sim \text{Exponential}\left(\frac{1}{\hat{y}_i}\right)$$

Eq. 20

The probability density of the exponential distribution in Eq. 18 is given by the following:

$$p(y_i|\lambda) = \lambda \exp(-\lambda y_i)$$

Eq. 21

So, if we plug $\lambda = 1/\hat{y}_i$ into Eq. 21 and take logs, we can write the following:

$$\log p(y_i \mid \hat{y}_i) = -\log \hat{y}_i - \frac{y_i}{\hat{y}_i}$$

Eq. 22

From Eq. 22, we can then write down the log-likelihood of N exponentially distributed independent observations as follows:

$$\log L = -\sum_{i=1}^{N}\left[\log \hat{y}(x_i \mid \beta) + \frac{y_i}{\hat{y}(x_i \mid \beta)}\right]$$

Eq. 23

From Eq. 18 to Eq. 23, we have gone from specifying what shape of random variation we think we will see in each observation through to deriving the log-likelihood in Eq. 23. We have done this by linking the parameters of the distribution, in this case, its mean λ, to our model. This is the general approach we take when building probabilistic models. This approach can be summarized as follows:

1. Specify the mathematical form of the distribution of the random component in the data.

2. Specify a predictive model from which we calculate the parameters of the distribution of the random component.

3. Calculate the likelihood of the data in terms of the distribution parameters and hence in terms of the predictive model.

Later on in this chapter, we will meet other ways of estimating the model parameters that make use of the likelihood but do not use maximum likelihood estimation. However, even for these alternative methods of parameter estimation, we will still follow the three broad steps outlined above.

At first sight, the process we went through to derive the log-likelihood in Eq. 23 looks very different to the process we used to derive the log-likelihood in Eq. 10 for independent observations with Gaussian additive noise. Actually, it isn't. We'll show this by re-deriving the log-likelihood in Eq. 10 for independent observations with Gaussian additive noise but using the steps outlined above.

Step 1: We have observations y_i that are continuous, and we think their random variation will cause them to follow a Gaussian (normal) distribution. So we will write this:

$$y_i \sim \text{Normal}(\mu_i, \sigma^2)$$

Eq. 24

Here, we have been explicit that the mean of each observation can be different, while we consider the standard deviation of the random variation to be the same for each observation.

Step 2: We specify a mathematical form for a predictive model $\hat{y}(x_i | \beta)$, which we will use as a model of the mean parameter μ_i. That is, we will put $\mu_i = \hat{y}(x_i | \beta)$, and so we have this:

$$y_i \sim \text{Normal}(\hat{y}(x_i | \beta), \sigma^2)$$

Eq. 25

We have not specified the precise details of the predictive model at this stage. That's because we don't need to yet. You are free to choose. You could use a simple linear model for $\hat{y}(x_i | \beta)$, or you could use a neural network if you felt that was justified. You can use your favorite machine learning model to model the relationship between the features x_i and the expectation $\mathbb{E}(y_i)$, as long as it is a suitably appropriate model.

Step 3: The log of the probability density of a Gaussian random variable is, in terms of μ_i and σ^2, as follows:

$$\log p(y_i | \mu_i, \sigma^2) = -\tfrac{1}{2}\log 2\pi\sigma^2 - \tfrac{1}{2\sigma^2}(y_i - \mu_i)^2$$

Eq. 26

Using Eq. 26 and plugging in our model $\hat{y}$ for μ_i, we get the log-likelihood of N Gaussian distributed independent observations:

$$\log L = -\frac{N}{2}\log 2\pi\sigma^2 - \frac{1}{2\sigma^2}\sum_{i=1}^{N}\left(y_i - \hat{y}\left(x_i\middle|\beta\right)\right)^2$$

Eq. 27

This is the same as Eq. 10.

Maximum likelihood for non-linear models

It is worth noting that for both of the preceding examples, our predictive model was directly a model of the expectation value of the observations. That is, we equated, $\mathbb{E}\left(y_i\middle|x_i\right) = \hat{y}\left(x_i\middle|\beta\right)$. This doesn't always have to be the case. The wording of step 2 was very deliberate – *"Specify a predictive model from which we calculate the parameters of the distribution of the random component."* Imagine we have a predictive model $f\left(x\middle|\beta\right)$. We can apply transformations to the output of our predictive model $f\left(x\middle|\beta\right)$ to calculate the expectation value of the observation variable y. In this case, we would equate the following:

$$\mathbb{E}(y\mid x) = g^{-1}\left(f\left(x\middle|\beta\right)\right)$$

Eq. 28

The function g is called a **link function**. It links the expectation of the observations to the predictive model. It is traditional to specify the transformation as the inverse of a function g. Now, you may ask, why not just include the final transformation g^{-1} in the definition of our predictive model f? This is because the terminology of "link functions" comes from classical statistics, where our predictive model f would typically be linear. The downstream application of a transformation g^{-1} was then a way of modeling non-linear relationships in classical statistics.

The best way to understand how link functions are used is to see one used in practice. We'll model shoppers' decisions about whether to buy a product or not. We have N shoppers in our dataset and we know whether they bought the product in question or not. We'll represent the outcome of the ith shopper's decision by the binary outcome variable n_i. So, $n_i = 0$ if the ith shopper didn't buy the product, and $n_i = 1$ if they did. The outcome n_i can be affected by several factors, such as the price of the product at the time, or the age demographic the shopper is in. We can put all the relevant factors into our feature vector x_i.

Now let's follow our 3-step recipe to derive the log-likelihood.

Step 1: The outcome variable n_i is a random variable. Why is this so? Well, even if the price and other factors in x_i are such that you'd think a particular shopper was ideally suited to the product, they may not buy it. The expectation of this random variable, $\mathbb{E}(n_i|x_i)$, is the probability p_i that the ith shopper will buy the product. The variable n_i is an example of a Bernoulli random variable that we introduced in *Chapter 2*. Modeling the shopper choice as a Bernoulli random variable is our choice for the mathematical form of the randomness in the observed data.

Step 2: Since n_i is a discrete random variable, unlike our previous two examples, we write down the probability distribution for n_i instead of a probability density. We need to express this probability distribution in terms of p_i. For a Bernoulli distribution, this is straightforward to do, and we get this:

$$P(n_i \mid p_i) = p_i^{n_i}(1 - p_i)^{1 - n_i}$$

Eq. 29

Plugging in values of $n_i = 0$ and $n_i = 1$ into Eq. 29 should convince you that we recover the correct probabilities. On the log scale, we have the following:

$$\log(P(n_i \mid p_i)) = n_i \log p_i + (1 - n_i)\log(1 - p_i)$$

Eq. 30

The probability p_i depends upon the feature vector x_i, and it is this relationship that we will build our predictive model for. We'll consider the simplest possible form for our predictive model, namely a linear one. Without loss of generality, we can write this linear model in the form $x^\top\beta$, where we have used our usual trick of including a constant by having a feature $x_0 = 1$ for all shoppers, with a corresponding parameter β_0. However, we have a problem! Our linear model can potentially take any value between $-\infty$ and $+\infty$, but a probability can only be between 0 and 1. So, we need a transformation that will map the whole real line $(-\infty, \infty)$ to the interval $[0,1]$. To do this, we use a simple sigmoid transformation called the **logistic function**:

$$p_i = p(x_i \mid \beta) = \frac{\exp(x_i^\top\beta)}{1 + \exp(x_i^\top\beta)} = \frac{1}{1 + \exp(-x_i^\top\beta)}$$

Eq. 31

The quantity $x^\top\beta$ is called the **linear predictor** because it is the linear part of the prediction model. It is often denoted by the symbol η. The plot in *Figure 5.1* shows how the logistic function transformation function in Eq. 31 varies with the linear predictor η.

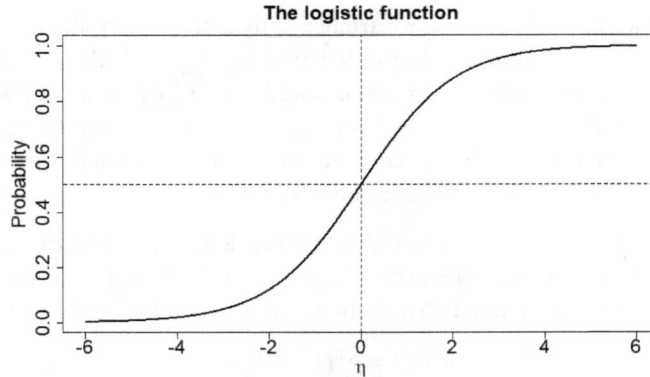

Figure 5.1: The logistic function

The plot in *Figure 5.1* is a plot of the function $g^{-1}(\eta)$, with $g^{-1}(\eta)$ given by the following:

$$g^{-1}(\eta) = \frac{1}{1+\exp(-\eta)}$$

Eq. 32

This function is the inverse of our link function, so what does the link function $g(p)$ look like? If we invert Eq. 32, we get this:

$$g(p) = \log\left(\frac{p}{1-p}\right)$$

Eq. 33

This is our link function. It is called the **logit function**.

Step 3: Now that we have our predictive model, we can calculate the log-likelihood. For each of our N shoppers, we have the observation n_i and the feature vector $\underline{x}_i$. Putting together Eq. 30 and Eq. 31, we get this:

$$\log L = \sum_{i=1}^{N}\left[n_i\log\left(\frac{\exp(\underline{x}_i^\top\underline{\beta})}{1+\exp(\underline{x}_i^\top\underline{\beta})}\right) + (1-n_i)\log\left(\frac{1}{1+\exp(\underline{x}_i^\top\underline{\beta})}\right)\right]$$

$$= \sum_{i=1}^{N}\left[n_i\underline{x}_i^\top\underline{\beta} - \log\left(1+\exp(\underline{x}_i^\top\underline{\beta})\right)\right]$$

Eq. 34

The log-likelihood in Eq. 34 can then be maximized with respect to the model parameters $\underline{\beta}$ to give the maximum likelihood estimate for $\underline{\beta}$. From this example, we can see how we can use maximum likelihood to train a model that includes a link function.

That is enough for now about likelihood, so let's finish this section by summarizing what we have learned.

What we have learned

In this section, we have learned the following:

- How the likelihood measures the probability of a dataset given the parameters of a predictive model and the parameters of a random process

- How maximum likelihood provides us with a way to use the likelihood to estimate the model parameters from data

- How maximum likelihood estimation of a predictive model's parameters corresponds to least squares estimation of the parameters, when we have data that contains i.i.d. Gaussian additive noise

- How to formulate a probabilistic model for any modeling problem using a short series of steps

Having learned about likelihood as the probability of the data given the model parameters, in the next section, we will move on to learning how to formulate the probability of the model parameters given the data.

Bayes' theorem

When we learned about maximum likelihood for estimating the parameters of a model, it felt like an intuitively sensible thing to do. Who can argue with the idea of choosing the model parameters β so that we have the highest possible probability of obtaining the data we have actually observed? But we didn't really derive maximum likelihood in any formal way. Yes, choosing parameters by maximizing $P(\text{Data} \mid \text{Model})$ seems reasonable, but aren't we really interested in the probability of the parameters given the data, that is $P(\text{Model} \mid \text{Data})$? Working with the likelihood $P(\text{Data} \mid \text{Model})$ seems close to what we want, but not quite there. If only there was a way we could calculate $P(\text{Model} \mid \text{Data})$ from $P(\text{Data} \mid \text{Model})$. There is. Enter **Bayes' theorem**.

This section will be relatively short as we will only introduce Bayes' theorem here. In the next two sections, we will explain how Bayes' theorem is used in practice.

Conditional probability and Bayes' theorem

Bayes' theorem, named after the Reverend Thomas Bayes, is about conditional probabilities. The probability $P(\text{Data} \mid \text{Model})$ is a conditional probability. It is the probability of the data **given** the model, or in other words the probability of the data conditional on having that particular model form with that particular set of model parameter values.

Bayes' theorem says that if we have two events A and B (things whose probabilities we are interested in knowing), then the conditional probabilities $P(A|B)$ and $P(B|A)$ are related via this formula:

$$P(B|A) = \frac{P(A|B)P(B)}{P(A)}$$

Eq. 35

You may wonder where Eq. 35 comes from. The form in Eq. 35 hides its simplicity. Consider, instead the joint probability $P(A, B)$. Now, from the rules of conditional probability, $P(A, B) = P(A|B)P(B)$. In words, the probability of A and B happening together is the probability of A happening given I know B has happened, multiplied by the probability that B has indeed actually happened. But here's the neat trick. The joint probability $P(A, B) = P(B, A)$ is just the probability of both A and B happening together, or both being true, so the ordering doesn't matter. This means we can also write $P(A, B) = P(B|A)P(A)$. Again, in words this make sense – it is the probability of B happening given I know A has happened, multiplied by the probability that A has indeed actually happened. If we now equate these two expressions for $P(A, B)$ we get this:

$$P(A, B) = P(A|B)P(B) = P(B|A)P(A)$$

Eq. 36

If we re-arrange Eq. 36, we get Bayes' theorem in Eq. 35. In fact, the expression in Eq. 36 is how I always remember Bayes' theorem when doing Bayesian modeling. If I have to derive or understand what looks like a complicated Bayesian formula, I just write out the joint probability of whatever events $(A, B, ...)$ I'm interested in (in the multiple ways involving conditional probabilities), and then I re-arrange it.

So, how does Bayes' theorem help us with working out $P(\text{Model} | \text{Data})$? Well, let's put $A = \text{Data}$ and $B = \text{Model}$, and plug them into Bayes' theorem in Eq. 35. We get this:

$$P(\text{Model} | \text{Data}) = \frac{P(\text{Data} | \text{Model}) \times P(\text{Model})}{P(\text{Data})}$$

Eq. 37

Excellent! We have a means of calculating $P(\text{Model} | \text{Data})$.

We have written Bayes' theorem in terms of probabilities. However, for many of our discussions on likelihood, we have been talking about probability densities. As you might expect, we can also apply Bayes' theorem to probability densities by replacing the probabilities in Eq. 37 with the corresponding densities. From now on, when discussing Bayes' theorem, we will for brevity refer to probabilities even when we mean probability densities. It will be clear from the context whether we are dealing with a probability or a probability density.

Finally, we should emphasize that when we refer to a model, for example, in Eq. 37, we mean both the parameters β for the predictive model, but also the parameters that control the random component of the data, such as the noise variance σ^2 in our model in Eq. 1. We will use the symbol $\underline{\theta}$ in general to represent the combined parameters from the predictive model and the random component, so $\underline{\theta} = (\underline{\beta}, \sigma^2)$ for our model in Eq. 1. We will use the word "model" and the symbol $\underline{\theta}$ interchangeably.

Priors

Let's unpack Eq. 37 in a bit more detail. Firstly, what is the quantity $P(\text{Model})$? At face value it is the probability of the model. But conditional on what? Well, nothing. Not on the data. So, it is just the probability of the model before, or prior, to us receiving the data. For this reason, it is called a prior probability, or simply a **prior**.

The symbol $\underline{\theta}$ summarizes the model, so $P(\text{Model})$ is $P(\underline{\theta})$ and is the probability we attach to the parameters $\underline{\theta}$ before we have seen any data. It therefore encapsulates our pre-defined beliefs about what values $\underline{\theta}$ should take. So, in most cases, $P(\underline{\theta})$ is subjective – the probability that you might attach to a particular value of $\underline{\theta}$ could be different from the probability that I attach to that same value of $\underline{\theta}$.

This subjective element of the prior is the reason why some people are not keen on Bayesian methods. Consequently, an alternative and often used approach is to use an **uninformative prior**; that is a prior that is not based on any subjective belief, but only on incontrovertible facts we know about $\underline{\theta}$, such as the upper and lower bounds for parameter values or based on the geometry of the space in which the parameters lie.

Whether we use a subjective prior or uninformative prior, in both cases we are constructing the prior from information. In the case of an uninformative prior, that information is the minimal properties that the parameters must satisfy, while in the case where I have constructed the prior from my own subjective beliefs, I am constructing the prior based upon information coming from my expert judgment, possibly based on years of domain experience. So, another way to think about the prior $P(\underline{\theta})$ is that it is the distribution of $\underline{\theta}$ based upon whatever information we have available to us before we have received the data for the current analysis. The prior information that we have available to us could even be from a previous analysis.

The posterior

We dealt with and explained the factor $P(\underline{\theta})$ in the Eq. 37 form of Bayes' theorem. What about the probability $P(\text{Data})$ in the denominator of Eq. 37? It looks like a prior. The more appropriate way to interpret $P(\text{Data})$ is as a normalizing factor for the probability $P(\text{Model} \mid \text{Data})$. From the rules of probability, we have this:

$$P(\text{Data}) = \int P(\text{Data}, \underline{\theta})\, d\underline{\theta} = \int P(\text{Data} \mid \underline{\theta})\, P(\underline{\theta})\, d\underline{\theta}$$

Eq. 38

This means we can write Bayes' theorem in Eq. 37 as follows:

$$P(\theta \mid \text{Data}) = \frac{P(\text{Data} \mid \theta)P(\theta)}{\int P(\text{Data} \mid \theta)P(\theta)d\theta}$$

Eq. 39

The denominator is just making sure the probabilities over all possible values of θ add up to 1.

Let's return to the numerator of Eq. 37. The probability $P(\text{Data} \mid \text{Model})$ is the likelihood. This means Bayes' theorem can be written as follows:

$$P(\text{Model} \mid \text{Data}) \propto \text{Likelihood} \times \text{Prior}$$

Eq. 40

We can just re-arrange that to write this:

$$P(\text{Model} \mid \text{Data}) \propto \text{Prior} \times \text{Likelihood}$$

Eq. 41

How does Eq. 41 help us? It tells us that the probability of the model θ given the data is our prior probability multiplied by the likelihood (and appropriately normalized). The likelihood has updated the distribution of θ from what we believed before we got the data – the prior $P(\theta)$ – to what we believe after we get the data – the distribution $P(\theta \mid \text{Data})$. Because of this, $P(\theta \mid \text{Data})$ is called the **posterior distribution** or simply the **posterior**. In simple terms, the prior is what we think is the distribution of θ **before** we get the data, while the posterior is what we think is the distribution of θ **after** we get the data – hence the name posterior.

We said this section would be short, so we'll now recap what we have learned.

What we have learned

In this section, we have learned about the following:

- Bayes' theorem and how we can use it to calculate the probability distribution of the model parameters given the data
- Prior distributions and how they encode the beliefs we already have about model parameters before we have received any new data or information
- The posterior distribution and how it is calculated by multiplying the likelihood of the data and the prior
- How the posterior distribution represents our belief of the distribution of the model parameters after we have received the new data or information

Having introduced Bayes' theorem and the concepts of prior and posterior distributions, in the next section, we're going to move onto how the posterior distribution is used in Bayesian modeling.

Bayesian modeling

The posterior $P(\underline{\theta} \mid \text{Data})$ encapsulates the philosophy of Bayesian modeling and changes how we view the model represented by $\underline{\theta}$. In Bayesian modeling, there is not a single "correct" underlying value of $\underline{\theta}$, for which we construct uncertain estimates. Instead, different values of $\underline{\theta}$ have different probabilities given the available data or evidence. $\underline{\theta}$ is a random variable, and we update what we think is the distribution of that random variable using Bayes' theorem and the additional data or information we receive.

With that statement about the philosophical interpretation of the posterior made, we now move on to how we use the posterior in a calculational sense. There are two potential ways in which we can use the posterior distribution:

- To evaluate expectation values. Here, we are using the posterior as it is intended, as a distribution. Here, the posterior is used to calculate predictions over lots of different models. This is called *Bayesian model averaging*.

- To identify a suitable single value, or point estimate, of the parameter vector $\underline{\theta}$ that we can use as a single model to make predictions with. The most obvious point value we can use is the value of $\underline{\theta}$ that is the most probable given the data. We can identify this value of $\underline{\theta}$ by maximizing the posterior $P(\underline{\theta} \mid \text{Data})$ with respect to $\underline{\theta}$. Since we are maximizing the posterior, this is known as the **maximum a posteriori** estimate, or **MAP** estimate for short. The MAP estimate of $\underline{\theta}$ satisfies this:

$$\underline{\theta}_{MAP} = \underset{\underline{\theta}}{\text{argmax}} \; P(\underline{\theta} \mid \text{Data})$$

Eq. 42

We will normally try and identify MAP estimates by solving the stationarity condition:

$$\frac{\partial P(\underline{\theta} \mid \text{Data})}{\partial \underline{\theta}} \bigg|_{\underline{\theta} = \underline{\theta}_{MAP}} = \underline{0}$$

Eq. 43

Bayesian model averaging and MAP estimation both have their advantages and disadvantages. It is worth covering those advantages and disadvantages in detail, so we will do so next.

Bayesian model averaging

Bayesian model averaging can be as simple as wanting to know what the typical value of the model parameters is given all the data or information available to us to date. In this case, we would calculate the expectation $\mathbb{E}(\theta \mid \text{Data})$ over the posterior distribution, so in other words, we calculate this:

$$\mathbb{E}(\theta \mid \text{Data}) = \int \theta P(\theta \mid \text{Data}) \, d\theta$$

Eq. 44

We can also use Bayesian model averaging to calculate the posterior covariance of θ. This would be calculated as follows:

$$\mathbb{E}(\theta \, \theta^{\mathsf{T}} \mid \text{Data}) - \mathbb{E}(\theta \mid \text{Data})\mathbb{E}(\theta^{\mathsf{T}} \mid \text{Data}) = \int \theta \theta^{\mathsf{T}} P(\theta \mid \text{Data}) \, d\theta - \mathbb{E}(\theta \mid \text{Data})\mathbb{E}(\theta^{\mathsf{T}} \mid \text{Data})$$

Eq. 45

Bayesian averaging gives us a convenient method for quantifying the spread of values that are possible for the model parameters θ given the data we have received.

In another situation, we might want to calculate the model prediction for some new feature value x. We want to calculate $\hat{y}(x \mid \beta)$. But what value of model parameters β do we use? Obviously, we want to use values of β that are guided by the data. However, remember in a Bayesian framework that β is a random variable, so this makes the prediction $\hat{y}(x \mid \beta)$ also a random variable. The solution is to calculate the expectation value of the prediction $\hat{y}$ over the posterior distribution. So, we calculate the following:

$$\mathbb{E}\left(\hat{y}(x \mid \beta) \mid \text{Data}\right) = \int \hat{y}(x \mid \beta) P(\theta \mid \text{Data}) \, d\theta$$

Eq. 46

You may ask why the expression in Eq. 46 is an average over the parameters θ and not over just β? After all, it is the average of the prediction we are interested in calculating, and the predictive model only depends on the parameter β. The answer is that there may be correlations between the parameter β and the parameters controlling the random component in the data. So when, say, σ^2 changes, we can't ignore the fact that this will change the probability of a given value of β. Consequently, we must average over both β and any parameters controlling the random component in the data, i.e., we average over θ.

Bayesian model averaging constructs a consensus from many models

The expression in Eq. 46 is useful because it allows us to calculate the typical value we would expect from our prediction, given the new input feature vector x. ·

One of the real benefits of Bayesian model averaging is the fact that we are taking an average over multiple models. The models that contribute most to the average in Eq. 46 will be those that have the highest probability given the data. There can be many models that are nearly all equally likely but some of which may give very different values for the prediction $\hat{y}(\underline{x} \mid \underline{\beta})$. Averaging over all these high-probability models ensures we have a suitable and sensible consensus for our prediction of what will happen at the new feature vector $\underline{x}$. In this way, we can think of Bayesian model averaging as a formal and rigorous way of doing model averaging, or committee voting, which are well-known techniques in machine learning.

The expression in Eq. 46 is a mathematical one. It doesn't tell us how to calculate the Bayesian average in practice. Doing so can be hard. We will learn about some computational techniques in the next section, such as **Markov Chain Monte Carlo (MCMC)** simulation, that approximate the calculation of the Bayesian average in Eq. 46.

Another way to approximate the calculation in Eq. 46 is to assume that a single model is representative of all the high probability models, and also representative in any downstream calculations, such as the calculation of the prediction $\hat{y}(\underline{x} \mid \underline{\beta})$. If we are comfortable with this assumption, then the most sensible single model to use is the MAP estimate $\underline{\theta}_{MAP}$.

MAP estimation

For our model in Eq. 1, our model parameters $\underline{\theta}$ are $(\underline{\beta}, \sigma^2)$, so $\underline{\theta}_{MAP} = (\underline{\beta}_{MAP}, \sigma^2_{MAP})$ and we would make predictions using the model $\hat{y}(\underline{x} \mid \underline{\beta}_{MAP})$. Obviously, when using just $\underline{\theta}_{MAP}$ we don't get the benefits of Bayesian model averaging, but sometimes the posterior distribution is so tightly distributed around its maximum at $\underline{\theta}_{MAP}$ that only a small range of models $\underline{\theta}$ have any reasonable posterior probability associated with them, and they are all reasonably well approximated by the MAP value $\underline{\theta}_{MAP}$.

This is equivalent to approximating the true posterior $P(\underline{\theta} \mid \text{Data})$ by a delta function. That is, we are approximating in the following way:

$$P(\underline{\theta} \mid \text{Data}) \approx \delta(\underline{\theta} - \underline{\theta}_{MAP})$$

Eq. 47

Now, remember from *Chapter 2* that we can loosely think of a Dirac delta function as an infinitely narrow, infinitely high spike. When would this be a good approximation of the true posterior $P(\underline{\theta} \mid \text{Data})$? To answer that, we need to understand when the posterior becomes a narrow, tall, distribution centered around the MAP estimate, $\underline{\theta}_{MAP}$.

Let's look at the definition of $\underline{\theta}_{MAP}$. It is the value of $\underline{\theta}$ that maximizes $P(\underline{\theta} \mid \text{Data})$. Again, applying our trick of maximizing the logarithm, $\underline{\theta}_{MAP}$ is also the value that maximizes $\log P(\underline{\theta} \mid \text{Data})$. Using Bayes' theorem to calculate $\log P(\underline{\theta} \mid \text{Data})$, we find that $\underline{\theta}_{MAP}$ is the value of $\underline{\theta}$ that maximizes the following:

$$\text{logLikelihood} + \text{logPrior} - \text{log}P(\text{Data})$$

Eq. 48

Since P(Data) doesn't depend on $\underline{\theta}$ we can ignore it when identifying $\underline{\theta}_{MAP}$, and so $\underline{\theta}_{MAP}$ is the value of $\underline{\theta}$ that maximizes,

$$logLikelihood + logPrior$$

Eq. 49

For large sample sizes, the likelihood dominates

Now, recall we said that the log-likelihood scales linearly with the number of data points N. So, the magnitude of the log-likelihood increases with N. In contrast, the prior does not, by definition, depend on the data, so the magnitude of the log-prior doesn't scale with N. Firstly, this means that as N increases, the expression in Eq. 49 is dominated by the log-likelihood. So, as $N \to \infty$, the MAP estimate $\underline{\theta}_{MAP}$ tends to the maximum-likelihood estimate $\underline{\theta}_{ML}$. Secondly, since the magnitude of the log-likelihood is increasing as $N \to \infty$, this means the magnitude of the posterior at $\underline{\theta} = \underline{\theta}_{MAP}$ is increasing indefinitely as $N \to \infty$. So, the posterior is becoming infinitely high. Since the posterior is a properly normalized probability density, it also becomes infinitely thin as $N \to \infty$. So, as $N \to \infty$ the approximation in Eq. 47 becomes accurate. While we have only given a hand-waving demonstration of this, it is still valid. A rigorous proof of this point is beyond the scope of this book.

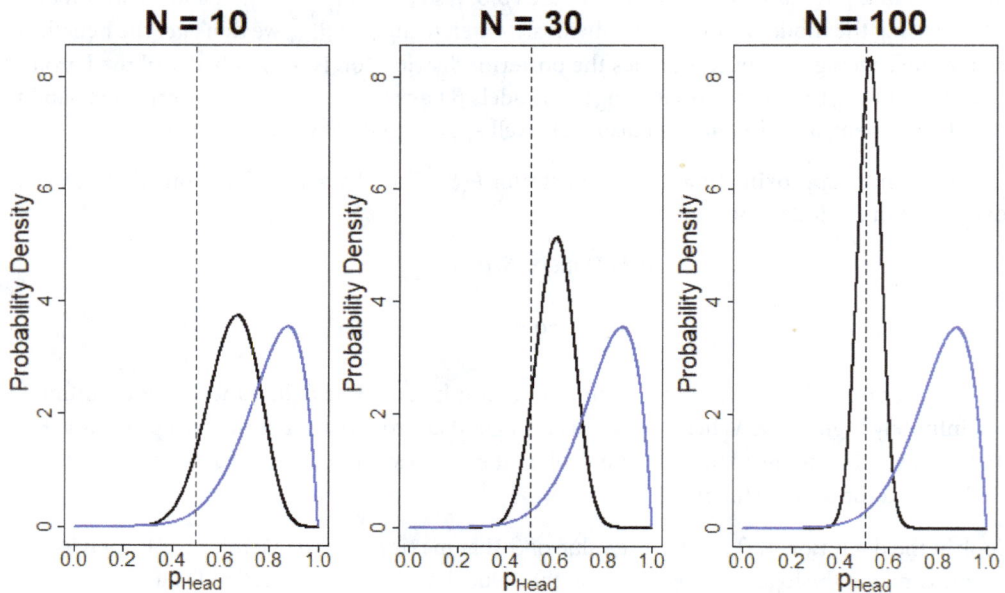

Figure 5.2: Plots of the prior and posterior distributions for three games of coin tossing with different sample sizes

Figure 5.2 shows a numerical demonstration of this. We have plotted the prior (the blue line) and posterior (the black line) distributions for a series of three experiments. In each plot, the sample data is the number of times a fair coin turns up heads when tossed N times. This is a chapter on Bayesian modeling, so we couldn't not have an example about tossing coins, could we? The plots are for different values of N with N = 10, 30, 100 in the plots from left to right, respectively. To make this interesting, I have set the coin tossing in the wild west of America in the 1840s. I'm betting against "cowboy Duke", a hardened gambler. Every time the coin is tossed and lands tails up, I win $1. I suspect that Duke is cheating and has an unfair coin. It turns out this isn't true, but in the wild west of the 1840s I'm on my guard and suspicious, so my prior is that the probability of heads, p_{Head}, for this coin is 0.8. I have used a Beta(8, 2) distribution for my prior so that the mean of my prior is 0.8, but it has some spread, as you can see by the blue line in the plots. The prior, the blue line, is the same in each of the plots, as you expect because the prior does not depend on the data and so can't depend on the sample size N.

For the left-hand plot, where N = 10, the number of heads was 5, but the influence of the prior on the posterior is clear. The prior has pulled the posterior away from being centered around the true value of p_{Head} = 0.5 (shown by the vertical dashed line), even though the observed proportion of heads was 5/10 = 0.5. In this case, the sample size of N = 10, isn't big enough for the likelihood to override the influence of a poorly chosen prior.

For the middle plot, with N = 30 and 16 heads observed, we can see the posterior distribution is closer to being centered around the true value of p_{Head} = 0.5, and it is narrower and taller compared to the left-hand plot, but the influence of the prior is still clear. The maximum of the posterior is still noticeably different from p_{Head} = 0.5.

In contrast, in the right-hand plot where N = 100 and 49 heads are observed, the likelihood dominates the posterior and the influence of the prior is small. The posterior distribution is very tall and narrow and centered almost exactly over p_{Head} = 0.5. The influence of my poorly chosen prior has been almost completely counteracted by the observation of the data.

Let's use the data from *Figure 5.2* in a MAP estimation code example.

MAP estimation code example

The following code example can be found in the `Code_Examples_Chap5.ipynb` notebook in the GitHub repository.

We'll calculate the MAP estimate for the binomial data we illustrated in the left-hand panel of *Figure 5.2*. We have data from a series of Bernoulli trials. The likelihood only depends on the number of trials, the number of successes (number of heads for the example in *Figure 5.2*), and the success probability p. We'll use the in-built optimization algorithms in the SciPy package to maximize the posterior. We do this by minimizing the negative of the log-posterior. We only need to calculate the sum of the log-likelihood and the log-prior since we can drop the normalizing constant from the log-posterior, as this does not depend upon the success probability. The parameter we maximize the log-posterior with respect to will actually be the logit of the success probability. Since the logistic function is a monotonic function, stationary points of the log-posterior with respect to p will also be stationary points of the log-posterior with respect to $\log(p/(1 - p))$. First, we'll need to define the log-posterior function:

```python
import pandas as pd
import numpy as np
import matplotlib.pyplot as plt

from scipy.special import loggamma
from scipy.optimize import minimize

def get_neg_log_binomial_posterior(n_trial, n_success, alpha, beta):
    '''
    A function to construct a callable that returns the negative of
    the log-posterior for binomially distributed data, with a Beta
    prior for the success probability of the Bernoulli trials.

    Returns a callable that returns the negative of the log-posterior
    (up to a global constant) and takes the logit of the success
    probability logit(p) as input.
    '''

    def neg_log_binomial_posterior(logit_p):
        '''
        A function to compute the negative log-posterior (up to a
        global constant) for binomially distributed data, with a Beta
        prior for the success probability of the Bernoulli trials.

        logit_p is the logit of the success probability.

        Returns the negative of the sum of the log-likelihood and the
        the log-prior
        '''

        # Compute the success probability p from logit(p)
        p = np.exp(logit_p)/ (1.0 + np.exp(logit_p))

        # Compute the log-prior
        log_prior = loggamma(alpha + beta) - loggamma(alpha) - \
            loggamma(beta)
        log_prior += ((alpha-1.0)*np.log(p)) + \
            ((beta-1.0)*np.log(1.0 - p))

        # Compute the log-likelihood
```

```
        log_likelihood = loggamma(n_trial +1.0)
        log_likelihood -= loggamma(n_trial - n_success +1.0)
        log_likelihood -= loggamma(n_success +1.0)
        log_likelihood += (n_success*np.log(p))
        log_likelihood += ((n_trial-n_success)*np.log(1.0-p))

        # Compute the log-posterior,
        #up to the global normalization factor,
        # as the sum of the log prior and log-likelihood
        log_posterior = log_likelihood + log_prior

        return -log_posterior

    return neg_log_binomial_posterior
```

Now we'll set the data. In this example, we'll use the values from the left-hand panel of *Figure 5.2*:

```
# Specify the sample size and the number of successes
n_trial = 10
n_success = 5

# Specify the parameters of the prior
alpha = 8
beta = 2

# Get the objective function to minimized
neg_log_posterior = get_neg_log_binomial_posterior(n_trial, n_success,
                                                   alpha, beta)

# Construct an initial estimate for the optimal parameter.
# We'll use the sample success proportion to do this
# (and take the logit)
p0 = float(n_success)/float(n_trial)
logit_p0 = np.log(p0/(1.0-p0))
x0 = np.array([logit_p0])
```

Now we'll do the minimization of the negative log-posterior using the SciPy implementation of the **Broyden-Fletcher-Goldfarb-Shanno (BFGS)** algorithm:

```
map_estimate = minimize(neg_log_posterior,
                        x0,
                        method='BFGS',
                        options={'disp': True})
# Convert from logit(p) to p
```

```
p_optimal = np.exp(map_estimate['x'][0])/ (
    1.0 + np.exp(map_estimate['x'][0]))
print("MAP estimate of success probability = ", p_optimal)
```

This gives the following output:

```
MAP estimate of success probability = 0.666666667917668
```

The MAP estimate of the success probability p corresponds to the position of the maximum of the posterior that we can see in the left-hand panel of *Figure 5.2*.

As *N* becomes large the prior becomes irrelevant

We have already illustrated the main consequence of the sample size N becoming large – the likelihood dominates the posterior – but let's unpack that conclusion a bit more. Effectively, the prior becomes irrelevant as the sample size increases. How quickly the prior becomes irrelevant depends upon the precise details of the likelihood – the predictive model and the nature of the random component in the data – but also the precise details of the prior. A narrow prior, $P(\theta)$, indicates a high degree of confidence in that prior belief, and this would take a greater amount of information (data) in the likelihood for the influence of the prior to be overruled. A high degree of confidence does not mean that the prior is correct. Instead, it means that we have a strong **a priori** belief that the only values of θ that are possible are those that are in the narrow range where $P(\theta)$ is high. But even the influence of a narrow poorly chosen prior can be overcome with sufficient data. With one exception. If our prior $P(\theta)$ is itself a delta function, then no finite amount of data can overcome the influence of the prior. In other words, if we are 100% certain in our **a priori** beliefs about the model θ, then no amount of data can convince us otherwise, even if those beliefs are wrong.

Least squares as an approximation to Bayesian modeling

As a final comment on the two main ways of using the posterior in calculations, we'll return to a comment we made in *Chapter 4* about least squares being a heuristic algorithm, but that it was possible to provide more formal under-pinning to least squares estimation of model parameters.

The delta function approximation in Eq. 47 highlights that we can view MAP estimation as an approximation to the correct posterior. We also know that MAP estimation becomes more appropriate when we have large sample sizes N. We also know that in the limit $N \to \infty$ the MAP estimate of θ becomes the maximum likelihood estimate of θ. And we know that when the random component in our data is of the form of additive Gaussian noise, then the model parameter β, from the maximum-likelihood estimate of $\theta = (\beta, \sigma^2)$, is the same as the least squares estimate of β. So, finally, we have the more rigorous justification for the least squares estimation of model parameters that we promised back in *Chapter 4*. We can view the least squares estimation of the model parameters β as the result of a chain of approximations applied to the posterior expectation of β.

We have covered a lot of the theory behind Bayesian modeling in this section, so it is time to recap what we have learned before we move on to how we put that theory into practice in numerical calculations in the next section.

What we have learned

In this section, we have learned about the following:

- Bayesian model averaging and how it uses the probabilities given by the posterior $P(\theta \mid \text{Data})$ to perform a weighted average calculated over a set of models θ

- **Maximum a posteriori (MAP)** estimation and how it approximates the posterior distribution by a single representative model, θ_{MAP}, that has the highest probability given the data

- How the MAP estimate for a model θ tends to the maximum likelihood estimate of the model θ, as the sample size $N \to \infty$

- How MAP estimation provides a justification for the least squares estimation of a model's parameters when the random component in the data is additive Gaussian noise

Having learned about the theory of Bayesian modeling, in the next section we'll learn about some of the practical aspects, and about a class of modeling tools called **Probabilistic Programming Languages (PPLs)**.

Bayesian modeling in practice

Bayesian model averaging, as encapsulated by Eq. 46, is a very powerful tool for any data scientist to have in their toolkit. In practice, it can take a bit more experience to fully make use of its potential. We haven't yet said how one goes about computing the expectation value in Eq. 46. This is the practice of Bayesian modeling.

To make Bayesian modeling averaging a practical tool, there are two main approaches we can take:

- Analytical calculation, whereby we approximate the posterior to the extent that calculation of the expectation in Eq. 46 can be done in closed-form or nearly in closed-form, and so we only need to perform a small number of numerical calculations

- Computationally intensive sampling, whereby we numerically approximate the integration in Eq. 46 by sampling many different model values of θ

We will now cover those two approaches in more detail.

Analytic approximation of the posterior

We have already introduced an analytic approximation to the posterior in Eq. 47. When we introduced the MAP estimate we explained that the approximation in Eq. 47 is appropriate if the posterior is tall and narrow. We know this happens as the dataset size N becomes large. But what happens if N is not large? What happens if the posterior is not tall and narrow? What analytic approximation can we use then? The next most obvious step is to approximate the posterior by a multi-variate Gaussian distribution centered around the maximum of the posterior, i.e., centered around $\underline{\theta}_{\text{MAP}}$. Effectively, we are approximating the log of $P(\underline{\theta}\,|\,\text{Data})$ by its second order Taylor-expansion about $\underline{\theta}_{\text{MAP}}$. Doing so gives us the following approximation:

$$P(\underline{\theta}\,|\,\text{Data}) \approx \frac{1}{(2\pi)^{\frac{|\underline{\theta}|}{2}}}\,\sqrt{\det(-\underline{H})}\,\exp\!\left(\tfrac{1}{2}\left(\underline{\theta}-\underline{\theta}_{\text{MAP}}\right)^{\top}\underline{H}\left(\underline{\theta}-\underline{\theta}_{\text{MAP}}\right)\right)$$

Eq. 50

Here, $|\underline{\theta}|$ denotes the cardinality of the vector $\underline{\theta}$, that is, the number of components in the vector $\underline{\theta}$. The matrix $\underline{H}$ is the Hessian of the log-posterior evaluated at the MAP point $\underline{\theta}_{\text{MAP}}$, and so is calculated as follows:

$$\underline{H} = \left.\frac{\partial^2 \log P(\underline{\theta}\,|\,\text{Data})}{\partial \underline{\theta}^2}\right|_{\underline{\theta}=\underline{\theta}_{\text{MAP}}}$$

Eq. 51

If we have a function, $f(\underline{x}, \underline{\theta})$, that depends upon the model $\underline{\theta}$ and we want to calculate its expectation, $\mathbb{E}_{\underline{\theta}}(f(\underline{x}, \underline{\theta}))$, over the posterior, then we can now approximate that expectation as follows:

$$\mathbb{E}_{\underline{\theta}}(f(\underline{x}, \underline{\theta})) \approx \frac{1}{(2\pi)^{\frac{|\underline{\theta}|}{2}}}\,\sqrt{\det(-\underline{H})}\,\int f(\underline{x},\underline{\theta})\,\exp\!\left(\tfrac{1}{2}\left(\underline{\theta}-\underline{\theta}_{\text{MAP}}\right)^{\top}\underline{H}\left(\underline{\theta}-\underline{\theta}_{\text{MAP}}\right)\right)d\underline{\theta}$$

Eq. 52

Many integrals with Gaussian distributions can be evaluated exactly, i.e., in closed-form, and this is one of the main benefits of the analytic approximation approach. When the integral in Eq. 52 cannot be evaluated exactly, a common approach is to also expand the function $f(\underline{x}, \underline{\theta})$ about $\underline{\theta}_{\text{MAP}}$, so we have this:

$$f(\underline{x}, \underline{\theta}) \approx f(\underline{x},\underline{\theta}_{\text{MAP}}) + \left(\underline{\theta}-\underline{\theta}_{\text{MAP}}\right)^{\top}\left.\frac{\partial f(\underline{x}, \underline{\theta})}{\partial \underline{\theta}}\right|_{\underline{\theta}=\underline{\theta}_{\text{MAP}}} + \tfrac{1}{2}\left(\underline{\theta}-\underline{\theta}_{\text{MAP}}\right)^{\top}\left.\frac{\partial^2 f(\underline{x}, \underline{\theta})}{\partial \underline{\theta}^2}\right|_{\underline{\theta}=\underline{\theta}_{\text{MAP}}}\left(\underline{\theta}-\underline{\theta}_{\text{MAP}}\right) + \cdots$$

Eq. 53

Plugging Eq. 53 into Eq. 52 and evaluating the integral in Eq. 52 then gives us this:

$$\mathbb{E}_{\underline{\theta}}(f(\underline{x}, \underline{\theta})) \approx f(\underline{x},\underline{\theta}_{\text{MAP}}) - \tfrac{1}{2}\text{tr}\!\left(\underline{H}^{-1}\left.\frac{\partial^2 f(\underline{x}, \underline{\theta})}{\partial \underline{\theta}^2}\right|_{\underline{\theta}=\underline{\theta}_{\text{MAP}}}\right) + \cdots$$

Eq. 54

We have now reduced the calculation down to an optimization problem – locating the value of $\underline{\theta}_{\text{MAP}}$ – and a linear algebra calculation in Eq. 54, both of which we have established software packages to do.

One of the benefits of using a Gaussian approximation centered on the MAP estimate θ_{MAP} is that we have already located the MAP value θ_{MAP} when we did MAP estimation. In fact, we can think of the delta function approximation in Eq. 47 as the first order approximation to the posterior $P(\theta \mid \text{Data})$, compared to the second-order approximation that Eq. 50 represents. One could construct higher-order approximations to the posterior by continuing the Taylor expansion of $\log P(\theta \mid \text{Data})$ about θ_{MAP} to third order or higher. However, this is very rarely done in machine learning or statistics.

Another benefit of the analytic approximation in Eq. 50 is that it is very general. We can apply the approximate posterior distribution given in Eq. 50 to calculate the expectation of any function $f(x, \theta)$. Likewise, the approximation in Eq. 54 can be applied to any function $f(x, \theta)$.

Computational sampling

The idea behind sampling approaches to Bayesian averaging is very simple. If we want to calculate the expectation, $\mathbb{E}_\theta(f(x, \theta))$, over a distribution of models, then we could approximate this population average by a sample average. In other words, we generate a random sample of values of θ from the distribution in question, plug those sample values of θ into $f(x, \theta)$ and calculate the sample mean. For our Bayesian averaging challenge the distribution in question from which we sample values of θ is the posterior $P(\theta \mid \text{Data})$. If we generate K sample values of θ from $P(\theta \mid \text{Data})$, then the expectation of $f(x, \theta)$ is approximated by this:

$$\mathbb{E}_\theta(f(x, \theta)) \approx \frac{1}{K} \sum_{i=1}^{K} f(x, \theta_i) \quad \theta_i \sim \text{Posterior(Data)}$$

Eq. 55

The larger K is, i.e., the larger the number of samples of θ we generate from the posterior, the more accurate the approximation in Eq. 55 will be. As with the analytic approximation techniques, this is a completely general approach – we haven't said what the function $f(x, \theta)$ is, and so this method can be applied to any function $f(x, \theta)$.

The only question that now remains is, how do we generate samples of θ from the posterior $P(\theta \mid \text{Data})$? We covered a bit about sampling from distributions in *Chapter 2*, but we commented that sampling from continuous distributions can sometimes be challenging. Fortunately, there is a general computational method that comes to our rescue – **Markov Chain Monte Carlo** (**MCMC**). MCMC is a Monte Carlo method, meaning we generate values of our random variable, θ in this case, at random. The Markov Chain part of the algorithm name means that when we generate our next value of θ at random, we do so conditionally on our current value of θ, and so we get a sequence or a **chain** of θ values. MCMC algorithms typically take the following form:

1.   Set the iteration number $i = 0$ and set θ to some initial value θ_0.

From the current value θ_i, propose a new value trial value $\theta_{\text{trial}} = \theta_i + \Delta\theta$, where $\Delta\theta$ is an adjustment that we sample from a simple fixed distribution, e.g. a uniform distribution with a small range.

1. Accept or reject the trial value $\underline{\theta}_{trial}$ by applying a stochastic comparison rule to the pair of values $P(\underline{\theta}_i|\,\text{Data})$ and $P(\underline{\theta}_{trial}|\,\text{Data})$.

2. If we accept the trial value $\underline{\theta}_{trial}$ then we set $\underline{\theta}_{i+1} = \underline{\theta}_{trial}$, else we set $\underline{\theta}_{i+1} = \underline{\theta}_i$. Record $\underline{\theta}_{i+1}$. Increment the iteration, $i \rightarrow i + 1$.

3. Repeat steps 2 – 4 until M iterations have been performed.

One of the simplest comparison rules is the **Metropolis-Hastings importance sampling** scheme, which takes the following form:

$$\text{Accept } \underline{\theta}_{trial} \text{ with probability } \min\left(\frac{P(\underline{\theta}_{trial}|\,\text{Data})}{P(\underline{\theta}_i|\,\text{Data})}, 1\right)$$

Eq. 56

Eq. 56 says that if our trial value $\underline{\theta}_{trial}$ has higher probability than our current value $\underline{\theta}_i$, then we move from $\underline{\theta}_i$ to $\underline{\theta}_{trial}$ with probability 1, i.e., we definitely move. While if $\underline{\theta}_{trial}$ has lower probability than $\underline{\theta}_i$ we can still move to $\underline{\theta}_{trial}$, but it is not guaranteed. The smaller the probability $P(\underline{\theta}_{trial}|\,\text{Data})$ is compared to $P(\underline{\theta}_i|\,\text{Data})$, the less likely we are to move to $\underline{\theta}_{trial}$ and so the more likely we are to stay in $\underline{\theta}_i$. Overall, this means we move, over time and in a stochastic fashion, to regions of higher and higher posterior probability.

You rightly ask how we choose the initial value $\underline{\theta}_0$. It doesn't matter. After we have performed a reasonable number of iterations, the values of $\underline{\theta}$ being generated will be correctly sampled from the posterior $P(\underline{\theta}|\,\text{Data})$. However, the early values of $\underline{\theta}$ will not be correctly sampled from the posterior, and therefore it is usual to discard these values of $\underline{\theta}$ from the early part of the chain. The period where we are just iterating the MCMC algorithm but not recording the generated values of $\underline{\theta}$ is called the **burn-in** period. After the burn-in period, we collect and use the generated values of $\underline{\theta}$ as our sample to plug into the sample average calculation in Eq. 55. Let's illustrate these concepts with a code example.

MCMC code example

We'll code up a very basic version of the Metropolis-Hastings algorithm and use it to sample logit(p) values from the posterior $P\left(\log\left(\frac{p}{1-p}\right)|\,\text{Data}\right)$ for the data in the left-hand plot of *Figure 5.2*. As with our earlier MAP estimation example, we are working with logit(p) rather than the success probability p so that we can sample a parameter that is unconstrained, i.e., it lies in the range $(-\infty, +\infty)$. This means we'll need to calculate the posterior for $\log\left(\frac{p}{1-p}\right)$, but we can do this using the rule for transforming probability distributions that we covered in *Chapter 2*. Doing so gives us this:

$$\log P\left(\log\left(\frac{p}{1-p}\right)\bigg|\,\text{Data}\right) = \log P(p\,|\,\text{Data}) + \log(p(1-p))$$

Eq. 57

We'll reuse the code from the previous MAP estimation code example to generate a callable Python function that returns $-\log P(p\,|\,\text{Data})$.

The following code example can be found in the `Code_Examples_Chap5.ipynb` notebook in the GitHub repository:

1. First, we'll define a function to perform a single trial Metropolis-Hastings move:

```python
import numpy as np

def perform_mh_trial(x, log_post, delta_x, neg_log_posterior):
    '''
    Function to perform a Metropolis-Hastings trial move.
    x is the current logit(p) value.
    log_post is the current log-posterior value.
    delta_x is the half-width of the range from which the trial
    adjustments to logit(p) are made.
    neg_log_posterior is a callable that returns the negative
    log-posterior.

    We return a tuple of the updated logit(p) value and updated
    log-posterior value.
    '''

    accept_trial = False
    x_trial = x + (delta_x*(2.0*np.random.rand(1) -0))
    p_trial = np.exp(x_trial)/(1.0 + np.exp(x_trial))

    # Calculate the log-posterior for the trial point.
    # Note we'll need to flip the sign of neg_log_posterior, as
    # our callable returns the negative of the log-posterior.
    log_post_trial = -neg_log_posterior(x_trial) + np.log(p_
    trial*(1.0 - p_trial))

    # Calculate the change in log-posterior if we move to the
    # trial point
    delta_log_post = log_post_trial - log_post

    # Work out if should accept the trial point
    if delta_log_post > 0.0:
        accept_trial = True
    else:
        if np.log(np.random.rand(1)) < delta_log_post:
            accept_trial = True

    # If we accept the trial point then update the current
    # value of the parameter
    # and the log-posterior
```

```
        if accept_trial==True:
            x = x_trial
            log_post = log_post_trial

        return x, log_post
```

2. Now, we can use the preceding function to define another function that runs the Markov chain for a user-specified number of burn-in iterations, followed by a user-specified number of sampling iterations:

```
def mh_mcmc(n_burnin, n_iter, x0, delta_x, neg_log_posterior):
    '''
    A function to run a simple Metropolis-Hastings MCMC
    calculation.

    n_burnin is the number of burnin iterations to be run.
    n_iter is the number of sampling iterations to be run.
    x0 is the starting value for logit(p).
    delta_x is the half-width of the range from which the trial
    adjustments to logit(p) are made.
    neg_log_posterior is a callable that returns the negative
    log-posterior.

    We return an array of the sampled logit(p) values
    '''

    #Calculate starting log_posterior
    x = x0
    p0 = np.exp(x0)/(1.0 + np.exp(x0))
    log_post = -neg_log_posterior(x) + np.log(p0*(1.0-p0))

    # Run the chain for the specified burn-in length
    for iter in range(n_burnin):
        x, log_post = perform_mh_trial(
            x, log_post, delta_x, neg_log_posterior)

    # Initialize an empty array to hold the sampled
    # parameter values
    x_chain = np.zeros(n_iter)

    # Continue the chain for the specified number of
    # sampling points
    # Store the sampled parameter values
    for iter in range(n_iter):
```

```
        x, log_post = perform_mh_trial(
            x, log_post, delta_x, neg_log_posterior)
        x_chain[iter] = x

    return x_chain
```

3. Now, we'll set the data. Again, in this example we'll use the values from the left-hand panel of *Figure 5.2*:

```
# Specify the sample size and the number of successes
n_trial = 10
n_success = 5

# Specify the parameters of the Beta prior
alpha = 8
beta = 2

# Construct a starting point for the MCMC calculation.
# We'll use the sample success proportion to do this
# (and take the logit)
p0 = float(n_success)/float(n_trial)
logit_p0 = np.log(p0/(1.0-p0))
```

4. Now, we'll run the MCMC calculation. We'll run a long burn-in period of 20,000 iterations to be sure, and then we'll take 1 million samples. Finally, we convert the sampled values of logit (*p*) to values of *p*:

```
# Run the MCMC calculation
x_chain = mh_mcmc(n_burnin=20000,
                  n_iter=1000000,
                  x0=logit_p0,
                  delta_x=0.1,
                  neg_log_posterior=neg_log_posterior)

# Convert the MCMC sampled logit(p) values back to values of
# the success probability
p_mcmc = np.exp(x_chain)/ (1.0 + np.exp(x_chain))
```

5. Now, let's look at the histogram of the MCMC sampled success probabilities and compare it to the true posterior curve. The true posterior curve will be the same posterior curve that is shown in the left-hand panel of *Figure 5.2*:

```
import matplotlib.pyplot as plt

# Plot the histogram of posterior sampled success probabilities
# and overlay the true posterior distribution
```

```
hist = plt.hist(p_mcmc, bins=100, density=True)
posterior = plt.plot(p_sequence, true_posterior_sequence)
plt.title('Histogram of MCMC samples', fontsize=24)
plt.xlabel(r'$p$', fontsize=20)
plt.ylabel('Probability Density', fontsize=20)
plt.xticks(fontsize=14)
plt.yticks(fontsize=14)
plt.show()
```

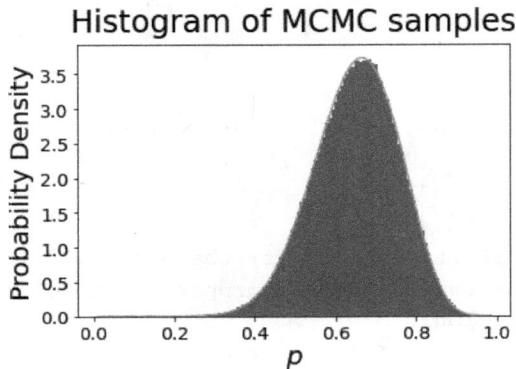

Figure 5.3: Histogram of MCMC samples and the true posterior

We can see in *Figure 5.3* how the values of *p* sampled using the Metropolis-Hastings algorithm match the true posterior distribution (the orange curve in *Figure 5.3*) exactly.

That was a lengthy code example, and it can be tedious to write the MCMC code yourself. If only it was easier. This brings us neatly on to our next topic, **probabilistic programming languages**.

Probabilistic programming languages

In practice, choosing the length of the burn-in period can be a bit of an art form. There are also other variants of MCMC algorithms that are more sophisticated than the basic Metropolis-Hastings rule. Again, these more sophisticated MCMC algorithms can be applied to evaluate the expectation of any function and for any posterior. Due to the wide applicability of these MCMC algorithms, many of them are coded into specialist languages called **Probabilistic Programming Languages**, **(PPLs)**. These programming languages allow the user to easily express a probabilistic model and configure the MCMC calculation, and then the PPL takes care of running the MCMC. The main benefit of PPLs is that the user does not have to write any MCMC calculation code. The user only writes code that expresses the model.

Since the goal of PPLs is to make Bayesian modeling and inference easy by abstracting away a lot of the technical code that runs the MCMC calculation, many PPLs also make MAP estimation easy. We have already mentioned that the way in which we use MAP estimates is very generic, such as through the analytic approximations to the posterior in Eq. 47 and Eq. 50. Consequently, most PPLs also provide functionality to obtain MAP estimates without the user having to write additional MAP estimation code. Again, the user only has to write code that expresses the model.

As you might have guessed, there are many different PPLs available. Some of the more well-established ones are listed here:

- **PyMC**: One of the most widely used Python open source PPLs that uses Theano as a backend. Details are available at `https://www.pymc.io/welcome.html`.

- **Stan**: Probably the other most widely used PPL. Stan has its own model specification language, which is then translated to C++ code that is then compiled to machine code. It is perhaps more widely used within the R programming community, but it has good interfaces to a number of languages, including Python. General details are available at `https://mc-stan.org/`. Details on the Python interface, PyStan, can be found at `https://pystan.readthedocs.io/en/latest/`.

- **Pyro**: A PPL created by Uber AI Labs with a PyTorch back-end. Details are available at `https://pyro.ai/`.

- **NumPyro**: A variant of Pyro that uses a backend based on NumPy and JAX, and so can provide a speedup over Pyro for a subset of model types. Details are available at `https://num.pyro.ai/en/stable/`.

You'll have spotted that each PPL either makes use of an existing computational backend, such as Theano, PyTorch, TensorFlow, or JAX, or has its own language that ultimately can be compiled to machine code. This computational power is necessary because of the computationally intensive nature of MCMC calculations, and also because PPLs have to be able to handle what can be very complex user-specified probabilistic models. Because of this need for PPLs to have access to a computational backend, they can be trickier to install and set up than, say, your average Python package. However, once you are comfortable with Bayesian modeling concepts and have gone through the pain of installing a PPL, they are very useful and fun to work with.

That finishes this subsection on PPLs, so we'll conclude by summarizing what we have learned about Bayesian modeling in practice, and then we'll summarize the chapter overall.

What we have learned

In this section, we have learned the following:

- How to make simple analytical approximations to the posterior to use in closed-form Bayesian averaging calculations
- About **Markov Chain Monte Carlo** (**MCMC**) algorithms and how they can be used to generate samples from a posterior distribution, which then allow us to approximate an expectation over the posterior via a sample mean
- About **Probabilistic Programming Languages** (**PPLs**) and how they automate away a lot of the cumbersome and repeated tasks involved in MCMC or MAP estimation calculations

Summary

This chapter has been a culmination of the many ideas and concepts we have introduced in the previous three chapters. At the heart of this chapter is the idea that because data is random, predictive models that attempt to explain and use that data should be probabilistic. To build probabilistic models, we have had to learn about the probability distributions that describe the data and the distributions that describe the models. Specifically, we have had to learn about the following:

- Likelihood as the probability of data given a model
- How to use the likelihood to estimate model parameters via maximum likelihood
- Bayes' theorem and about prior and posterior distributions
- How the posterior distribution quantifies the probability of the model parameters given the data or information we have received
- How we can use the posterior in Bayesian model averaging, or MAP estimation calculations
- How to perform those Bayesian model averaging and MAP estimation calculations in practice
- **Probabilistic Programming Languages** (**PPLs**) and how they automate a lot of the practical tasks in probabilistic modeling

This chapter represents the last of the core math concepts and techniques we cover in this book. This chapter and the preceding three chapters cover what I consider to be the absolute minimum core math concepts and techniques that any data scientist should be familiar with. For the rest of the book, we will get more specialized, covering individual math concepts that tend to be focused on a particular type of data or a particular domain. To start, in the next chapter, we move onto time series data.

Exercises

Here is a series of exercises. Answers to all the exercises are given in the Jupyter `Answers_to_ Exercises_Chap5.ipynb` notebook in the GitHub repository:

1. For the MAP estimation code example in the text, we used the `scipy.optimize.minimize` function to do the optimization of the log-posterior. The `minimize` function has the option for the user to supply a callable function that calculates the gradient of the objective function with respect to the objective function parameters. Work out on paper the gradient of the log-posterior and implement a function that returns the gradient of the log-posterior. Re-run the MAP estimation process using `scipy.optimize.minimize`, but when you pass in your log-posterior gradient function, you'll need to look at the online documentation for the `scipy.optimize.minimize` function to see how your callable gradient function should be passed in.

2. The `Data/coffee_or_tea.csv` file in the GitHub repository contains two columns of data, corresponding to 250 days' worth of observations on what drink I had each morning (tea or coffee) when I was at home, and also whether it rained the night before. The two columns are called `rained`, with a value of `1` indicating that it rained the night before (`0` indicating it did not), and `coffee`, with a value of `1` indicating that I drank coffee that morning (`0` indicating I drank tea instead). Formulate a probabilistic model (hint – look at the binomial shopper decision example in the text) that models my decision to drink coffee or not, and that uses the `rained` variable as a predictive feature. The linear predictor for the model will take the form, $\eta_{coffee} = \beta_0 + \beta_1 \times$ rained. Use the `scipy.optimize.minimize` function to obtain maximum likelihood estimates for the model parameters, β_0, β_1.

3. Using a $N(0, 1)$ prior for both β_0 and β_1 in the model you formulated in Q2, obtain MAP estimates for β_0 and β_1 using the same dataset used in Q2.

4. Using the posterior you developed in Q3, adapt the MCMC code example in the text and obtain 10,000 samples of the tuple (β_0, β_1). Plot separate histograms of β_0 and β_1 from those 10,000 samples.

Part 2:
Intermediate Concepts

In this part, we will introduce you to more math concepts that you are very likely to encounter the longer you work in data science. In contrast to Part 1, each chapter is focused on a standalone data science task, modeling technique, or type of data. By the end of Part 2, you will have gained a solid understanding of time series data, how to run a hypothesis test, model complexity, how to build up a function from a set of simpler parts, and network data.

This section contains the following chapters:

- *Chapter 6, Time Series and Forecasting*
- *Chapter 7, Hypothesis Testing*
- *Chapter 8, Model Complexity*
- *Chapter 9, Function Decomposition*
- *Chapter 10, Network Analysis*

Time Series and Forecasting

Time series data is very common. This makes time series data analysis an important and highly relevant topic. It also means that time series analysis is an enormous topic and one that we can't fully cover here – see point 1 in the *Notes and further reading* section at the end of the chapter. Consequently, our focus in this chapter will be on introducing and explaining the key math concepts that are at the heart of any time series data. These concepts are also what underpin the dominant classical approach to time series modeling – **ARIMA modeling**. Therefore, we will unapologetically focus overwhelmingly on introducing and explaining ARIMA models. This is not to say that there aren't alternative time series modeling techniques out there. However, any well-founded time series modeling technique has to deal with the core concepts that ARIMA modeling focuses on. Therefore, ARIMA modeling serves as a concise and useful way of explaining those concepts. We do this by covering the following topics:

- *What is time series data?*: In this section, we will explain what makes time series data different from other data you will encounter, and what the ramifications of those differences are

- *ARIMA modeling*: In this section, we will explain the classical approach to time series modeling

- *ARIMA modeling in practice*: In this section, we will highlight some of the practical subtleties and nuances of building ARIMA models

- *Machine learning approaches*: In this section, we will give a brief summary of some of the modern alternatives to ARIMA for time series modeling

Technical requirements

All code examples given in this chapter (and additional examples) can be found in the GitHub repository, https://github.com/PacktPublishing/15-Math-Concepts-Every-Data-Scientist-Should-Know/tree/main/Chapter06. To run the Jupyter notebooks you will need a full Python installation, including the following packages:

- pandas (>=2.0.3)

- numpy (>=1.24.3)

- `matplotlib` (>=3.7.2)
- `statsmodels` (>=0.14.0)
- `pmdarima` (>=2.0.4)

What is time series data?

The title of this section may seem a little odd. Is time series data somehow different from other sorts of data? The answer is yes. Look at the time series plotted in *Figure 6.1*. It is a plot of the UK's monthly seasonally adjusted unemployment rate from February 1971 to December 2019 inclusive. The unemployment rate (percentage) is for all individuals aged 16 and over. I have taken the data from UK's Office of National Statistics website – see point 2 in the *Notes and further reading* section. The data is publicly available and regularly updated.

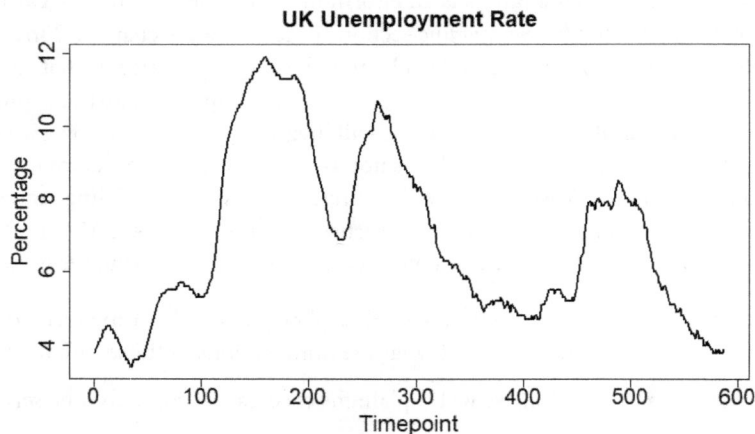

Figure 6.1: Monthly UK seasonally-adjusted unemployment rate (percentage)

The key here is the time aspect of the data. Each data point in a time series dataset is associated with a particular time point. This introduces two new key aspects to the data:

- There is a natural ordering to the data, that is, we can order the data points by the timepoint, and this ordering makes sense. Consequently, it is also natural to plot time series against the time point, i.e., to almost always plot time series data with time t as our x-axis.

- We would expect in many time series datasets for data points close together in time to show some sort of similarity in value, i.e., there is a correlation between any two nearby data point values, with the strength of the correlation decreasing as we increase the separation in time between the two data points. Since this correlation is across time, we call it *temporal correlation*, and secondly, since this correlation is a correlation between the response variable and itself (at a later time point), it is called *auto-correlation*.

What does auto-correlation mean for modeling time series data?

The presence of auto-correlation in time series data means that any approach to modeling that time series data should ideally take the auto-correlation into account. That is, we should ideally use a model form that has auto-correlation baked in from the start. We'll learn about the most common classical way of doing this in the next section.

We have said that modeling of time series data should ideally bake in the auto-correlative nature of time series into our mathematical model. This would appear to imply that sometimes we don't take the auto-correlation into account, or that we can leave it out of our model. Is this ok to do? The short answer is no. But, as with any piece of mathematical modeling, there are choices and trade-offs to be made. We can choose to ignore the auto-correlation present within a time series dataset, but that is a conscious choice, and we should be aware of the implications of that choice. Ignoring the auto-correlation present in a time series dataset adds an extra assumption to our mathematical model. We assume that the auto-correlation is not important. We may have reached that conclusion via a number of routes:

- We have assessed or quantified the level of auto-correlation present in the data and decided that it is sufficiently weak that it can be ignored to first order approximation.

- The presence of the auto-correlation in the data does not materially affect the conclusions we draw from the mathematical model. This will depend upon what we want to use the model for. For example, we may be confident that a classifier built from the data is still sufficiently accurate for our needs that we can ignore the auto-correlation. Taking into account the auto-correlation will produce a better model, but we have chosen to make a trade-off between accuracy and development effort.

- We will attempt to take account of the auto-correlation in an ad hoc way via appropriate pre-processing of the data or post-processing of the output from the model. Consequently, we do not bake auto-correlation into our main mathematical model.

From this, you will have gathered that ignoring auto-correlation in time series data does happen. Yes, virtually every experienced data scientist will have done it. But to emphasize, you should always be aware of when you are doing it, and what the potential ramifications of doing it are.

The first stage in understanding auto-correlation is quantifying it. To do that we need to introduce the **auto-correlation function**.

The auto-correlation function (ACF)

To start, we should make clear what we mean by correlation. The **Pearson correlation coefficient**, ρ_{XY}, between two random variables, X and Y is defined as follows:

$$\rho_{XY} = \frac{\text{Cov}(X, Y)}{\sqrt{\text{Var}(X)\text{Var}(Y)}}$$

Eq. 1

Here, Cov(X, Y) is the covariance between X and Y. From a sample of data, where we have measurements, x_i, y_i of the variables X, Y made on the same objects (i.e., datapoints) $i = 1,2,\ldots,N$, we can estimate ρ_{XY} from various sample quantities using the following formula:

$$\hat{\rho}_{XY} = \frac{\frac{1}{N-1}\sum_{i=1}^{N}(x_i - m_X)(y_i - m_Y)}{\sqrt{s_X^2 s_Y^2}}$$

Eq. 2

m_X and s_X^2 are the sample mean and sample variance calculated from the set of observations for X, like this:

$$m_X = \frac{1}{N}\sum_{i=1}^{N}x_i \qquad s_X^2 = \frac{1}{N-1}\sum_{i=1}^{N}(x_i - m_X)^2$$

Eq. 3

Similarly, m_Y and s_Y^2 in Eq. 2 are the sample mean and sample variance calculated from the set of observations for Y, using similar formula to those in Eq. 3.

We can think of the correlation coefficient in Eq. 1 as a normalized covariance, normalized by the product of the scales of the variation in each of the variables X and Y. It measures how much X and Y go up and down together, irrespective of the scale of their own individual variation.

Since we are defining the sample variances s_X^2, s_Y^2 in Eq. 3 and the sample covariance in Eq. 2 with a denominator of $N - 1$, to ensure unbiased estimators (see *Chapter 2*), then we can cancel the factors of $1/(N - 1)$ in the numerator and denominator of Eq. 2 to simplify it a bit. Doing so, we get this:

$$\hat{\rho}_{XY} = \frac{\sum_{i=1}^{N}(x_i - m_X)(y_i - m_Y)}{\sqrt{\left(\sum_{i=1}^{N}(x_i - m_X)^2\right)\left(\sum_{i'=1}^{N}(y_{i'} - m_Y)^2\right)}}$$

Eq. 4

Now, for a time series dataset, we have observations, y_t, with $t = 1,2,\ldots,N$. We think of this as being a set of single observations from a series of random variables, Y_t, $t = 1,2,\ldots,N$, so that y_1 is an observation of the random variable Y_1, y_2 is an observation of the random variable Y_2, and so on. Given this, we can now attempt to quantify the correlation between a random variable Y_t and one that is k time points behind, i.e., Y_{t-k}. As we have lots of observations (well, N of them, at least), we have lots of pairs of values of (Y_t, Y_{t-k}). For example, if we chose $k = 3$, then we could use the pairs of random variables $(Y_4, Y_1), (Y_5, Y_2), \ldots, (Y_N, Y_{N-3})$ to quantify this $k = 3$ auto-correlation. As we want to quantify an auto-correlation, we simply follow the idea we used to define ordinary correlation and define autocorrelation as the covariance between the random variable at time t and one k time points earlier but normalized by the variation over the whole series. Unsurprisingly, we use the corresponding paired observations $(y_4, y_1), \ldots, (y_N, y_{N-3})$ to estimate this quantity. The formula to do so is analogous to that in Eq. 4 but with pairs of observations (x_i, y_i) replaced by pairs of observations (y_t, y_{t-k}). The formula is as follows:

$$r_k = \frac{\sum_{t=k+1}^{N}(y_t - m_Y)(y_{t-k} - m_Y)}{\sum_{t=1}^{N}(y_t - m_Y)^2}$$

Eq. 5

Here, m_Y is the sample mean calculated from the whole set of observations $y_1, y_2, \ldots, y_N$.

This auto-correlation estimate in Eq. 5 is called the lag k auto-correlation, since it measures the auto-correlation between a time point t and one lagging k steps behind.

Clearly, we can calculate the auto-correlation for any positive value of k we want, for example, for $k = 1, 2, 3, \ldots$. Doing so gives us a series of auto-correlation estimates that we can plot against k. This is the **auto-correlation function**, or **ACF** for short. Many time series analysis packages will provide functionality to compute and plot the ACF from an input time series. Let's look at a code example using the statsmodels package. We'll compute the ACF for our UK unemployment time series data plotted in *Figure 6.1*.

ACF code example

The following code example can also be found in the Code_Examples_Chap6.ipynb Jupyter notebook in the GitHub repository. First, we'll read in the raw data:

```
import pandas as pd
import matplotlib.pyplot as plt
from statsmodels.graphics.tsaplots import plot_acf, plot_pacf

df_unemployment = pd.read_csv("../Data/uk_unemployment_rate_monthly.
csv")
```

Now, we'll use the statsmodels.graphics.tsaplots.plot_acf function. We can just pass it the pandas series we want to calculate the ACF for:

```
plot_acf(df_unemployment['Unemployment'], auto_ylims=True)
plt.title('ACF vs lag', fontsize=24)
plt.xlabel(r'lag $k$', fontsize=20)
plt.ylabel('ACF', fontsize=20)
plt.xticks(fontsize=14)
plt.yticks(fontsize=14)
plt.show()
```

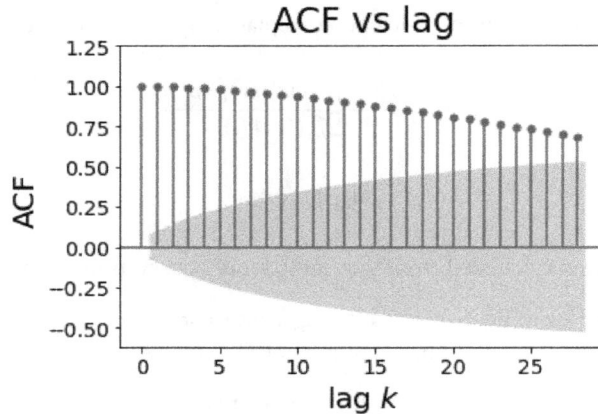

Figure 6.2: ACF plot for the unemployment time series

At lag zero ($k = 0$), the auto-correlation is 1 because the numerator and denominator are both just the sample variance. The shaded blue area represents the 95% confidence interval (centered on zero), so if the auto-correlation value is outside the 95% confidence value we can conclude that the auto-correlation is statistically significant. This ACF plot suggests that there are statistically significant auto-correlation values at multiple different lags. However, because auto-correlation means values at nearby time points are correlated, we get a domino-like effect of correlations being propagated all the way forward in time. If we have auto-correlation at lag k, then we will have auto-correlation at a lag $k' > k$. This brings us neatly to the concept of the **partial auto-correlation function (PACF)**.

The partial auto-correlation function (PACF)

Since long-term auto-correlation may just be a consequence of a short-term auto-correlation being propagated from one time point to the next, it would be ideal if we could assess the level of auto-correlation at lag k having first corrected or adjusted for any auto-correlation present in the data at shorter lags, i.e. at $k' < k$. This is what the PACF attempts to do. There are many ways in which we can perform this adjustment. Some make more explicit assumptions (in the form of a model) about how the data was generated but lead to unbiased estimates of the PACF if those assumptions are true, while other methods make fewer assumptions but produce biased estimates of the PACF. Reviewing all the possible PACF estimation methods is impractical here, so we will just give a code example of one and show how to interpret the resulting plot.

PACF code example

We'll use the UK unemployment time series example again, and we'll assume we have the appropriate packages loaded and data held as a `pandas` dataframe, following from our ACF example. To plot the PACF of the unemployment time series we use the `statsmodels.graphics.tsaplots.`

plot_pacf function. Again, the following code example can also be found in the Code_Examples_ Chap6.ipynb Jupyter notebook:

```
plot_pacf(df_unemployment['Unemployment'],
          method='ywm', auto_ylims=True)
plt.title('PACF vs lag', fontsize=24)
plt.xlabel(r'lag $k$', fontsize=20)
plt.ylabel('PACF', fontsize=20)
plt.xticks(fontsize=14)
plt.yticks(fontsize=14)
plt.show()
```

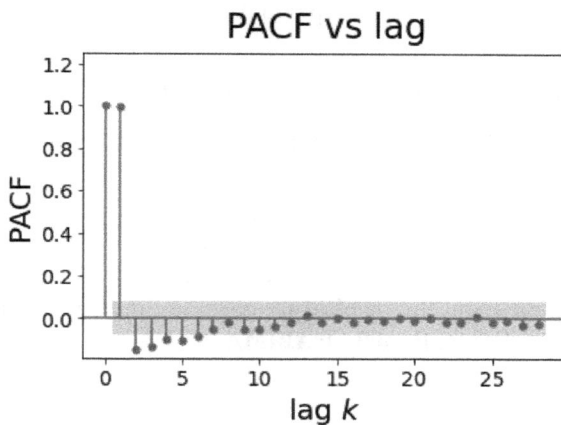

Figure 6.3: PACF plot for the unemployment time series

Again, the blue shaded region in *Figure 6.3* shows the width of the 95% confidence interval, so any partial auto-correlation function values outside the shaded region indicate a statistically significant PACF value. The first value is at lag $k = 0$ and so automatically has a PACF value of 1. We can see that there is a strong partial auto-correlation at lag $k = 1$, with some weaker partial auto-correlation at lags $k = 2$ and $k = 3$.

Other data science implications of time series data

As well as introducing the concept of auto-correlation, time series data also introduces some practical implications from a machine learning perspective. Attention must be paid to how one splits data in training, validation, and test datasets. The presence of auto-correlation means that held-out data points that are in between training points are likely to be predicted well simply because of the constraint that the held-out value must be similar to its neighbors due to the auto-correlation. Testing a time series model using holdout data that is interspersed with the training data is not a rigorous enough test of the predictive accuracy of the time series model. Instead, we must always split our time series data temporally into train and test periods. For example, for the UK unemployment data, I could choose to split the data into a dataset for developing the model, say, from January 1971 to December 2016, and then use the remaining data from January 2017 to December 2019 to test the model.

Having introduced how the ACF and PACF quantify auto-correlation in a time series, we are at the end of this section. So, we'll recap what we have learned so far about time series data.

What we have learned

In this section, we have learned the following:

- How time series data differs from other types of data due to the natural temporal ordering of the data points

- How the temporal ordering of the data introduces the concept of auto-correlation

- How we can quantity the auto-correlation present in a time series dataset using the ACF

- How the PACF quantifies the auto-correlation structure present in a time series dataset in a more concise way

Having learned about the unique aspects of time series data and the important concept of auto-correlation, we will now learn about the classical approach to time series modeling that bakes in auto-correlation into the models from the outset: **ARIMA models**.

ARIMA models

Until the advent of the ideas underpinning ARIMA modeling, time series analysis largely lacked a rigorous foundation. A large part of the uptake, popularity, and hence success of ARIMA models is the rigorous foundations that have been developed. Those foundations were primarily developed by the famous statisticians George Box and Gwilym Jenkins in the late 1960s and 1970s.

ARIMA modeling provides us with a mathematical framework to generate auto-correlation in a time series using very simple equations that specify how the time series evolves. Because of its simplicity and power, ARIMA modeling has for many years been considered the classic and only way to approach time series modeling. Only recently have modern machine learning methods such as deep learning neural networks begun to rival these classic ARIMA methods. Therefore, even if your preference is for more modern techniques, there is still huge value in learning and understanding these classic methods.

The name **ARIMA** stands for **A**uto-**R**egressive **I**ntegrated **M**oving **A**verage models. The three components of ARIMA models attempt to capture the main auto-correlative mechanisms that can contribute to a time series. These components are as follows:

- **Auto-regressive**: Future values of the time series depend on previous, that is, lagged, values of the response variable.

- **Integrated**: Future values of the time series depend on previous, that is, lagged, values of the response variable to such an extent that the variance-covariance structure of the time series is non-stationary, that is, it changes with time.

- **Moving average**: Future values of the time series depend upon lagged values of the additive noise.

Integrated refers to a property of a time series, while **auto-regressive** and **moving average** refer to types of simple models that are combined in an ARIMA model. Those short explanations do not really do justice to the beauty and flexibility of ARIMA modeling, so we will explain each of these three components in detail. We'll start with the concept of an integrated time series. Whether a time series is integrated or not is one of the first things we typically want to establish when we're analyzing it.

Integrated

The easiest way to explain the concept of an integrated time series is to give an example of one. We'll use the classic 1D random walk, also known as **Brownian motion**, and colloquially as the "drunkard's walk." As the last name suggests, it consists of very random movement. We build up the random walk by adding Gaussian random noise to our last observation, so we generate the random walk time series of T observations with the following equation:

$$y_t = y_{t-1} + \varepsilon_t \quad t = 1, 2, \ldots, T.$$

Eq. 6

The additive noise values $\varepsilon_1, \varepsilon_2, \ldots, \varepsilon_T$ are taken as i.i.d. random variables with mean zero and variance σ^2. So, we have this:

$$\mathbb{E}(\varepsilon_t) = 0 \quad \text{for all } t \quad \mathbb{E}(\varepsilon_t^2) = \sigma^2 \quad \text{for all } t \quad \mathbb{E}(\varepsilon_t \varepsilon_{t'}) = 0 \quad \text{if } t' \neq t$$

Eq. 7

Usually, we will take the noise values to be Gaussian distributed, so $\varepsilon_t \sim N(0, \sigma^2)$. We start the time series off at some given value y_0. For convenience, but without loss of generality, we'll take $y_0 = 0$ from now on.

Eq. 6 is an example of a 1D random walk because at each time step, we move a random amount up or down, that is, along a single vertical direction. Random walks, or Brownian motion, can be generalized to higher dimensions, whereby at each time step a random Gaussian move is made in a D-dimensional space. However, for classic time series, we need only concern ourselves with the 1D random walk.

Figure 6.4 shows an example of a random walk generated using the relationship in Eq. 6. We have set $\sigma^2 = 1$ and $T = 300$. The example in *Figure 6.4* is random because the noise values ε_t are random. I have sampled them from a Gaussian distribution Normal(0, σ^2). If I repeated the process, I'd get a different set of noise values and hence a different shape to the random walk. The particular random walk shown in *Figure 6.4* is a single sample from an infinite number of possible random walks that can all be generated using the relationship in Eq. 6. The example shown in *Figure 6.4* is called a **sample path** because it is a single path or walk, sampled from the generating relationship represented by Eq. 6.

Example Random Walk

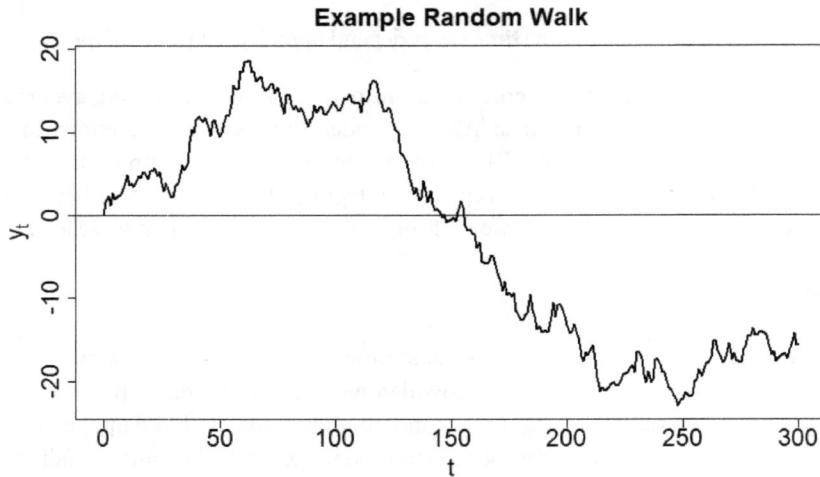

Figure 6.4: An example random walk

Integrated time series have non-stationary statistical properties

The evolution rule in Eq. 6 is very simple but has some interesting properties. The example in *Figure 6.4* might give the impression that the mean of each observation y_t is non-zero, but this is not the case. We can use Eq. 6 to calculate the expectation value $\mathbb{E}(y_t)$ for each time point t. Taking the expectation of both sides of Eq. 6, we get this:

$$\mathbb{E}(y_t) = \mathbb{E}(y_{t-1})$$

Eq. 8

This is because the expectation of the noise ε_t is zero. Because y_0 is a fixed value of zero, it means that $\mathbb{E}(y_0) = 0$, and so applying Eq. 8 repeatedly we get $\mathbb{E}(y_t) = 0$ for all t. This looks boring. However, when we look at the variance it gets more interesting. Applying the same approach to calculate the variance of each observation, we get this:

$$\mathrm{Var}(y_t) = \mathrm{Var}(y_{t-1}) + \sigma^2$$

Eq. 9

So, if we apply Eq. 9 repeatedly, we get $\text{Var}(y_t) = t\sigma^2$. From this, we can see that the variance of the observations y_t increases the further along the time series we are. The statistical properties of the observations, such as the variance, do not settle down to a well-defined stable value, but instead increase indefinitely. Because the statistical properties do not become static, we say the time series is **non-stationary**. The distribution from which y_t is drawn is not the same as the distribution from which y_{t+k} is drawn (for $k > 0$). To emphasize the non-stationary nature of the random walk, we have plotted several sample paths in *Figure 6.5* that have all been generated using Eq. 6 (with $\sigma^2 = 1$ And $T = 300$).

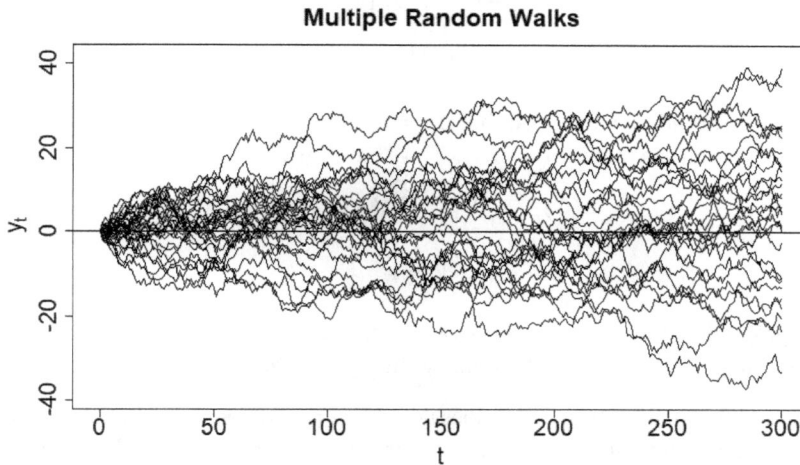

Figure 6.5: Multiple random walks

In *Figure 6.5*, the vertical range of values observed clearly increases as the timepoint increases, highlighting the increasing variance. The approximately symmetrical vertical scatter about the horizontal zero line also highlights the zero expectation value at every time point.

Differencing as a way to remove integration

Non-stationarity is a problem. When we are building probabilistic models, we prefer for the statistical properties of the various probability distributions to be static. It makes them easier to estimate. We would prefer it if we didn't have to work with non-stationary time series. But what has caused the non-stationarity here?

The non-stationary nature of the 1D random walk arises from the fact that at each timepoint we are adding a new random noise value ε_t. This means the tth observation y_t is the sum of all those random noise values:

$$y_t = \sum_{t'=1}^{t} \varepsilon_{t'}$$

Eq. 10

In mathematical language, y_t is the integration of the noise series up until the time point t. We say that the time series y is **integrated**, or it is an **integrated series**.

Writing the observation y_t in its integrated form in Eq. 10 also gives a hint at how we may remove the non-stationarity from the time series y. We perform the opposite of integration. In this case, we take the difference between consecutive observations. So, let's define a new time series of observations as follows:

$$\Delta y_t = y_t - y_{t-1}$$

Eq. 11

A quick look at Eq. 6 should convince you that $\Delta y_t = \varepsilon_t$ for $t = 1, \ldots, T$ (and because we have fixed $y_0 = 0$). From this, we can see that for $t = 1, \ldots, T$, we have this:

$$\mathbb{E}(\Delta y_t) = 0 \qquad \text{Var}(\Delta y_t) = \sigma^2$$

Eq. 12

So, the statistical properties of this new time series are static. This makes this new time series **stationary**. We have removed the problematic non-stationarity simply by **differencing**.

But what happens if differencing a time series does not yield a new times series that is stationary? Well, take the difference again. The number of times, d, that we have to apply this differencing process to the original time series y, before we get a new time series that is stationary, is called the **integration order** of the original time series y.

In fact, we'll define a differencing operator D to be the process of taking lag 1 differences:

$$D y_t = y_t - y_{t-1}$$

Eq. 13

Then, if y is of integration order d, we will have $D^d y$ is non-stationary when $d' < d$, and stationary when $d' \geq d$. That is, we have to apply the differencing process at least d times to get a stationary time series. If a time series is integrated to order d, we denote this as $I(d)$.

So, now we know how to remove the non-stationary aspect of an integrated time series, we can assume from now on that all the time series we will deal with are indeed stationary. Now, we move on to how we incorporate the auto-correlative aspect into our time series models.

Auto-regression

Explaining how we incorporate an auto-regressive element into our time series model is again best explained with an explicit example. Let's consider the time series given by the following relationship:

$$y_t = \phi y_{t-1} + \varepsilon_t$$

Eq. 14

Again, well take the additive noise values ε_t to be i.i.d. Gaussian with zero mean and variance σ^2. The term auto-regressive means we are predicting or regressing the response variable y using earlier values of the response variable itself. But wait a minute, weren't we effectively doing that with the random walk model in Eq. 6? Eq. 14 looks remarkably similar to Eq. 6. In fact, setting $\phi = 1$ in Eq. 14 gives us the random walk relationship in Eq. 6. This suggests that $\phi = 1$ might be an interesting boundary in the behaviour of the auto-regressive model in Eq. 14. Let's analyze the properties of the relationship in Eq. 14 in the same way that we did for Eq. 6. To start, taking expectations, we have this:

$$\mathbb{E}(y_t) = \phi \mathbb{E}(y_{t-1})$$

Eq. 15

Again, if we have $y_0 = 0$ then iterating the relation in Eq. 15 tells us that $\mathbb{E}(y_t) = 0$ for all t. This is the same as for the 1D random walk. Now, let's look at the variance. Taking the variance of both sides of Eq. 14, we get this:

$$\mathrm{Var}(y_t) = \phi^2 \mathrm{Var}(y_{t-1}) + \sigma^2$$

Eq. 16

Again, clearly, the variance is a dynamic quantity, that is, it varies with t. But does the variance grow indefinitely as it did for the 1D random walk, or does it settle down to a stable limiting value? Let's assume the latter and see what conditions we need to hold for that assumption to be true. We'll denote the (assumed) finite limiting variance as V_∞. By definition, $V_\infty = \lim_{t \to \infty} \mathrm{Var}(y_t) = \lim_{t \to \infty} \mathrm{Var}(y_{t-1})$. This means we can take the limit $t \to \infty$ in Eq. 16 and replace the variance on the left-hand and right-hand sides with V_∞. Doing so gives us this:

$$V_\infty = \phi^2 V_\infty + \sigma^2 \quad \Rightarrow V_\infty = \frac{\sigma^2}{1 - \phi^2}$$

Eq. 17

The right-hand result in Eq. 17 shows that we can indeed get a sensible and finite limiting variance V_∞ if we have $\phi^2 < 1$. So, for $\phi^2 < 1$ the auto-regressive model in Eq. 14 has stable statistical properties and so is stationary – a desirable property for our time series model to have. We can see that the 1D random walk, which corresponds to $\phi = 1$, is non-stationary according to this analysis, confirming what we found previously.

So, the simple auto-regressive model in Eq. 14 can produce models with stationary statistical properties. But what do the sample paths look like? *Figure 6.6* shows an example sample path generated from Eq. 14 with $\phi = 0.35$, $\sigma^2 = 1$ and $T = 300$. In contrast to the random walk sample paths shown in *Figure 6.5*, the sample path in *Figure 6.6* stays approximately within a fixed range at all time points t.

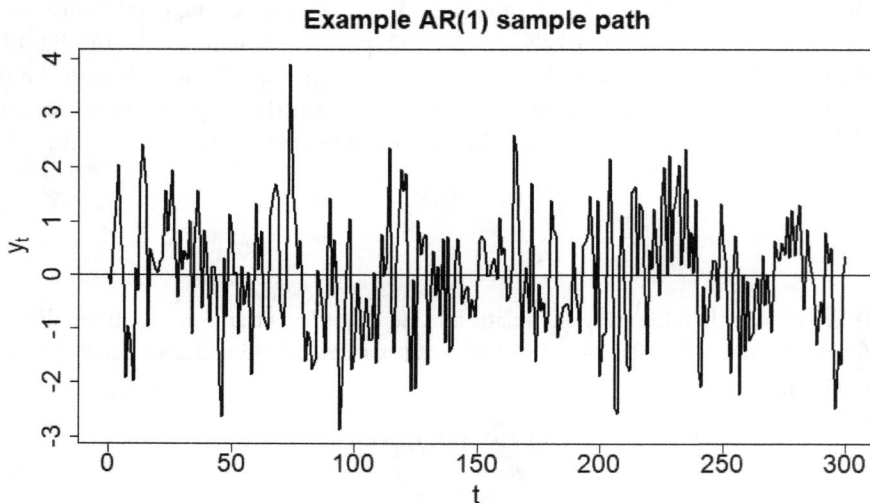

Figure 6.6: Example sample path from an AR(1) process

So far, we have been starting our sample paths from $y_0 = 0$. Are any of our conclusions changed if we start from a non-zero value of y_0? The short answer is no. The longer answer is that we leave this as one of the exercises at the end of the chapter.

Likewise, what would happen if we wanted an auto-regressive model that varied about some non-zero mean μ? Easy! We can just modify our auto-regressive model in Eq. 14 as follows:

$$y_t = (1 - \phi)\mu + \phi y_{t-1} + \varepsilon_t$$

Eq. 18

We have left it as one of the exercises at the end of the chapter to show that for the model in Eq. 18 we have $\mathbb{E}(y_t) \to \mu$ as $t \to \infty$ and $\mathrm{Var}(y_t) \to \sigma^2/(1 - \phi^2)$ as $t \to \infty$. This means we still have stationary statistical properties when $\phi^2 < 1$, but now with the desired mean μ.

The AR(p) model

The auto-regressive model in Eq. 14 has a single parameter ϕ. The value of y_t depends only the preceeding, lag 1, value of the time series. Consequently, it is referred to as a **lag 1 auto-regressive model**, or an **AR(1)** model for short. As you might have guessed, we can generalize the model in Eq. 14 to a situation where y_t depends on multiple lagged values by writing this relationship:

$$y_t = \phi_1 y_{t-1} + \phi_2 y_{t-2} + \cdots + \phi_p y_{t-p} + \varepsilon_t$$

Eq. 19

Unsurprisingly, the relationship in Eq. 19 is called a lag p auto-regressive model and is denoted as an **AR(p)** model. The integer p is referred to as the order of the auto-regressive model.

AR(p) models as infinite impulse response filters

So far, we have been starting our sample paths from zero. That is, we have set y_0 equal to the long-run mean of $\mu = 0$. What happens if for some reason we had a starting value of y_0 that was not equal to our mean $\mu = 0$? What if some large perturbation occurred that meant that we started with a large non-zero value for y_0. We'll look at an AR(1) model to understand what would happen. Clearly, the relationship in Eq. 14 means that the next value y_1 would also be affected by this initial perturbation, but what is the ongoing effect of that perturbation? We can do a simplified experiment with our AR(1) model. We'll start our AR(1) process in Eq. 14 from a non-zero value of y_0. In fact, we'll set $y_0 = 1$ and see what happens. *Figure 6.7* shows an example sample path where we have set $\phi = 0.9, \sigma = 0.1$ and $T = 300$.

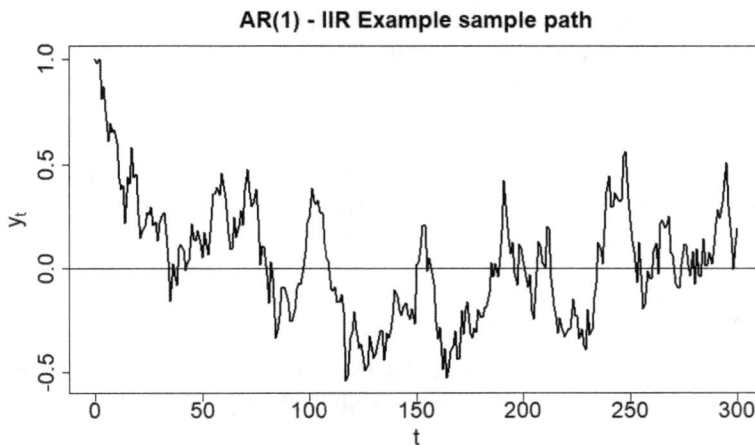

Figure 6.7: Example sample path for an AR(1) process starting from 1

The starting value of $y_0 = 1$ is clear from the left-hand end of the plot in *Figure 6.7*, but after that, the sample path decays towards its long-run mean of zero. What is the shape of that decay? It is difficult to see from a single sample path, as the presence of the noise hides any systematic behavior once y_t gets close to zero. It is easier if we look at the behavior of the expectation $\mathbb{E}(y_t)$ as t increases, rather than looking at a single sample path. We can calculate the mathematical form of $\mathbb{E}(y_t)$ given the relation in Eq. 14 by iterating the relationship in Eq. 15 starting with $\mathbb{E}(y_0) = 1$. Doing this gives us this:

$$\mathbb{E}(y_t) = \phi^t = e^{t\log\phi}$$

Eq. 20

Since we have $0 < \phi < 1$ for our example stationary AR(1) model, $e^{t\log\phi}$ represents exponential decay. In general, we can write this:

$$\mathbb{E}(y_t) = \phi^t = \text{sign}(\phi)^t e^{t\log|\phi|}$$

Eq. 21

We still have exponential decay of the magnitude of $\mathbb{E}(y_t)$ whenever $|\phi| < 1$, i.e. whenever we have a stationary model.

What does this exponential decay mean? It means the effect of our initial perturbation reduces over time, as we would expect and as the example sample path in *Figure 6.7* suggested. More importantly, though, it tells us that the effect of the initial perturbation at $y_0 = 1$ takes an infinite amount of time to die out. At any finite value of t, no matter how large, the value of $\mathbb{E}(y_t)$ is non-zero. It may be small, but there is still some impact of the initial perturbation. Its effect lasts forever. This is also true for an AR(p) model when $p > 1$. Because of this an AR model is said to have an **infinite impulse response** (**IIR**) – its response to an impulse (the perturbation at $t = 0$) lasts for an infinitely long time. You may also see an AR model described as being an **IIR filter**.

As an AR(p) model has an exponential decay response to a perturbation, it is a useful model if we want to capture an exponentially decaying pattern within a time series dataset. But what if we wanted to capture a pattern in a time series that was of finite duration? That is where **moving average** models come in.

Moving average

As with the other components of an ARIMA model, the moving average component is most easily explained by an example. Consider the model given by the following relationship:

$$y_t = \varepsilon_t + \theta_1 \varepsilon_{t-1}$$

Eq. 22

Here, the response variable at time t is a combination of the noise ε_t at t and the lagged noise ε_{t-1}. As usual, the noise values are taken to be i.i.d. Gaussian with zero mean and variance of σ^2. Although we have called the values ε_t "noise," we can more correctly think of them as perturbations (from the long-run mean of zero) that are being introduced into the time series. As such, Eq. 22 says that y_t is a weighted combination of the perturbations at the current and lag 1 time points. In other words, y_t is a weighted average applied to a selected window of the ε series. As the time point t is incremented by one so that $t \rightarrow t + 1$, the window moves along one place. The result is that the entire series y is obtained as a moving average calculation applied to the series, ε, of perturbations. Consequently, the model in Eq. 22 is called a moving average model. It has one parameter, θ_1, which multiplies the lagged 1 perturbation ε_{t-1}. The model in Eq. 22 is called a lag 1 moving average model, or an **MA(1)** model for short.

We can generalize Eq. 22 to a moving average over q lagged values of the perturbations using the following relationship:

$$y_t = \varepsilon_t + \theta_1 \varepsilon_{t-1} + \theta_2 \varepsilon_{t-2} + \cdots + \theta_q \varepsilon_{t-q}$$

Eq. 23

The model in Eq. 23 uses q lagged values and is referred to as a lag q moving average model, or an **MA(q)** model.

What should be clear from Eq. 23 is that after time point $t + q$ the perturbation ε_t has no effect on the response variable y. In contrast to an AR(p) model, the impact of a perturbation in an MA(q) model lasts a finite amount of time. Thus, moving average models are said to have a **finite impulse response (FIR)** and are referred to as **FIR filters**. As such, they are ideal for capturing the effect of a pattern that is of finite duration.

Combining the AR(p), I(d), and MA(q) into an ARIMA model

Now that we have explained the three different component concepts of an ARIMA model, we can piece them back together. In words, an ARIMA model of order (p, d, q) for a time series y consists of an AR component of order p and an MA component of order q applied to a time series y that has been differenced d times. Mathematically, we write this as follows:

$$\tilde{y}_t = \phi_1 \tilde{y}_{t-1} + \phi_2 \tilde{y}_{t-2} + \cdots + \phi_p \tilde{y}_{t-p} + \varepsilon_t + \theta_1 \varepsilon_{t-1} + \cdots + \theta_q \theta_{t-q}$$

Eq. 24

$\tilde{y}$ is the time series obtained from y by differencing d times.

Differencing is just the process of subtracting the lag 1 values of a series from itself. To represent this mathematically, we'll introduce the concept of the **lag operator** L. The lag operator L applied to y_t extracts the value at the preceding time point, or in other words, $Ly_t = y_{t-1}$. You will also see the lag operator referred to as the **backshift operator**, denoted by B.

Using the lag operator L means we can represent the process of differencing as applying the operator $1 - L$ to the time series y. Differencing y d times is the same as applying the operator $(1 - L)^d$ to y. So, the time series $\tilde{y}$ can be defined as follows:

$$\tilde{y}_t = (1 - L)^d y_t$$

Eq. 25

You may also have also spotted that in our definition of an AR(p) model and our definition of an MA(q) model, we are working either with lagged values of the times series variable y, or with lagged versions of the perturbation variable ε_t. This means we can write the ARIMA model in Eq. 24 in a more succinct format using the lag operator L:

$$\tilde{y}_t = \phi_1 L\tilde{y}_t + \phi_2 L^2 \tilde{y}_t + \cdots + \phi_p L^p \tilde{y}_t + \varepsilon_t + \theta_1 L\varepsilon_t + \theta_2 L^2 \varepsilon_t + \cdots + \theta_q L^q \varepsilon_t$$

Eq. 26

It is traditional to write this in the compact form:

$$\left(1 - \sum_{i=1}^{p} \phi_i L^i\right) \tilde{y}_t = \left(1 + \sum_{k=1}^{q} \theta_k L^k\right) \varepsilon_t$$

Eq. 27

If we plug the expression for $\tilde{y}_t$ in Eq. 25 into Eq. 27, we finally get the following succinct form for our ARIMA model:

$$\left(1 - \sum_{i=1}^{p} \phi_i L^i\right) (1 - L)^d y_t = \left(1 + \sum_{k=1}^{q} \theta_k L^k\right) \varepsilon_t$$

Eq. 28

The mathematical form in Eq. 28 is useful for performing formal manipulations of ARIMA models to rigorously prove certain time series results. However, I personally prefer the longer form of Eq. 24 as it shows more explicitly how the observation y_t is built up.

Variants of ARIMA modeling

Having explained ARIMA models, it is natural to ask what they have omitted. As we said at the beginning of this section on ARIMA modeling, the focus is very much on how to capture auto-correlation through very simple mechanisms. The question then arises, what happens if we want to do some of the usual things we do with predictive models, such as include other predictive features? Or how do we cope with patterns such as seasonality, which we would naturally expect to see in many time series? ARIMA modeling can be extended in many directions. Now, we will explain just a few of those common extensions.

ARIMA modeling with a non-zero mean

If we want to model a non-integrated time series $\tilde{y}$ that has a non-zero mean μ, we can simply subtract μ from each observation $\tilde{y}_t$, and then work with the resulting time series that has zero mean by construction.

If we want to represent this mathematically, we can replace $\tilde{y}_t$ by $\tilde{y}_t - \mu$ in the model in Eq. 24. If we do that, we get this:

$$\tilde{y}_t - \mu = \sum_{i=1}^{p} \phi_i(\tilde{y}_{t-i} - \mu) + \varepsilon_t + \sum_{k=1}^{q} \theta_k \varepsilon_{t-k}$$

Eq. 29

We can re-write this as follows:

$$\tilde{y}_t = \sum_{i=1}^{p} \phi_i \tilde{y}_{t-i} + \mu\left(1 - \sum_{i=1}^{p} \phi_i\right) + \varepsilon_t + \sum_{k=1}^{q} \theta_k \varepsilon_{t-k}$$

Eq. 30

Then, using the relationship in Eq. 25, we can write our non-zero mean ARIMA model as follows:

$$\left(1 - \sum_{i=1}^{p} \phi_i L^i\right)(1 - L)^d y_t = \mu\left(1 - \sum_{i=1}^{p} \phi_i\right) + \left(1 + \sum_{k=1}^{q} \theta_k L^k\right)\varepsilon_t$$

Eq. 31

Eq. 31 makes it clearer that μ is the long-run mean of $(1 - L)^d y_t$.

ARIMA modeling with other predictors

So far, we have been modeling the time series y using just the time series itself (using lagged observations) or the random perturbation series ε. But what happens if we know or suspect that the observations y_t are influenced by other factors that are external to the time series y? We might have a set of features $\underline{x}$ that we want to include in our predictive time series model. Such factors are called **exogenous** factors since they are external or outside of the time series itself. We include them in our time series model, just as we do in most of our probabilistic modeling; we just multiply them by some weights or coefficients and add to the predictive part of the model. So, our model now takes the following form:

$$\left(1 - \sum_{i=1}^{p}\phi_i L^i\right)(1 - L)^d y_t = \mu\left(1 - \sum_{i=1}^{p}\phi_i\right) + \left(1 + \sum_{k=1}^{q}\theta_k L^k\right)\varepsilon_t + \underline{x}_t^\top \underline{\beta}$$

Eq. 32

Note that in Eq. 32, we have added the linear predictor term $\underline{x}_t^\top \underline{\beta}$ to the right-hand side. It is the right-hand side that contains all the terms that express how new perturbations, signal, or information get into the time series evolution.

Since this approach consists of just adding the effect of exogenous predictors to the existing ARIMA model in Eq. 31, it is usually called **ARIMA+X** or **ARIMAX**.

ARIMA modeling with seasonality

When we have seasonality present in a time series, we are saying that a certain proportion of the observation y_t will have the same value as in the preceding season. If the season is of length S time points, then we are saying that y_t is predicted in part by y_{t-S}. How would we include y_{t-S} as a predictor in our time series model? Well, y_{t-S} is just a lagged observation, and we know how to include lagged observations in an ARIMA model. We can do that by introducing the concept of the **seasonal lag operator** L_s, which shifts observations back by one season so that $L_s y_t = y_{t-S}$. Since shifting by a season is equivalent to shifting by S time steps. This means we can write $L_S = L^S$, where L is our usual 1 timestep lag operator.

To include a seasonal lag of y_t in our ARIMA model we simply replace the auto-regressive part of our ARIMA model with the following substitution:

$$\left(1 - \sum_{i=1}^{p}\phi_i L^i\right)(1 - L)^d y_t \rightarrow \left(1 - \sum_{i=1}^{p}\phi_i L^i\right)\left(1 - \sum_{j=1}^{p}\Phi_j L_S^j\right)(1 - L)^d y_t$$

Eq. 33

The right-hand side of Eq. 33 now has an extra set of lag operations applied to y_t. These lag operations are applications of various powers of the seasonal lag operator L_S with coefficients $\Phi_1, \Phi_2, \ldots, \Phi_P$. This is the seasonal part of the seasonal ARIMA model. Sometimes, it is referred to as an **SAR**(P) model, meaning it is a seasonal auto-regressive model of order P. Note the use of capital letter for the order P of the seasonal auto-regressive model as well as the use of capital Greek letters $\Phi_1, \Phi_2, \ldots, \Phi_P$ for the coefficients. This is common when specifying seasonal ARIMA models.

The substitution in Eq. 33 tells us how to modify the auto-regressive part of the ARIMA equation to include seasonality, but what about the integration and moving average parts? Since those parts can also be expressed in terms of lag operators, extending them to include seasonal components proceeds with similar substitutions to that in Eq. 33. Our seasonal ARIMA model is finally specified as follows:

$$\left(1 - \sum_{i=1}^{p} \phi_i L^i\right)\left(1 - \sum_{j=1}^{P} \Phi_i L_S^j\right)(1-L)^d (1 - L_S)^D y_t = \left(1 + \sum_{k=1}^{q} \theta_k L^k\right)\left(1 + \sum_{l=1}^{Q} \Theta L_S^l\right)\varepsilon_t$$

Eq. 34

Note that now Eq. 34 has a seasonal auto-regressive component of order P, a seasonal integrated component of order D, and a seasonal moving average component of order Q. This model is referred to as a SARIMA$(p, d, q)(P, D, Q)S$ model.

The SARIMA model in Eq. 34 can also be easily extended to include a non-zero mean, as in Eq. 31, and the effect of exogenous influences, as in Eq. 32.

Okay, we have learned a lot about ARIMA models. We will use them for real in the next section, but for now, this is a good place to stop and recap what we have learned about ARIMA modeling.

What we have learned

In this section, we have learned the following:

- The concept of an integrated time series and how it has non-stationary statistical properties
- The concept of an auto-regressive AR(p) model and how it can capture IIR patterns in a time series
- The concept of a moving average MA(q) model and how it can capture FIR patterns in a time series
- How the concepts of integration, an AR(p) model, and an MA(q) model can be combined into a single ARIMA(p,d,q) model
- How ARIMA models can be extended to include exogenous effects and seasonality

Having explained the mathematical concepts behind ARIMA models, we'll now explain some of the practical issues involved with ARIMA time series modeling.

ARIMA modeling in practice

As with any theoretical framework, there are subtleties and nuances to getting ARIMA to work in practice. We cannot cover all the tips and tricks of ARIMA modeling in practice in this short section. Instead, we will focus on the practical issues related to what we introduced in the previous section, namely, how to identify the order (p, d, q) that we should use when building an ARIMA model of a given time series dataset.

Unit root testing

The first stage in deciding the appropriate orders of an ARIMA model is to determine whether the time series data you have represents an integrated series or not. This is usually done by testing for the presence of what is called a **unit root**. We won't go into the full details of what a unit root is, other than to say it is a root of a polynomial equation. A unit root has magnitude 1 and the polynomial equation is constructed from coefficients of an AR(p) model estimated from the time series. The presence of a unit root implies that the time series is non-stationary and integrated to some order d.

There are various statistical hypothesis tests that test for the presence of a unit root given the time series data – we will cover hypothesis tests in *Chapter 7*. Two of the most common hypothesis tests for testing for the presence of a unit root are the **Augmented-Dickey-Fuller (ADF)** test, and the **Kwiatkowski-Phillips-Schmidt-Shin (KPSS)** test.

The statsmodels package provides easy-to-use functions to run both the ADF and KPSS tests. The functions are statsmodels.tsa.stattools.adfuller for the ADF test and statsmodels.tsa.stattools.kpss for the KPSS test. Using these functions is as straightforward as passing in a 1D NumPy array or pandas series of response variable values, and leaving all the other arguments at their default settings, for example, using a syntax in the following form:

```
from statsmodels.tsa.stattools import adfuller, kpss
adf_test = adfuller(my_time_series)
kpss_test = kpss(my_time_series)
```

Interpreting the output from the ADF and KPSS tests requires knowledge of hypothesis testing, which we won't cover until the next chapter. However, we will give you a warning about something that can trip up the inexperienced time series modeler. The ADF and KPSS tests make different starting assumptions. The ADF test has a **null hypothesis** that a unit root is present, that is, it starts with the assumption that the time series is integrated, so is I(1), and then tests whether that assumption can be rejected. In contrast, the KPSS test has a null hypothesis that a unit root is absent, that is, it starts with the assumption that the time series is not integrated, so is I(0), and then tests whether that assumption can be rejected. I will typically run both tests when testing if a time series is integrated. It is possible for the two tests to give inconsistent conclusions depending upon the data and the false positive rate thresholds we apply to the two tests. However, if this happens, it indicates that the evidence from the data is not overwhelmingly conclusive for either the presence or absence of a unit root. If there is strong evidence for the presence (or absence) of a unit root you will typically see consistent conclusions from ADF and KPSS tests.

Interpreting ACF and PACF plots

Determining the order of the auto-regressive and moving average components of an ARIMA model can be done by looking at the ACF and PACF plots. The key point to bear in mind is that since an $MA(q)$ process is an FIR process, any auto-correlation will be zero for a lag of more than q timesteps, and so for an $MA(q)$ process, we will see the ACF plot reduce significantly after lag q. In contrast, an $AR(p)$ process will still show structure, not necessarily significant, in the ACF plot at lags beyond p.

When we look at the PACF plot if we have an $AR(p)$ process, then we expect to see no significant structure beyond lag p. For an $MA(q)$ process the PACF plot can show significant values beyond lag q.

Overall, this is best illustrated graphically. In *Figure 6.8*, we have the ACF and PACF plot for a time series of 1,000 data points. The time series has been generated using an AR(2) process with $\phi_1 = 0.6$, $\phi_2 = 0.3$ and $\sigma = 0.1$. The first two time point values have been sampled from Normal$(0, \sigma^2)$. The blue horizontal dashed lines in *Figure 6.8* are the 95% confidence intervals.

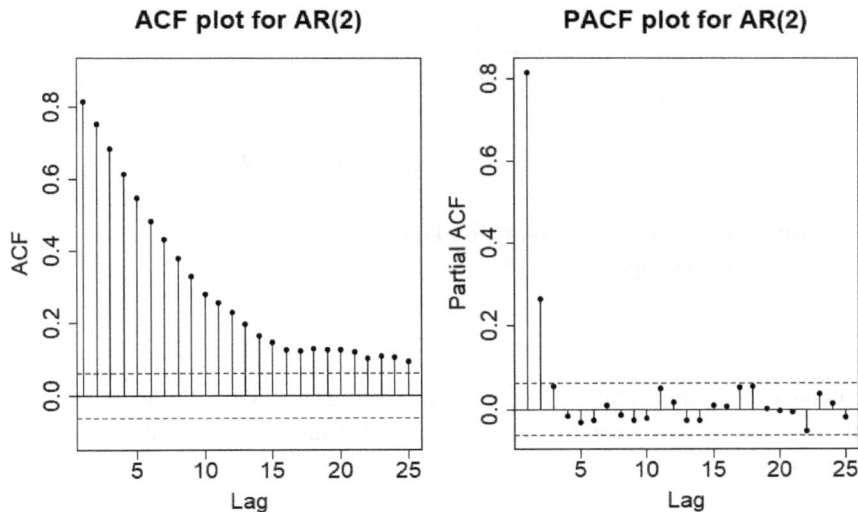

Figure 6.8: ACF and PACF plot for an example AR(2) process

The extended auto-correlations at lags > 2 are clearly visible in the ACF plot in *Figure 6.8*, but the PACF plot in *Figure 6.8* reveals that there are only two auto-regressive terms that we should really include in any ARIMA model of this time series.

We can contrast *Figure 6.8* with what the ACF and PACF plots look like for an example MA(2) process. This is shown in *Figure 6.9*, where we have plotted the ACF and PACF for a time series of 1,000 time points generated using an MA(2) process with $\theta_1 = 0.6$, $\theta_2 = 0.3$ and $\sigma = 0.1$. The first two time point values have been set equal to the noise values ε_1 and ε_2.

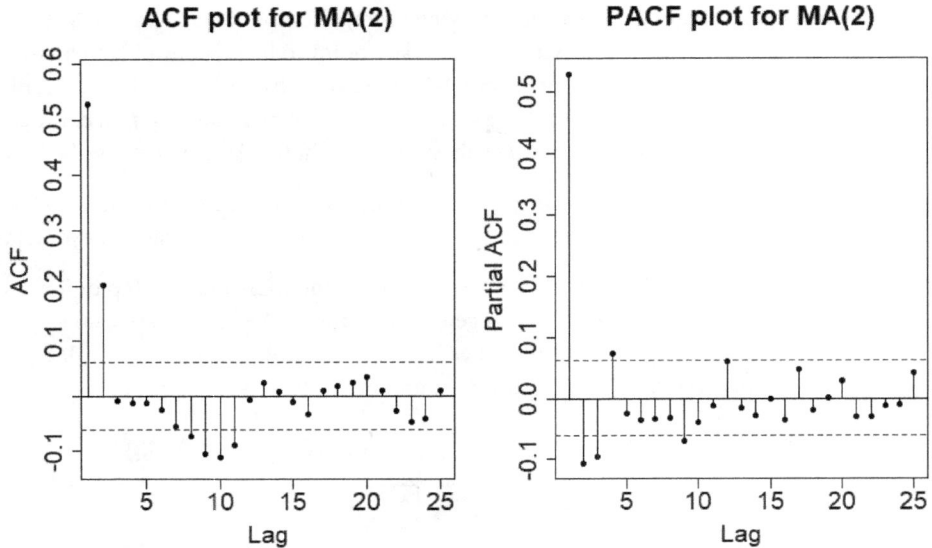

Figure 6.9: ACF and PACF plot for an example MA(2) process

The expected sharp drop-off in the ACF plot after lag=2 is clearly present in *Figure 6.9*, whilst the PACF plot still appears to have significant PACF values at lags beyond lag 2.

auto.arima

Testing for unit roots and interpreting ACF/PACF plots sounds like both hard work and something of an art form. It is. That is why most practitioners will in the first instance make use of some form of **auto.arima** tool.

The processes of unit root testing and selecting values for *p* and *q* can largely be automated. The term "auto.arima" refers to a function that will take a time series as input, select the optimal values of p, d, q for that time series, and possibly also estimate the model coefficients $\phi_1, \ldots, \phi_p, \theta_1, \ldots, \theta_q$. A function called `auto.arima` is part of the popular `forecast` package in R. This functionality has been ported to a number of Python packages, such as the `Pmdarima` package and the `statsforecast` package. We'll demonstrate the `auto.arima` functionality of the `Pmdarima` package in the following code example using a new dataset. We'll use the solar activity dataset that is part of the `statsmodels` package. For simplicity, we have created a de-seasonalized and centered version of this data. The data is in the `sunactivity_deseasonalized.csv` file in the `Data` folder of the GitHub repository.

auto.arima code example

The following code example, and more, is also available in the `Code_Examples_Chap6.ipynb` Jupyter notebook in the GitHub repository. The first thing we need to do is read in the data:

```
import numpy as np
import pandas as pd
import matplotlib.pyplot as plt
from statsmodels.graphics.tsaplots import plot_acf, plot_pacf
from statsmodels.tsa.arima.model import ARIMA
from pmdarima.arima import auto_arima
# Read in the de-seasonalized and centered sun activity data
df_sunactivity = pd.read_csv('../Data/sunactivity_deseasonalized.csv')
# We'll split the data into a training segment and a hold-out period
# at the end, which we'll forecast using our ARIMA model
y = df_sunactivity['sunactivity_deseasonalized'].to_numpy()

# Set the number of forecast steps
n_steps = 6

# Calculate the length of the training data and extract
n_train = y.shape[0] - n_steps
y_train = y[0:n_train]
```

Let's look at the ACF plot:

```
plot_acf(y_train, auto_ylims=True)
plt.title('SUN ACTIVITY ACF ', fontsize=24)
plt.xlabel(r'lag $k$', fontsize=20)
plt.ylabel('ACF', fontsize=20)
plt.xticks(fontsize=14)
plt.yticks(fontsize=14)
plt.show()
```

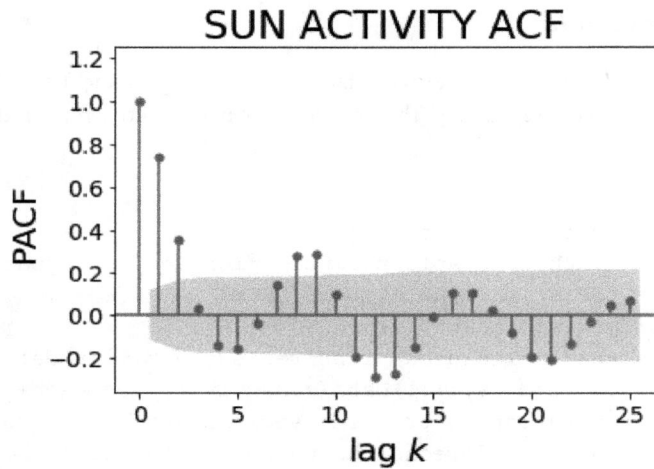

Figure 6.10: ACF plot for the sun activity data

The ACF plot hints at three auto-regressive parameters. Let's check the PACF plot:

```
plot_pacf(y_train, method='ldb', auto_ylims=True)
plt.title('SUN ACTIVITY PACF ', fontsize=24)
plt.xlabel(r'lag $k$', fontsize=20)
plt.ylabel('PACF', fontsize=20)
plt.xticks(fontsize=14)
plt.yticks(fontsize=14)
plt.show()
```

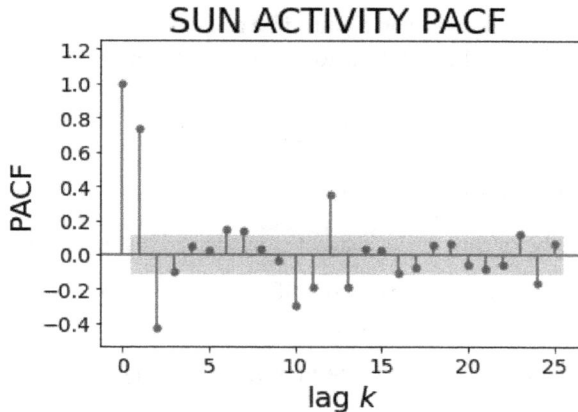

Figure 6.11: PACF plot for the sun activity data

The PACF plot still suggests an AR(3) model. There are some contributions at lags 10 and 12, but we'll keep the model simple and try an AR(3) model. Let's see what `auto.arima` suggests. We'll use the `auto_arima` function from the `Pmdarima` package:

```
# Run the auto_arima function
auto_arima(y_train)
```

This gives the following output:

```
ARIMA(3,0,0)(0,0,0)[0]
```

`auto_arima` also suggests an AR(3) model, so we can then use the ARIMA model fitting functionality of the `statsmodels` package to do this. This is illustrated in the `Code_Examples_Chap6. ipynb` notebook.

You may ask that if we have `auto.arima` functions available to us, why do we even bother learning about unit root tests and how to interpret ACF and PACF plots? The answer is that `auto.arima` functions are not infallible. Sometimes, the choices of p, d, q from an `auto.arima` function can be inappropriate, or sometimes they need adjusting slightly to yield a more appropriate model. The key skill here is the ability to understand what p, d, q represent and the ability to check, where necessary, the underlying evidence such as ACF and PACF plots to see if the values recommended by the `auto. arima` function stack up. In fact, if you are building relatively small-scale models, for example, models of just one or two time series, I would always recommend checking the output of an `auto. arima` function to check whether it is sensible. Where you are building time series models at scale, for example, as part of an automated model building pipeline, checking the `auto.arima` output can be more tricky. However, even in these latter cases, understanding how to determine the order (p, d, q) is still useful, e.g. for pipeline debugging purposes.

Practicing building time series models is worth your time as a data scientist, and there are many more tips and tricks that we could discuss. But it is time to finish this section and recap what we have learned.

What we have learned

In this section, we have learned the following:

- How to identify whether a time series dataset is integrated, using unit root tests
- How to interpret ACF and PACF plots to identify the orders of the auto-regressive and moving average components of an ARIMA model
- How to use an `auto.arima` function to identify appropriate values for the orders (p, d, q) of a time series dataset

Having covered some of the practical issues involved with ARIMA time series modeling, we will finish the chapter with a short summary of some of the modern alternatives to ARIMA for time series modeling.

Machine learning approaches to time series analysis

Everything we have explained so far in this chapter may give the impression that ARIMA modeling is the only approach to time series modeling available. This is certainly not true. The rapid development and success of machine learning techniques over the last two decades has inevitably meant that machine learning algorithms have been applied to time series datasets.

However, the field of machine learning algorithms applied to time series is probably comparable to or bigger than that of ARIMA modeling, and it is still a rapidly developing field. As with other branches of machine learning, it is a very applied field. There are not simple clear-cut mathematical concepts or principles that define the field of machine learning for time series analysis, like there are for ARIMA analysis of time series. Therefore, in this section, we focus on giving you a short review of how the field has developed and its current state. Since it is a very short review section, we will refer you to many of the original sources and key articles.

Routine application of machine learning to time series analysis

Machine learning techniques such as feed-forward neural networks have been used for time series analysis for many years. Other common machine learning algorithms, such as Random Forests and XGBoost, have also been applied to time series data. However, the approach often taken has been to treat time series data as an ordinary dataset, that is, as a training set of response variable observations that need to be predicted in terms of various predictive features. Consequently, the auto-correlation elements of the time series data are ignored. Given that the auto-correlation element present in a time series dataset is essentially what makes a time series, it is not surprising that many practitioners reported that simply applying standard machine learning techniques to time series data did not lead to improved predictive accuracy compared to the classical ARIMA methods. Combined with the simplicity, familiarity, and easy interpretability of the classical ARIMA methods, it is not surprising that ARIMA is still a widely used time series analysis technique and that machine learning methods have not had the uptake they have had in other fields.

This does not mean that machine learning approaches haven't been successful for time series analysis of specific problems, specific datasets, or specific domains. Indeed, it is interesting to note that in the recent M5 forecasting competition (see point 3 in the *Notes and further reading* section) the top 50 performing entries all made use of LightGBM, a machine learning algorithm that uses gradient boosted trees. The preceding iteration of the M forecasting competitions, M4, was also won by a neural network-based solution (see point 4 in the *Notes and further reading* section), albeit a hybrid solution that combined the neural network with a classical exponential smoothing approach. At the same time, the M5 competition still highlighted the power and utility of simple ARIMA models. Entries to the M5 competition consisted of other ARIMA-based models, machine learning models, and hybrid approaches, yet 92.5% of entries failed to beat an exponential smoothing benchmark algorithm – see point 5 in the *Notes and further reading* section. A linear exponential smoothing algorithm is equivalent to an ARIMA model, that is, usually either an ARIMA(0,1,1) or an ARIMA (0,2,2) model.

Overall, we would conclude that proclaiming either the death of ARIMA modeling, or the failure of standard machine learning approaches when applied to time series analysis, is somewhat naïve and incorrect.

Deep learning approaches to time series analysis

The success of deep learning neural networks on such tasks as image recognition and classification has opened a new sub-field of machine learning for time series analysis. Specifically, the advent of **deep recurrent neural networks (deep-RNNs)** and modular structures such as **long short-term memory (LSTM)** units, has meant that it is possible to construct machine learning models that also directly incorporate auto-regressive components in the model and so directly address the auto-correlation present in the time series data. The DeepAR model from Amazon Research is one example of a successful deep-RNN based time series analysis/forecasting solution that reportedly outperforms classical ARIMA based models fitted to the same data – see point 6 in the *Notes and further reading* section. The DeepAR model achieves this performance by sharing information across multiple time series, in this case learning a set of common parameters of the RNN part of the model and the common parameters of the likelihood model. An implementation of the DeepAR algorithm is now available through the **Amazon Web Services (AWS)** SageMaker AI platform (see point 7 in the *Notes and further reading* section). The literature of this sub-field of deep learning for time series forecasting has been recently reviewed by the same group that produced the DeepAR model – see point 8 in the *Notes and further reading* section.

AutoML approaches to time series analysis

An alternative machine learning approach to time series analysis is to automate the model-building process as much as possible but using relatively simple model forms and architectures, instead of more complex models such as deep learning neural networks or gradient boosted trees. This is in the same spirit as the `auto.arima` functions we demonstrated earlier. The model forms used can be other classical statistical techniques, such as **generalized additive models** (**GAMs**), which give more flexibility than basic ARIMA models but are still simple enough to be incorporated into a robust AutoML solution.

Several of the large tech firms, such as Meta and Uber have released their own open-source AutoML tools for time series analysis and forecasting. Probably the most popular of these is Prophet (`https://facebook.github.io/prophet/`) from Facebook (Meta). Prophet uses GAMs for modeling and uses the Stan PPL (see *Chapter 5*) under the hood to do the model fitting. Prophet is intended to be an easy-to-use tool. As with any automated tool, there is the possibility for mistakes to creep in, but this is where what you have learned in the previous sections will help you check and interpret the output from an AutoML time series tool such as Prophet.

Orbit (`https://uber.com/en-GB/blog/orbit/`) from Uber is another open source AutoML tool for time series analysis from one of the big tech companies. Orbit also uses a PPL as its backend calculation engine, but there is some flexibility about which PPL you can choose to use. Orbit focuses on Bayesian time series modeling but provides more flexibility in the choice of model form you can use, in contrast to Prophet, which only uses GAMs. Despite this increase in flexibility in model form, Orbit aims to still maintain ease of specification, which is the key selling point of AutoML tools.

The use of AutoML tools for time series data analysis divides opinions. Tools such as Prophet are popular but can also attract criticism from experienced practitioners. Overall, the application of machine learning techniques to time series data is still a rapidly evolving field, hence why we have only given a brief and high-level overview. Having given that high-level overview it is time to summarize what we have learned in this section and in the chapter overall.

What we have learned

In this section, we have learned the following:

- How standard machine learning algorithms have been applied to time series analysis but have not replaced ARIMA-based modeling approaches

- How deep learning neural network models that incorporate an auto-regressive component, such as the DeepAR model from Amazon Research, can combine the power of deep learning with the simple auto-correlative focused approaches of ARIMA modeling

- About AutoML tools for time series analysis/forecasting, such as Prophet from Meta and Orbit from Uber

Summary

The focus of this chapter has been heavily on ARIMA modeling. This is because ARIMA models, by their very definition, focus on the simple core mathematical concepts that effectively define what a time series is. Focusing on ARIMA modeling allowed us to easily introduce and understand those math concepts. Specifically, we have had to learn about the following:

- How the time element of time series data provides a natural ordering to the data and introduces auto-correlation between the values from nearby time points

- How ARIMA models are built up from simple mechanisms that capture auto-correlation patterns in data

- How ARIMA models can be expressed in simple math terms using the lag operator

- How ARIMA models can be extended to include exogenous factors and seasonality

- How ARIMA models are built in practice

- How machine learning algorithms can also be applied to time series data; in particular, how deep learning auto-regressive neural networks incorporate aspects of both classical auto-regressive models and machine learning techniques

- How AutoML forecasting tools can provide automated machine learning approaches to time series analysis

In the next chapter, we cover another core but self-contained topic, that of hypothesis testing. Like the ARIMA models we focused on in this chapter, the bulk of what we will explain about hypothesis testing relates to classical statistical analysis techniques. A good understanding of these classical techniques is invaluable for any data scientist.

Exercises

Here is a series of exercises. The answers to all the exercises are given in the `Answers_to_Exercises_Chap6.ipynb` Jupyter notebook in the GitHub repository.

1. For the AR(1) process defined by the following equation,

$$y_t = \phi y_{t-1} + \varepsilon_t \quad \varepsilon_t \sim \text{Normal}(0, \sigma^2)$$

Eq. 35

show that $\mathbb{E}(y_t) \to 0$ and $\text{Var}(y_t) \to \sigma^2/(1 - \phi^2)$ as $t \to \infty$, for any starting value y_0. For this exercise you can assume that $0 < \phi < 1$.

2. For the AR(1) process defined by the following equation,

$$y_t = (1-\phi)\mu + \phi y_{t-1} + \varepsilon_t \qquad \varepsilon_t \sim \text{Normal}(0, \sigma^2)$$

Eq. 36

show that $\mathbb{E}(y_t) \to \mu$ and $\text{Var}(y_t) \to \sigma^2/(1-\phi^2)$ as $t \to \infty$, for any starting value y_0. For this exercise, you can assume that $0 < \phi < 1$. See if you can derive the values of $\mathbb{E}(y_t)$ and $\text{Var}(y_t)$ for any value of t, not just the asymptotically limiting values.

3. Use the ARIMA model form in Eq. 28 to generate a sample time series of length $T = 500$ timepoints, that is of order $p = 1, d = 1, q = 1$, with coefficients $\phi_1 = 0.6, \theta_1 = 0.7$. The noise values should be i.i.d. Normal$(0, 0.1^2)$. You can set the first value, y_0, of the generated series to zero. Use the `statsmodels.tsa.arima.model.ARIMA` function from the `statsmodels` package to fit an ARIMA(1,1,1) model to the simulated data you have just generated. Look at how close the estimated AR(1) and MA(1) parameters are to the values $\phi_1 = 0.6, \theta_1 = 0.7$ you used to generate the data.

Notes and further reading

1. The recent article `https://www.sciencedirect.com/science/article/pii/S0169207021001758` in the *International Journal of Forecasting* attempts to summarize the majority of the time series/forecasting field. A preprint version of the article can be found in the arXiv archive at `https://arxiv.org/pdf/2012.03854.pdf`

2. The unemployment data is ONS series MGSX, and can be accessed via `https://www.ons.gov.uk/employmentandlabourmarket/peoplenotinwork/unemployment/timeseries/mgsx/lms`. The data plotted in *Figure 6.1* was accessed on 29th March 2023.

3. The M series of forecasting competitions is perhaps the main Kaggle-like competition within the forecasting community and has been run intermittently over the last 40 years.

4. Details of the winning M4 entry can be found at `https://www.uber.com/en-GB/blog/m4-forecasting-competition/`.

5. For a summary of the M5 competition outcomes and further discussion of the exponential smoothing benchmark, see, "*M5 accuracy competition: Results, findings, and conclusions*", S. Makridakis, E. Spiliotis and V. Assimakopoulos, International Journal of Forecasting, 38(4):1346-1364, 2022, `https://www.sciencedirect.com/science/article/pii/S0169207021001874`.

6. "*DeepAR: Probabilistic forecasting with autoregressive recurrent networks*", D. Salinas et al., International Journal of Forecasting 36(3):1181-1191, 2020. The arXiv pre-print version can be found at `https://arxiv.org/pdf/1704.04110.pdf`.

7.   Documentation of the SageMaker implementation can be found at `https://docs.aws.amazon.com/sagemaker/latest/dg/deepar.html`.

8.   The review and tutorial article, *"Deep Learning for Time Series Forecasting: Tutorial and Literature Survey"*, Benidis et al., ACM Computing Surveys 55(6), 2023, article No.: 12, can be found at `https://dl.acm.org/doi/10.1145/3533382`. The arXiv pre-print version can be found at `https://arxiv.org/pdf/2004.10240.pdf`.

7
Hypothesis Testing

Hypothesis tests are a ubiquitous part of classical statistics. They often have a very simple objective, such as testing whether two samples of data indicate there is a difference in the means of the underlying populations from which those samples were taken. Despite the simplicity of these aims and questions, hypothesis tests have very practical applications. The question of whether two populations have different means is precisely what we ask when running an A/B test to decide whether the A variant of an e-commerce site has a higher click-through rate, compared to the B variant. As such, hypothesis testing is an important skill to master for any data scientist working with real-world data. Despite the simplicity of the question that a hypothesis test asks, the mathematical machinery needed to run a hypothesis test is full of concepts and nuances that can trip up a new data scientist – concepts such as p-values, degrees of freedom, confidence intervals, Type-I and Type-II errors, and power. To master hypothesis testing, it is essential to get a basic understanding of these concepts. That is what we aim to do in this chapter. We do this by covering the following topics:

- *What a hypothesis test is*: In this section, we will explore how hypothesis tests are structured and how they follow a common pattern

- *Confidence intervals*: In this section, we will see how confidence intervals quantify the uncertainty associated with estimates of parameters

- *Type I and Type II errors, and power*: Finally, will will learn how to pre-compute the sample size necessary to achieve a specified false negative rate

Technical requirements

All code examples given in this chapter (and additional examples) can be found at the GitHub repository: `https://github.com/PacktPublishing/15-Math-Concepts-Every-Data-Scientist-Should-Know/tree/main/Chapter07`. To run the Jupyter notebooks, you will need a full Python installation that includes the following packages:

- `pandas` (>=2.0.3)
- `numpy` (>=1.24.3)
- `scipy` (>=1.11.1)
- `statsmodels` (>=0.14.0)

What is a hypothesis test?

A hypothesis test does exactly what it says it does. It tests a hypothesis. Typically, the hypothesis we test is well formulated and can be simply and precisely expressed mathematically. For example, in a scientific experiment, we might want to know whether the average weight of one group of animals is different from that of another group of animals. The hypothesis here can be precisely expressed mathematically as,

$$\mathbb{E}(\text{weight of animals in group A}) \neq \mathbb{E}(\text{weights of animals in group B})$$

Eq. 1

Alternatively, we might be interested in knowing whether website visitors complete purchases at the same rate on variant A of an e-commerce site compared to variant B of the website. In this case, the hypothesis can be expressed as,

$$\mathbb{E}(\text{Click-Through Rate for site A}) \neq \mathbb{E}(\text{Click-Through Rate for site B})$$

Eq. 2

Note that in both these hypotheses, the mathematical expression is in terms of the expectation. In other words, we want to test the hypothesis at the level of populations, not for a finite sample of the A and B populations. In the e-commerce example, we want to know that if everybody in our target population were to visit variant A of our website, would we get the same volume of purchases compared to if we had deployed variant B of the site. If we use μ_A and μ_B to denote the population mean click-through rate for site A and site B, respectively, then the hypothesis in Eq. 2 is the same as asking,

$$\mu_A \neq \mu_B?$$

Eq. 3

However, we usually only have data from a sample of the A and B populations available to us, so we can only calculate sample means. If we denote the sample means m_A and m_B, then the equivalent comparison to Eq. 3 is,

$$m_A \neq m_B?$$

Eq. 4

The problem is that m_A and m_B are calculated from data, so they have random components. Those random components mean that m_A could be different from m_B just by chance, due to the random variation of the sample means. Consequently, we could have Eq. 4 as true even if Eq. 3 isn't.

What do we do? Do we just give up on using the sample means m_A and m_B to test the hypothesis in Eq. 3? No. Since we can't ask the question in Eq. 3 directly, we will embrace the uncertainty inherent in m_A and m_B and take a statistical approach to asking the question in Eq. 3.

We will use our understanding from *Chapter 2* to quantify how frequently Eq. 4 is true when Eq. 3 isn't. Doing so will allow us to assess whether an observation that $m_A \neq m_B$ provides evidence for the hypothesis $\mu_A \neq \mu_B$.

Example

We'll give a slightly contrived example, but one which will help unpack how we use the statistical ideas from *Chapter 2* to assess the evidence contained in the values of m_A and m_B.

Imagine we have two samples of data, an A sample and a B sample, of sizes N_A and N_B, respectively. These could be the two datasets of observations of whether visitors to our e-commerce website made a purchase. The following provides more details on each of those two samples:

- **Sample A:** N_A = 100 observations. Each observation is from a Gaussian random variable that has a mean μ_A = 0.5 and variance σ_A^2 = 1 – that is, sample A is of 100 i.i.d. *Normal*(0.5, 1) random variables. The sample mean of those 100 observations is m_A = 0.529.

- **Sample B:** N_B = 100 observations. Each observation is from a Gaussian random variable that has a mean μ_B = 0.1 and variance σ_B^2 = 1 – that is, sample B is of 100 i.i.d. *Normal*(0.1, 1) random variables. The sample mean of those 100 observations is m_B = 0.176.

Let's highlight a couple of things about our two samples:

- Firstly, to simplify the explanation, we have said our observations come from Gaussian random variables, so those observations could be anywhere between $-\infty$ and $+\infty$. This is not realistic for a click-through rate, but we mainly used the e-commerce example as a way of demonstrating why hypothesis testing is a relevant concept to know about.

- Secondly, we know what the values of the population variances σ_A^2 and σ_B^2 are. In this example, they have a common value, σ^2, which we have set equal to 1. In real life, we won't know the values of σ_A^2 and σ_B^2. In real life, we can use the sample variances, s_A^2 and s_B^2, as estimates of σ_A^2 and σ_B^2. Alternatively, for simplicity, we can still assume that σ_A^2 and σ_B^2 have a common value σ^2, and we can use the sample variances s_A^2 and s_B^2 to construct an estimate of this common value, σ^2. We will return later to explain the implications of only knowing the sample variances.

Now, if we didn't know the values of μ_A and μ_B we could start our analysis by assuming that $\mu_A = \mu_B$ and proceed from there. If we compare the sample means by calculating $m_A - m_B$, we get $0.529 - 0.176 = 0.353$. Superficially, it looks like the mean of group A is different from the mean of group B. This is surprising, particularly since we started with the assumption that $\mu_A = \mu_B$. Does this provide evidence that our starting assumption is incorrect and in fact $\mu_A \neq \mu_B$? We also note that it is the magnitude of the difference between m_A and m_B that surprises us. We would have been equally surprised if $m_A - m_B = -0.353$. What we want to know is how probable is a value of 0.353 or larger for $\left| m_A - m_B \right|$ if there is, in fact, no difference between μ_A and μ_B. To calculate this, we need to work out the probability distribution of $m_A - m_B$ when $\mu_A = \mu_B$. How do we do this? Read on.

Both m_A and m_B are sums of Gaussian random variables because they are means of our data values, which we have said are i.i.d. Normal. A key fact about Gaussian random variables is that when you add two or more Gaussian random variables, you get another Gaussian random variable. This makes both m_A and m_B Gaussian random variables. Since $m_A - m_B$ is simply another sum, this means that $m_A - m_B$ is also a Gaussian random variable. Remember from *Chapter 2* that a Gaussian random variable is characterized entirely by its mean and variance. That implies that once we know $\mathbb{E}(m_A - m_B)$ and $\text{Var}(m_A - m_B)$, we know everything there is to know about the distribution of $m_A - m_B$.

What are the values of $\mathbb{E}(m_A - m_B)$ and $\text{Var}(m_A - m_B)$? We can use the rules established in *Chapter 2* to do this. If we write out $m_A - m_B$ in full we have,

$$m_A - m_B = \frac{1}{N_A}\sum_{i=1}^{N_A} x_i - \frac{1}{N_B}\sum_{i=1}^{N_B} y_i$$

$$x_i \sim Normal(\mu_A, \sigma_A^2) \text{ for } i = 1,\ldots,N_A \qquad y_i \sim Normal(\mu_B, \sigma_B^2) \text{ for } i = 1,\ldots,N_B$$

Eq. 5

Using this long-hand definition of $m_A - m_B$ and the distributions of x_i and y_i, we have,

$$\mathbb{E}(m_A - m_B) = \frac{1}{N_A}\sum_{i=1}^{N_A} \mathbb{E}(x_i) - \frac{1}{N_B}\sum_{i=1}^{N_B} \mathbb{E}(y_i) = \frac{N_A}{N_A}\mu_A - \frac{N_B}{N_B}\mu_B = \mu_A - \mu_B$$

Eq. 6

With our starting assumption that $\mu_A = \mu_B$, Eq. 6 tells us that $\mathbb{E}(m_A - m_B) = 0$. Since the observations are independent of each other, a similar calculation gives,

$$\text{Var}(m_A - m_B) = \frac{1}{N_A^2} \sum_{i=1}^{N_A} \text{Var}(x_i) + \frac{1}{N_B^2} \sum_{i=1}^{N_B} \text{Var}(y_i) = \frac{1}{N_A} \sigma_A^2 + \frac{1}{N_B} \sigma_B^2$$

Eq. 7

Note how the variance of $m_A - m_B$ gets smaller as we increase either N_A or N_B in Eq. 7.

Now, we want to calculate the probability of getting a value of 0.353 or larger for $|m_A - m_B|$ – that is, to calculate the probability $P(|m_A - m_B| \geq 0.353)$. We can split this into two separate contributions:

$$P(|m_A - m_B| \geq 0.353) = P(m_A - m_B \leq -0.353) + P(m_A - m_B \geq 0.353)$$

Eq. 8

Since we know the mathematical expression for the probability density of a Gaussian random variable, this probability can be written as the integral:

$$P(|m_A - m_B| \geq 0.353) = \frac{2}{\sqrt{2\pi\text{Var}(m_A - m_B)}} \int_{0.353}^{\infty} \exp\left(-\frac{u^2}{2\text{Var}(m_A - m_B)}\right) du$$

Eq. 9

The integral in Eq. 9 just adds up (integrates) all the probability contributions where $|m_A - m_B| \geq 0.353$. The probability contributions are just the probability density function of a Gaussian distribution, with a mean of zero and variance $\text{Var}(m_A - m_B)$. The integral in Eq. 9 is such a common one that we have a special mathematical function or symbol for it. Eq. 9 can be evaluated in terms of the **error function** erf(z), which is defined as,

$$\text{erf}(z) = \frac{2}{\sqrt{\pi}} \int_{-\infty}^{z} e^{-t^2} dt$$

Eq. 10

The shape of the error function is plotted in *Figure 7.1*. For visual guidance, we have added dashed horizontal lines at 0, 0.5, and 1 and a dashed vertical line at 0.5 to the plot.

The Error Function

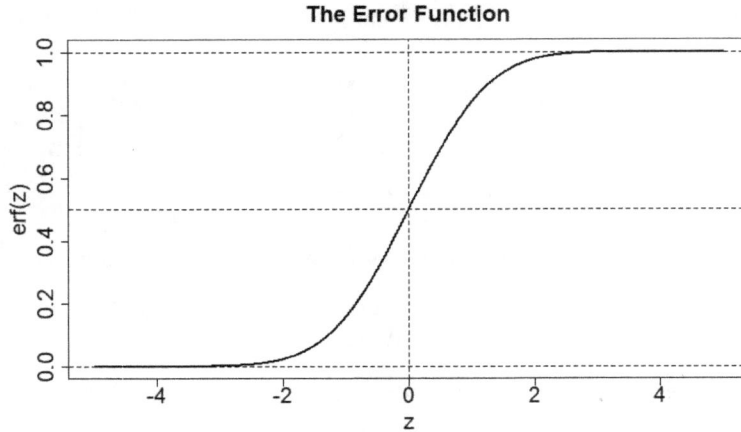

Figure 7.1: The shape of the error function, erf(z).

Using this definition of the error function, we can write *Eq. 9*, and hence the probability of getting a value of $|m_A - m_B|$ of 0.353 or larger, as,

$$P\left(|m_A - m_B| \geq 0.353\right) = 1 - \mathrm{erf}\left(\frac{0.353}{\sqrt{2\mathrm{Var}(m_A - m_B)}}\right)$$

Eq. 11

Using the expression in *Eq. 7* and the fact that $\sigma_A^2 = \sigma_B^2 = \sigma^2$, we can simplify *Eq. 11* to get,

$$P\left(|m_A - m_B| \geq 0.353\right) = 1 - \mathrm{erf}\left(\frac{0.353}{2\sigma\sqrt{\frac{1}{N_A} + \frac{1}{N_B}}}\right)$$

Eq. 12

Plugging the values $N_A = N_B = 100$ and $\sigma^2 = 1$ into *Eq. 12* and using the SciPy implementation of erf(z), we get the value 0.01256. It looks like we get a value of 0.353 or larger for $|m_A - m_B|$ less that 1.3% of the time if $\mu_A = \mu_B$. This is a low percentage, so while it is possible that the value of 0.353 that we observed for $m_A - m_B$ is just due to chance, can we believe we've been that unlucky and were part of that 1.3%? In this case, we don't we can. I think 0.01256 is a sufficiently low enough probability for me to doubt the starting assumption that $\mu_A = \mu_B$. Consequently, I reject the starting assumption that $\mu_A = \mu_B$, and I believe that my data provides evidence that $\mu_A \neq \mu_B$.

For the preceding calculation, we used the fact that $m_A - m_B$ is Gaussian-distributed to transform the calculation to one involving the standard Normal distribution $N(0,1)$. Since the z symbol is commonly used for a random variable that is distributed according to the standard Normal distribution, the type of hypothesis test we have just performed is called a **z-test**.

The wrinkle here is that usually we don't know the population variance of our measured quantity $m_A - m_B$, and we use sample variances instead, which complicates the matter. Don't worry, we'll explain how we solve this complication when we introduce the **t-test**. While using a z-test and assuming we know the population variances $\sigma_A^2 = \sigma_B^2 = \sigma^2$ is unrealistic, it has allowed us to simplify the introductory example and explanation of what a hypothesis test is. We shall make use of the z-test again for explanatory purposes later on in the chapter; however, be aware that, in reality, it is unlikely that you would ever perform a z-test on a real dataset.

From the preceding discussion, it should be becoming clear that what we have are two hypotheses – one, such as that in *Eq. 3*, that we are interested in understanding whether there is any evidence for from the data, and an opposite hypothesis that we use to process that evidence. In fact, all hypothesis tests essentially take this form – so much so that we can layout a general "recipe" to perform any hypothesis test. We will do this next.

The general form of a hypothesis test

We can boil any hypothesis test down to a small number of steps, as follows:

Recipe for a hypothesis test

1. Define the mutually exclusive hypotheses about the population that we want to decide between. These hypotheses are called the **null hypothesis** (denoted $\mathcal{H}_0$) and the **alternative hypothesis** (denoted $\mathcal{H}_1$). Usually, the alternative hypothesis is the one we're interested in and represents an interesting outcome. By contrast, the null hypothesis corresponds to uninteresting behavior.

2. Define a metric that, when applied to population quantities, can distinguish the null hypothesis from the alternative hypothesis.

3. Calculate the value of the metric using the corresponding sample quantities. This is called the test statistic.

4. Calculate the probability of getting a value of the test statistic equal to or larger than the observed test-statistic value, if the null hypothesis were true.

5. If the calculated probability is sufficiently small, then reject the null hypothesis and accept the alternative hypothesis. Conversely, if the calculated probability is not sufficiently small, then accept the null hypothesis.

We'll leave what we mean by "sufficiently small" in *step 5* until the next section. For now, we'll expand upon *steps 1–4*.

In our previous explanatory example, the metric we used was the difference between the means of the two A and B groups. When applied to the population means, this metric is equal to $\mu_A - \mu_B$, and so is zero if the null hypothesis is true and non-zero if the alternative hypothesis is true. It is common to define a test-statistic in this way, in which it has a value of zero for the null hypothesis and is non-zero for the alternative hypothesis. For our example, the metric applied to sample quantities then becomes $m_A - m_B$, and being calculated from sample quantities, it is now a random variable and has a probability distribution.

Step 4 is often the trickiest part of a hypothesis test. We can calculate the probability in *step 4* if we know the probability distribution of the test statistic, but working out the mathematical equation for the distribution of a test statistic can often be an advanced mathematical calculation. That is why we used a very simplified and contrived example previously, making *step 4* achievable using only the math that we've introduced so far in the book. Until the advent of intensive computational methods, the only way that the probability distribution for a given test statistic could be worked out was if some clever mathematician/statistician managed to do so. This meant that the number of different test statistics available for use was limited, and new test statistics and their corresponding probability distributions were topics of academic research. Later, we'll touch briefly upon computational methods to calculate the probabilities associated with any test statistic. For now, let's return to *step 5*.

The p-value

The probability we calculate in *step 4* is called the **p-value**. The *p*-value is the probability of getting a value of the test statistic equal to or larger than that observed in the data, if the null hypothesis were true. For our example in *Eq. 12*, when carrying out *step 5*, we decided that the calculated *p*-value was sufficiently small enough for us to reject the null hypothesis.

But how did we decide that the *p*-value in this instance was sufficiently small? What threshold did we compare it to, and who gets to choose that threshold? The threshold we compare the *p*-value to is usually denoted by α. The value chosen for α is subjective. You are free to set it to whatever value between 0 and 1 you think is appropriate. Commonly used values are 0.05, 0.01, and 0.001. Sometimes, α is referred to in percentage terms, so the commonly used values are 5%, 1%, and 0.1%. Of these commonly used values for α, 0.05 is the most prevalent, and that is the threshold I decided to use in my example.

When the *p*-value is below our chosen threshold, we say that the test statistic is "statistically significant." For our little example, we would say that there is "a statistically significant difference between the sample means of the two groups." Because a statistically significant finding can represent a potentially important scientific discovery, there is in many fields an unhealthy fixation on just the *p*-value (and whether it is statistically significant) when performing hypothesis tests, leading to both misunderstanding of what *p*-values represent and also an unethical data analysis practice, known as "*p*-hacking," whereby the analysis process is continually adjusted and changed until a statistically significant result is obtained. At the heart of these problems is a poor understanding of what the *p*-value and the threshold α represent. To counter this and to help you avoid these common mistakes, we'll dig further into them.

What the p-value does and does not represent

We use the calculated *p*-value to reject or accept the null hypothesis. This does not mean the *p*-value is the probability that the null hypothesis is true. Similarly, $1 - p$-value is not the probability that the alternative hypothesis is true. Both these viewpoints are common misconceptions of what the *p*-value represents. Nor is the *p*-value the probability that the data was generated by chance alone.

Misconceptions about p-values are so prevalent and problematic that the **American Statistical Association (ASA)** felt it necessary in 2016 to take the highly unusual step of issuing explicit statements and guidance about p-values in the form of six principles. The principles are well worth a read. You can find a link to the principles in the *Notes and further reading* section at the end of this chapter. We can't cover all the principles here, but the overall ethos of the ASA statement can be summarized as follows:

- You need to understand what a p-value represents

- Decisions made on applying arbitrary and subjective thresholds to p-values will always have problems, particularly when a p-value is close to a threshold

- Don't only use the p-value to assess the support that a dataset gives toward a hypothesis or conclusion

Despite the problems and issues with p-values, they are so entwined with the whole process of hypothesis testing that they are not going to disappear anytime soon. It is, therefore, important that you have a good understanding of what they are, what they are not, what affects them, and how we control them.

What the α threshold represents

The p-value definition means that if the null hypothesis is true, the probability of getting a p-value less than p is p. So, even when the null hypothesis is true, there is a probability of α of getting a p-value below α and consequently rejecting the null hypothesis. Firstly, this should emphasize to you the point that a hypothesis test can produce a false positive. Secondly, the false positive rate is, in fact, α. Since we set α, we control the false positive rate. Now that we have a feel for what α represents, we have a better feel for how to set its value. A value of $\alpha = 0.05$ corresponds to a 1-in-20 chance of a false positive. If the downside consequences of a false positive are severe (e.g., life-changing), you might not think a 1-in-20 chance is that low. Typically, when the ramifications of rejecting the null hypothesis are substantial or noteworthy, we set the value of α very small. For example, the discovery of new sub-atomic particles is based upon a hypothesis test. The threshold for declaring that a new particle has been discovered is set at what is termed "five-sigma." This corresponds to setting $\alpha \cong 2.87 \times 10^{-7}$.

If we do set a very low value for the threshold α, are we in danger of setting too stringent a threshold that it is impossible to pass? At this stage, it is worth looking at what factors affect the p-value.

The effect of increasing sample size

Let's return to *Eq. 12*, as it helps us to highlight another important aspect of hypothesis testing. You'll see from *Eq. 7* that the variance of the test statistic in our example decreases as we increase either sample size N_A or N_B. In general, this should be true of any appropriately designed test statistic. For our example in *Eq. 12*, the argument of the error function is our test statistic $m_A - m_B$ divided by the square-root of the variance in *Eq. 7*. With that variance decreasing as we increase either sample size, we can see that increasing the sample size has the effect of increasing the size of the argument in the error function. From the plot in *Figure 7.1*, you can see that $\text{erf}(z) \to 1^-$ monotonically as $z \to \infty$. This has the effect that the p-value in our example decreases monotonically as either N_A or N_B increase.

Again, this is true in general for any (appropriate) hypothesis test – larger sample sizes lead to smaller *p*-values. What are the consequences of this?

- For a given test-statistic value, the larger the sample size from which that test-statistic value has been calculated, the more likely it is we will reject the null hypothesis and declare the test-statistic value as being statistically significant. Increasing sample size is like turning up the resolution on a microscope. The larger the sample size, the smaller the differences between the two groups we can detect.

- If we want to be confident in detecting a difference between two groups, we know that we can do this simply by ensuring a sufficiently large enough sample size. How large is large enough is something we will explain in the *Type-I and Type-II errors, and power* section later in this chapter.

- However, be careful. Given a large enough sample size, any test-statistic value, no matter how small, will become statistically significant. This highlights that "statistically significant" does not necessarily mean impactful. We may have a large enough sample size in our e-commerce A/B test example to detect even very small differences in the click-through rate between our website variants. This does not mean those very small differences in the click-through rate are big enough to change a bottom-line metric that the business cares about. This can be particularly relevant in e-commerce A/B testing, where large sample sizes are relatively easy to come by for the more popular websites, due to potentially global scale traffic volumes coming to those websites.

The fact that even a very small test-statistic value can be statistically significant, or conversely that a large value of a test statistic may not be statistically significant, illustrates the problems with just focusing on the *p*-value in isolation and relates to one of the principles emphasized in the ASA statement on *p*-values. It highlights that whenever we perform a hypothesis test, it is good practice to report the size of any effect estimate (e.g., the observed difference in sample means, which is $m_A - m_B$ in our simple example), as well as report the *p*-value that results from that difference. Reporting just the *p*-value as the only outcome of a hypothesis should be considered bad data science practice and avoided.

The effect of decreasing noise

In a similar fashion, we can see from *Eq. 7* that the variance of the test statistic, $m_A - m_B$, also decreases if the variances, σ_A^2, σ_B^2, of the noise in the observations from either group decrease. The implications of this are obvious. The more accurate (less noise) we can make our observations, the more we can detect genuine differences between the groups, for a given sample size. Performing accurate measurements or observations allows us to make more efficient use of the data we have collected.

This is true in general, even when the variation we see in the sample data does not come from a measurement error. In general, the smaller the intrinsic sampling variation in the individual data points, the more able our hypothesis test is to reject the null hypothesis if it is indeed false.

However, note that the intrinsic sampling variation is not always within our control. Where variation is due to a measurement error, then, yes, we may be able to design and build more sensitive measuring equipment. Where variation is due to humans making decisions or choices, we have less control over the magnitude of the variation.

Having explained what the *p*-value represents and what factors affect it, we'll now briefly return to one aspect of the *p*-value calculation in our simple example that may have been puzzling you.

One-tailed and two-tailed tests

For the simple example, we have been using to illustrate a hypothesis test our test statistic, $m_A - m_B$ which was 0.353. We calculated the *p*-value as the probability of getting a value $|m_A - m_B| \geq 0.353$ if the null hypothesis were true. But why did we use the absolute value $|m_A - m_B|$ in our calculation of the *p*-value? We said earlier that this was because the *p*-value gives us a measure of how unusual or surprising the observed test-statistic value is under the null hypothesis, and we would have been just as surprised if we had seen a test-statistic value of -0.353 or lower. In other words, the calculated *p*-value reflects the fact that our alternative hypothesis is $\mu_A - \mu_B \neq 0$, so there would be evidence for the alternative hypothesis being true if we had seen a large positive value of the test statistic $m_A - m_B$, or if we had seen a large negative value of the test statistic. Because we are calculating the *p*-value as the probability of unusual values from either end or tail of the test-statistic probability distribution, it is called a **two-tailed** test. The part of the test-statistic probability distribution that contributes to the *p*-value is shown schematically by the gray area in the left-hand plot of *Figure 7.2*.

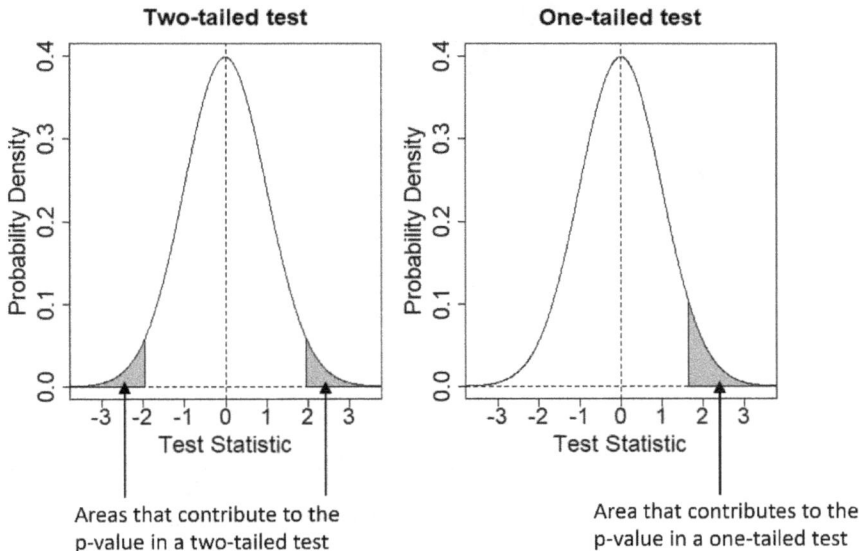

Figure 7.2: Areas contributing to the p-value for one-tailed and two-tailed hypothesis tests

Of course, this raises the question of whether we sometimes want to perform a **one-tailed** hypothesis test. The answer is yes. An example of a one-tailed test would be when we want to test an alternative hypothesis $\mathscr{H}_1 : \mu_A > \mu_B$. Our null hypothesis would still be the same (i.e., $\mathscr{H}_0 : \mu_A = \mu_B$). Now, we are saying we would only be surprised if we saw the sample mean m_A much larger than the sample mean m_B. We would only be surprised if the probability of getting $m_A - m_B > 0.353$ with the null hypothesis being true was very small. The part of the test-statistic probability distribution that contributes to the p-value for this one-tailed test is shown schematically in the gray area in the right-hand plot of *Figure 7.2*. Obviously, there may be situations where we would want to perform a one-tailed hypothesis, where the alternative hypothesis is $\mathscr{H}_1 : \mu_A < \mu_B$. For such a one-tailed test, the p-value would come from the left-hand tail of the test-statistic probability distribution.

When should we perform a two-tailed test, and when should we perform a one-tailed test? This depends upon what **a priori** knowledge we have about what might happen if the null hypothesis is not true. If we know beforehand that only situations where $\mu_A > \mu_B$ represent genuinely interesting scenarios, then we would probably want to run a one-tailed test with $\mathscr{H}_1 : \mu_A > \mu_B$. If, however, we considered both $\mu_A < \mu_B$ and $\mu_A > \mu_B$ to represent interesting scenarios or findings, then we would want to run a two-tailed test.

The difference between a one-tailed test and a two-tailed test may look like a philosophical one, but the difference is important to understand, as you may be tempted to use a one-tailed test just to get a smaller p-value. How so? Well, imagine we chose a significance threshold of $\alpha = 0.05$. What would be the test-statistic value that corresponds to precisely this threshold? In other words, for a two-tailed test, what test-statistic value would have a p-value of $p = 0.05$? For a two-tailed test, the left-hand schematic plot in *Figure 7.2* indicates that both the left-hand and right-hand gray areas contribute equally to the p-value and, therefore, contribute a probability of 0.025. For the standard Normal distribution, we can use the mathematical form of the **probability density function** (**PDF**) to find that this corresponds to a value of $\sqrt{2} \times \mathrm{erf}^{-1}(1 - 2 \times 0.025) \approx 1.96$. This means that for our simple example, if the null hypothesis were true, then,

$$P\left(\frac{|m_A - m_B|}{\sqrt{\mathrm{Var}(m_A - m_B)}} \geq 1.96 \right) = 0.05$$

Eq. 13

Now, let's repeat that calculation for a one-tailed test, as represented by the schematic in the right-hand plot of *Figure 7.2*. Now, instead, it is just the gray area in the right-hand tail that gives us the threshold value of 0.05. This now corresponds to a threshold of $\sqrt{2} \times \mathrm{erf}^{-1}(1 - 2 \times 0.05) \approx 1.64$. This means that if the null hypothesis were true for our simple example, then we would have,

$$P\left(\frac{m_A - m_B}{\sqrt{\mathrm{Var}(m_A - m_B)}} \geq 1.64 \right) = 0.05$$

Eq. 14

We can immediately see that, for the same p-value threshold value of $\alpha = 0.05$, the equivalent threshold we apply to the test statistic is smaller in magnitude for a one-tailed test than a two-tailed test, 1.64 versus 1.96. Naively, it appears that the one-tailed test is less stringent, since the threshold that we apply to determine whether the difference $m_A - m_B$ is statistically significant is lower. A one-tailed test is not less stringent. Instead, we utilized prior information to select a more appropriate alternative hypothesis that then led us to run a one-tailed test. However, I have seen some data scientists choose to run a one-tailed test after seeing what the value of $m_A - m_B$ was. They were hoping to get a statistically significant outcome from a relatively low value of $m_A - m_B$. Do not be tempted to do this! This is p-hacking – modifying how a test is run after the data has been obtained just to get a smaller p-value. The conclusions you reach from doing this will not be robust. You should decide whether you are running a one-tailed or two-tailed test before actually running it.

Using samples variances in the test statistic – the t-test

In our simple example, we started out by assuming we knew the population variances σ_A^2 and σ_B^2. We did so because it made the calculation of the p-value simpler to derive. In real life, we won't know the values of σ_A^2 and σ_B^2, but we can, of course, use the sample variances s_A^2 and s_B^2 as estimates for σ_A^2 and σ_B^2. However, doing so introduces a complexity in calculating the p-value. We'll now explain what that complexity is and how we tackle it.

The p-value calculation in our example in $Eq.\ 12$ is equivalent to using a scaled test statistic, whereby we divide $m_A - m_B$ by its standard deviation to give,

$$z = \frac{m_A - m_B}{\sqrt{\frac{\sigma_A^2}{N_A} + \frac{\sigma_B^2}{N_B}}} = \frac{m_A - m_B}{\sigma\sqrt{\frac{1}{N_A} + \frac{1}{N_B}}}$$

Eq. 15

We get the results on the far right-hand side of $Eq.\ 15$ because σ_A^2 and σ_B^2 have a common value, which is σ^2. If we now replace the population variances with sample variances, our new test statistic is,

$$t = \frac{m_A - m_B}{\hat{\sigma}\sqrt{\frac{1}{N_A} + \frac{1}{N_B}}}$$

Eq. 16

For simplicity, we have assumed in $Eq.\ 16$ that the underlying population variances σ_A^2 and σ_B^2 still have a common value σ^2, and the denominator in $Eq.\ 16$ reflects the fact that we have combined the sample variances s_A^2 and s_B^2 to construct an estimate of σ^2. That estimate, denoted by $\hat{\sigma}^2$, is calculated as,

$$\hat{\sigma}^2 = \frac{(N_A - 1)s_A^2 + (N_B - 1)s_B^2}{N_A + N_B - 2}$$

Eq. 17

The denominator in $Eq.\ 17$ has been set so that $\hat{\sigma}^2$ is an unbiased estimator of σ^2.

All we now need to do is work out the probability distribution for this new test statistic under the assumption that the null hypothesis, $\mu_A = \mu_B$, is true. The temptation here is to assume that because the individual data points come from a Gaussian distribution, then the distribution of t is still Gaussian because it looks like we are still just taking the difference between two sums of Gaussian random variables, as we did in *Eq. 15* in our simple example. This was true when we used σ_A^2 and σ_B^2 in the denominator of our scaled test statistic in *Eq. 15*, because σ_A^2 and σ_B^2 are just fixed constants whose values we know. But for the t test statistic in *Eq. 16*, the denominator is now also a function of the data, so we need to account for the fact that the denominator varies as the random variables that make up the data vary.

To calculate a *p*-value for the test statistic t, we need to see how t varies under the assumption that the null hypothesis is true. This means hypothetically varying the data that goes into the calculation of t but in a way that is consistent with the null hypothesis. To do this, we need to constrain the hypothetical data so that there is no difference in the sample means between the two groups, A and B. This imposes two constraints on the hypothetical data. In our hypothetical data, we do not have $N_A + N_B$ freely varying data values, but instead, we effectively have $N_A + N_B - 2$ freely varying values. This number is called the **degrees of freedom**, sometimes abbreviated to **DF**, or **DOF**, and usually denoted by the symbol v. In general, the number of DOF is the total sample size minus the number of parameters specified in our hypotheses $\mathscr{H}_0$ and $\mathscr{H}_1$, which we have had to estimate. In our example, we had a total sample size of $N_A + N_B$, and we had to estimate the two means, μ_A and μ_B, so the DOF $v = N_A + N_B - 2$.

As well as reducing the effective number of DOF, the presence of the constraints also changes the shape of the distribution of our test statistic. Whilst our test statistic in *Eq. 15* has a Gaussian distribution if the null distribution is true, our new test statistic t in *Eq. 16* follows the more complicated probability density function shape, given by the following formula,

$$p(t) = \frac{\Gamma(\frac{v+1}{2})}{\Gamma(\frac{v}{2})\sqrt{v\pi}} \left(1 + \frac{t^2}{v}\right)^{-\frac{(v+1)}{2}}$$

Eq. 18

Since it is the distribution of our test statistic t, *Eq. 18* is called the **t-distribution**. The formula in *Eq. 18* looks complicated. It isn't, and you don't have to remember it. We can use the `SciPy` implementation when we need to calculate it. What you do need to understand is what its shape looks like. We have plotted its shape in *Figure 7.3* for a few different values of the DOF v. For comparison, we have also plotted the density function of the standard Normal distribution $N(0,1)$. The standard Normal distribution is the probability density function of the test statistic z in *Eq. 15* that we used in our simple example.

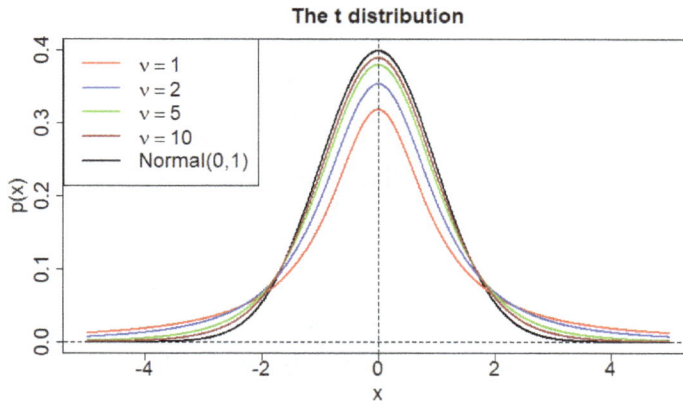

Figure 7.3: The shape of the t distribution

From *Figure 7.3*, we can see that for the smaller values of v, the t-distribution has heavier tails (the shape does not drop to zero as fast) than the standard Normal distribution, while for the larger values of v, there is very little difference between the t-distribution and the standard Normal distribution. In fact, as $v \rightarrow \infty$, the t-distribution becomes identical to the standard Normal distribution.

The comparison of two sample means, using the t-distribution to calculate the *p*-value, is known as a **t-test**. There are several variants of a t-test. All of the t-test variants are concerned with comparing sample means to assess whether the underlying populations from which the data came are the same in all regards, except for the population means, μ_A and μ_B. The different variants of t-test are as follows:

- **A two-sample t-test**: The particular variant we have explained previously is the "two sample" t-test, in which we have two independent samples, one from each group

- **A one-sample t-test**: Here, we have a sample of data from a single group, and we test the hypothesis that the underlying population mean of that group is different from zero

- **A paired-sample t-test**: This is where we have two samples of data, but they are related or matched (paired) in some way (e.g., observations from the same set of individuals before and after some treatment has been applied)

- **Welch's t-test**: This variant of the two-sample t-test drops the assumption that the variances of the two underlying populations are the same

Because comparing the means of two groups is such a common task – think of the website A/B test example we mentioned earlier – several Python packages have made performing a t-test easier for us. We will illustrate this next with a simple code example.

A t-test code example

The following code example can also be found in the `Code_Examples_Chap7.ipynb` Jupyter notebook in the GitHub repository. The data we'll use can be found in the `hypothesis_test_example.csv` file in the `Data` folder of the GitHub repository. The data corresponds to the simple example we have used for illustration purposes throughout this section. First, we'll read in the data:

```python
import pandas as pd
import numpy as np
from scipy.stats import ttest_ind
# Read in t-test example dataset
df_simple_example = pd.read_csv("../Data/hypothesis_test_example.csv")
```

We can take a quick look at the properties of the data using the `pandas describe` function:

```python
# Extract summary statistics of the data
df_simple_example.describe()
```

This gives the following output:

```
            x_A         x_B
count  100.000000  100.000000
mean     0.529000    0.176000
std      0.938470    0.983872
min     -1.036634   -1.957430
25%     -0.241617   -0.448830
50%      0.544784    0.250698
75%      1.108813    0.676324
max      3.126451    2.612088
```

From the summary statistics, we can see there is a difference in the sample means, with the B group sample data having a mean of 0.176, whilst the A group sample data has a mean of 0.529. But does this provide evidence for the underlying population means of group A and group B being different? Let's run the t-test to test this. We'll use the `ttest_ind` function from `scipy.stats`:

```python
# We pass in the two columns of data. We are assuming
# the underlying population variances are the same in each group.
ttest_ind(df_simple_example['x_A'], df_simple_example['x_B'])
```

The output gives the *t*-value (the test-statistic value) and the associated *p*-value.

```
Ttest_indResult(statistic=2.5961983095998966,
pvalue=0.010132851609223453, df=198.0)
```

From this t-test, we get a test statistic (a t-value) of 2.596 and a p-value of 0.0101. If we use an α threshold of $\alpha = 0.05$, then we would reject the null hypothesis and conclude that there is evidence (not proof) that the underlying population means of the two groups are different.

We have been able to perform this two-sample t-test because we can calculate the p-value using the t-distribution density function in *Eq. 18*. We have only been able to do so because the formula for the t-distribution PDF was worked out by statisticians over 100 years ago. But what happens if we don't have any clever statisticians or mathematicians available to us to work out the PDF of our test statistic or, even worse, if we've designed a test statistic that no statistician or mathematician is ever likely to be able to work out the PDF for? Fear not – the general form of hypothesis test that we laid out earlier is like a recipe, and in the age of computers, we can always follow that recipe computationally, even when some of the ingredients (the test statistic) are challenging. In the next subsection, we will briefly explore how computationally intensive techniques can be used to perform hypothesis tests.

Computationally intensive methods for p-value estimation

Calculating the p-value associated with a test-statistic value is *step 4* of our five-step recipe for hypothesis testing. Computationally intensive approaches rely upon the fact that we have specified what the null hypothesis is. This means we can just generate a load of simulated samples from that known null hypothesis and count how many times we get a test-statistic value as large as that observed. It is as easy as that. Again, we follow a short recipe, given as follows (for a two-tailed test):

Recipe for computationally intensive estimation of the p-value

1. Set the number of simulated datasets to be generated (e.g. $N_{sets} = 10,000$).

2. Set count = 0. This is to count how many times we see a test statistic in the simulated data that is larger than the test-statistic value we got from the real data.

3. Generate a simulated dataset from the known null hypothesis. The simulated dataset should be of the same sample size as the real data.

4. Calculate the test statistic for the simulated dataset. If the magnitude of the simulated test statistic is greater than or equal to the magnitude of the test statistic for the real dataset, then set count = count + 1.

5. Repeat *steps 3* and *4*, N_{sets} times.

6. Calculate the estimate of the p-value as $\hat{p} = \text{count}/N_{sets}$. Since the real data also gave us a test-statistic value equal to or larger to than that observed, it is more common to estimate the p-value as $\hat{p} = (1 + \text{count})/(1 + N_{sets})$.

Before we run a demonstration of this approach, there are a couple of subtleties to point out:

- Firstly, the value of N_{sets} determines the granularity with which we can estimate the p-value. A value of N_{sets} = 100 means we estimate the p-value to a granularity of 1 part in 100. To produce a more precise estimate of the p-value means, we must use larger and larger values of N_{sets}, thereby increasing the overall runtime – because we must generate more simulated datasets and calculate more test-statistic values. Hence why methods such as this are referred to as computationally intensive methods.

- Secondly, *step 3* involves generating data from the known null hypothesis. However, the null hypothesis may have been specified only in vague terms. For example, the null hypothesis may be that data from the two groups, A and B, is drawn from the same distribution. This doesn't tell us which distribution. But wait – didn't we assume that the data was Gaussian-distributed when calculating the t distribution in *Eq. 18*? Yes, we did, but the t distribution in *Eq. 18* is only correct if our data is Gaussian. What happens if the real data is not Gaussian-distributed, or we are not willing to make that assumption? We still want to generate simulated datasets assuming the null hypothesis to be true, (i.e., that all the data points came from the same distribution), but now we're going to have to use the exact and true distribution from which the data came. The trouble is, we're unlikely to know what the exact true distribution was. We've got a problem, right? It turns out we can still generate simulated datasets from the true underlying distribution and assume that the null hypothesis is true, even if we don't know what the true underlying distribution is. Let's imagine I just take a copy of the original data and randomize which group, A or B, each data point belongs to. I can do this by simply permuting the labels on the original data points. I now have a copy of the data where there is no meaningful difference in the group means. If there was to begin with, I certainly destroyed that difference when I randomized the labels. So now, the null hypothesis is true for my randomized copy of the data. However, because the data values (not the labels) of my copy are the same as the original data, all the statistical properties of my copy are the same as the statistical properties of the original. This means my copy dataset has effectively come from the same underlying distribution as my original dataset. My objective is complete – by a simple permutation of the labels attached to each data point, I have generated a simulated dataset for which the null hypothesis is true, and which comes from the same underlying distribution as my original data. All I need to do now is to repeat this permutation process N_{sets} times.

The astute among you will have spotted that the number of possible permutations of the original dataset I can generate is finite, so given what we have said in the preceding bullet point, there is a finite granularity to which I can estimate the p-value using this permutation approach. This is true, but for an original dataset consisting of N_A data points in group A and N_B data points in group B, the total number of distinct possible permutations is $\frac{(N_A + N_B)!}{N_A! N_B!}$. For any medium or large value of $N = N_A + N_B$, this is an astronomical number of possible permutations, so in practice, we aren't really restricted in the kinds of values for N_{sets} we can choose.

Let's look at this permutation-based approach in action in a code example.

A permutation-based p-value code example

The following code example can also be found in the `Code_Examples_Chap7.ipynb` Jupyter notebook in the GitHub repository.

First, we'll set the seed for the NumPy random number generator:

```
import numpy as np
np.random.seed(1869)
```

Next, we'll calculate the test-statistic value (the *t*-value) for the actual real dataset:

```
# Set a large, but reasonable number of permutations to run.
# In this case I've chosen to generate 100000 permuted datasets.
# This may take a couple of minutes to run.
n_permutations = 100000


# First I'll combine the original data into a single array. This makes
# performing the permutation easier.
x_All = np.concatenate((df_simple_example['x_A'].to_numpy(),
                        df_simple_example['x_B'].to_numpy()))

## Next I'll calculate the observed test-statistic value and store it
## in a variable called t_observed

# Create arrays to hold the indices of the datapoints
# belonging to the A group and the B group. To start, the
# A group datapoints are at indices 0:99. The B group datapoints
# are at indices 100:199
nA = df_simple_example.shape[0]
nB = nA


A_indices = np.arange(0, nA)
B_indices = np.arange(nA, (nA+nB))

# Calculate the mean of each sample group
m_A = np.mean(x_All[A_indices])
m_B = np.mean(x_All[B_indices])

# Calculate the sample variances of each sample group
# The ddof=1 means we are using unbiased estimators for
# the sample variance calculations
s2_A = np.var(x_All[A_indices], ddof=1)
s2_B = np.var(x_All[B_indices], ddof=1)
```

```
# Calculate the t-value test-statistic for the original data
sigma2_observed = (((nA-1)*s2_A) + ((nB-1)*s2_B))/(nA+nB-2)
t_observed = (m_A - m_B)/ (np.sqrt(sigma2_observed) * np.sqrt(2.0/nA))

print("Observed t-value is = ", t_observed)
```

Now, we can repeat the process of calculating the *t*-value but do it for a sequence of permuted datasets:

```
# Set our p-value estimate count to zero
p_count = 0.0

# Loop over the permutations
for i in range(n_permutations):
    #Generate the permutation
    permuted_indices = np.random.permutation(nA+nB)
    A_indices = permuted_indices[0:nA]
    B_indices = permuted_indices[nA:(nA+nB)]

    # Calculate the mean of each sample group
    # for the permuted dataset
    m_A = np.mean(x_All[A_indices])
    m_B = np.mean(x_All[B_indices])

    # Calculate the sample variances of each sample group
    # for the permuted dataset
    s2_A = np.var(x_All[A_indices], ddof=1)
    s2_B = np.var(x_All[B_indices], ddof=1)

    # Calculate the t-value for the permuted dataset
    sigma2_permuted = (((nA-1)*s2_A) + ((nB-1)*s2_B))/(nA+nB-2)
    t_permuted = (m_A - m_B)/ (
        np.sqrt(sigma2_permuted) * np.sqrt(2.0/nA))

    # Update our count if the t-value for the permuted dataset
    # exceeds (in magnitude) that for the real dataset
    if np.abs(t_permuted) >= np.abs(t_observed):
        p_count += 1.0

# Now estimate the p-value
p_value_permutation = (1.0+p_count)/(1.0+n_permutations)
print("Permutation estimated p-value = ", p_value_permutation)
```

This gives the following output:

```
Observed t-value is =   2.5961983095998966
Permutation estimated p-value =   0.01032989670103299
```

The permutation-based p-value estimate is very close to that from the t-test run using the `scipy.stats.ttest_ind` function. Obviously, since the permutation-based p-value estimate is based on a random generation of permutations, we would expect to see differences. As we increase the number of permutations used, we would expect the differences in the p-value estimates between the two methods to decrease.

Parametric versus non-parametric hypothesis tests

The permutation-based approach illustrated in the preceding code example is known as a **permutation test.** As we saw, one of its main advantages was that we did not have to assume in the null hypothesis what the shape of the distribution was from which the data came. We did not assume any **parameterized** equation for the probability density of the data. Consequently, this permutation test is an example of a **non-parametric hypothesis test**. In contrast, our t-test probability density in *Eq. 18* was derived from the null hypothesis assumption that each datapoint was drawn from the same Gaussian distribution – that is, we assumed in the t-test a specific parameterized mathematical form for the probability density of the data. Consequently, the t-test example we illustrated earlier is an example of a **parametric hypothesis test**.

There are many other non-parametric hypothesis tests. The aspect they all have in common is that they do not make any assumptions for the null hypothesis data distribution. Usually, this is done by constructing a test statistic that is calculated from the ranks of the data values, not the data values directly. This means that for a given ranking of the observations, the test statistic is invariant of the probability distribution from which the data is drawn. There are rank-based non-parametric hypothesis tests corresponding to both the one-sample and two-sample parametric t-tests we illustrated earlier. These are listed in *Table 7.1*.

Parametric t-test	Corresponding rank-based test
One-sample or paired-sample	Wilcoxon signed-rank test
Two-sample	Mann-Whitney U test

Table 7.1: Parametric t-tests and their corresponding non-parametric hypothesis test equivalents

For the two-sample non-parametric case, you could also use Mood's median test, which tests whether the samples came from distributions that have the same median.

When should we use a non-parametric test, and when should we use a parametric test? There are pros and cons for both. Because a parametric test makes an assumption about the form of the data distribution, it is sensitive to deviations away from that assumption. If the real data you were testing came from a distribution with a markedly different shape to the one assumed, the parametric test may reach the wrong conclusion. In contrast, a non-parametric hypothesis test is robust against the different distributions that the real data may have come from, as a non-parametric test does not make any assumptions about that underlying distribution. However, if the real data has come from a distribution identical to or very close to that assumed by a parametric test, then the correctness of that assumption allows us to identify differences between populations of data more accurately, and we have a more sensitive test. Parametric tests also have the advantage that part of the testing usually involves estimating a quantity that we are interested in. For example, the two-sample t-test estimates the difference in the population means, $\mu_A - \mu_B$, so if the null hypothesis is rejected and we conclude that there is evidence for a difference between the population means, we get an estimate of that difference automatically as part of the test. In contrast, a Mann-Whitney U test may reject the null hypothesis and conclude there is a difference in population medians, but we do not know how big that difference is.

We will meet non-parametric statistics again in *Chapter 14* when we learn about Bayesian non-parametric methods, but for the remainder of this chapter, we will discuss and use only parametric tests. If you want to learn more about non-parametric hypothesis tests, check out the second, third and fourth references in the *Notes and further reading* section at the end of the chapter for suggested extra material.

This discussion of parametric and non-parametric tests completes what has been a long section. We have learned a lot of new ideas and concepts, so let's end this section by summarizing what we learned.

What we learned

In this section, we learned the following:

- What a hypothesis test is
- A five-step recipe to perform any hypothesis test
- What the p-value represents and what the significance threshold α represents
- What factors affect the p-value
- About one-tailed and two-tailed hypothesis tests and the difference between them
- How to run a t-test
- How to use computationally intensive methods to calculate the p-value when you have a complex hypothesis test
- About parametric and non-parametric hypothesis tests

Having learned about hypothesis tests and how to perform them, in the next section, we'll look at how to calculate confidence intervals for some of the quantities we calculate as part of a hypothesis test, as well as what those confidence intervals represent.

Confidence intervals

Our explanation of hypothesis testing in the previous section focused a lot on how to calculate the p-value. We explained that this focus on p-values can be dangerous, as well as that reporting results from a hypothesis test should include not just the p-value but also the estimate of the effect (e.g., the estimate of $\mu_A - \mu_B$ for our difference in means example). The only problem is that our estimate is based on sample data, so it has an inherent uncertainty associated with it. Ideally, we should also report the degree of uncertainty associated with our estimate.

To summarize our estimate of the effect and also the uncertainty of that estimate, we'll introduce a new concept, a **confidence interval**. To explain what a confidence interval is, we'll first recap some results from the previous section.

Suppose we have run the two-sample z-test hypothesis test of the previous section. We used the calculated value of $m_A - m_B$ to reject the null hypothesis that $\mu_A - \mu_B = 0$, so we believe that $\mu_A - \mu_B \neq 0$. But what should be our estimate of this non-zero quantity? We can use the value of $m_A - m_B$ as our estimate of $\mu_A - \mu_B$. We can write this as,

$$\hat{\mu}_A - \hat{\mu}_B = m_A - m_B$$

Eq. 19

The hat symbols on the left-hand side of *Eq. 19* mean that this quantity is our estimate of $\mu_A - \mu_B$, but it is not actually $\mu_A - \mu_B$ itself. Due to the uncertainty in our estimate arising from the noise in the data, it is almost certainly true that $\hat{\mu}_A - \hat{\mu}_B \neq \mu_A - \mu_B$. The right-hand side of *Eq. 19* just says that the numerical value of our estimate $\hat{\mu}_A - \hat{\mu}_B$ is given by the numerical value of $m_A - m_B$. Currently, our estimate $\hat{\mu}_A - \hat{\mu}_B$ is just a single number. This is called a **point estimate**. The true value $\mu_A - \mu_B$ could be close to our point estimate $\hat{\mu}_A - \hat{\mu}_B$, or it could be a long way from it. Is it possible to construct a range in which we think it is likely that the true value $\mu_A - \mu_B$ lies? Yes – this is what the confidence interval quantifies.

The formal definition of a 95% confidence interval for $\mu_A - \mu_B$ would be a formula to calculate an interval $[l, u]$, with l and u calculated from the sample data, such that for 95% of cases, the true value $\mu_A - \mu_B$ would in lie in the calculated interval.

For our simple example in the previous section, we had sample sizes $N_A = N_B = 100$. From now on, we'll assume the sample sizes are always the same for the A and B samples, and we'll use N to denote this common sample size. If we know the common value σ^2 of the population variances σ_A^2, σ_B^2, then the 95% confidence interval is given by the formula,

$$m_A - m_B - 1.96 \times \frac{\sigma\sqrt{2}}{\sqrt{N}} \leq \mu_A - \mu_B \leq m_A - m_B + 1.96 \times \frac{\sigma\sqrt{2}}{\sqrt{N}}$$

Eq. 20

There are several things to note about this formula:

- First, note how the confidence interval depends upon the sample data through $m_A - m_B$.

- Secondly, note how the width of the confidence interval decreases as we increase the sample size N. This makes intuitive sense – the more observations we have, the more we expect the sample means m_A, m_B will be accurate estimates of their population counterparts μ_A, μ_B, so the narrower our uncertainty about the true value of $\mu_A - \mu_B$ should be.

- Thirdly, note the value 1.96 in the formula for the confidence value. This looks like the test-statistic threshold in *Eq. 13* that we derived from setting $\alpha = 0.05$ in the z-test. It is. But wasn't our choice of α subjective (i.e., someone else might have chosen to set $\alpha = 0.01$), and if so, does this mean that the confidence interval is subjective as well? Yes, to some degree. You are free to calculate whatever level of confidence interval you wish. We could calculate an 80% confidence interval if we wished, or a 99% confidence interval. For our simple example, the formula for a $(1 - \alpha) \times 100\%$ confidence interval is given by,

$$m_A - m_B - \sqrt{2}\ \text{erf}^{-1}(1 - \alpha) \times \frac{\sigma\sqrt{2}}{\sqrt{N}} \leq \mu_A - \mu_B \leq m_A - m_B + \sqrt{2}\ \text{erf}^{-1}(1 - \alpha) \times \frac{\sigma\sqrt{2}}{\sqrt{N}}$$

Eq. 21

From this formula and the plot in *Figure 7.1*, we can see that the higher the value of $1 - \alpha$, the wider the confidence interval is. Again, this makes intuitive sense – if we want a higher degree of confidence, we will need a wider interval. If we wanted to be 100% confident (i.e., absolutely certain that the true value of $\mu_A - \mu_B$ was in the interval), that interval would have to be infinitely wide.

- The formula to calculate the confidence interval depends upon what we know about the distribution of the data. In our simple example, we started by assuming (for simplicity purposes) that we knew the value of σ^2, the value of the population variance for both the A and B populations. As with the z-test hypothesis test, we can replace the population variance σ^2 with its estimate $\hat{\sigma}^2$ in *Eq. 17*, based upon the sample variances s_A^2 and s_B^2. The formula for the confidence interval then also changes and becomes,

$$m_A - m_B - \text{CDF}^{-1}_{t,2N-2}\left(1 - \frac{\alpha}{2}\right) \times \frac{\hat{\sigma}\sqrt{2}}{\sqrt{N}} \leq \mu_A - \mu_B \leq m_A - m_B + \text{CDF}^{-1}_{t,2N-2}\left(1 - \frac{\alpha}{2}\right) \times \frac{\hat{\sigma}\sqrt{2}}{\sqrt{N}}$$

Eq. 22

The changes in the formula are the replacement of σ with its estimate, $\hat{\sigma}$, and the replacement of the z test-statistic threshold, $\sqrt{2}\,\text{erf}^{-1}(1 - \alpha)$, with its t-distribution equivalent, which is the inverse of the cumulative distribution function of the t-distribution for $v = 2N - 2$ DOF, and which we have denoted with $\text{CDF}^{-1}_{t,2N-2}$. All the previous qualitative statements we have made about the confidence interval still apply.

What does a confidence interval really represent?

The comments we made about the confidence interval formulae in *Eq. 20*, *Eq. 21*, and *Eq. 22*, were specific to those formulae, although some aspects, such as the confidence interval width widening with the increase in the level of confidence and decreasing with the sample size, are more general. However, there are also some comments we can make about any confidence interval.

The most important of these more general comments is to emphasize that when we calculate a confidence interval from a single dataset using, say, *Eq. 22*, there is not a $(1 - \alpha)$ probability that the true value of $\mu_A - \mu_B$ lies in the interval calculated from this particular dataset. It may appear that is what the formal definition says, but the formal definition is deliberately more nuanced than that. When we talk about a confidence interval, we make statements about the true value of some parameter (e.g., $\mu_A - \mu_B$ in our example). Because we have taken a non-Bayesian approach, the true value of $\mu_A - \mu_B$ is a fixed number. When we calculate a confidence interval from a single dataset using *Eq. 22*, we get a single interval. The true value of $\mu_A - \mu_B$ is either in this interval or it is not. It is binary – yes/no, or 0/1. To attach a probability to a confidence interval, we must look at repeating the process multiple times with many different datasets under identical conditions. That is why we gave the formal definition of the 95% confidence interval as "a formula to calculate an interval $[l, u]$, with l and u calculated from the sample data, such that for 95% of cases, the true value $\mu_A - \mu_B$ would in lie in the calculated interval." This means that if our formula to calculate the confidence interval is valid, then if we repeated the process of sampling N datapoints from both the A and B populations 100 times and calculated a confidence interval for each of those 100 datasets, we would expect to see the true value inside the confidence interval approximately 95 times. It is important to realize that for each of those 100 datasets, we would get a slightly different confidence interval (because the value of $m_A - m_B$ would be different each time). If we repeated the process 1,000 times, we would expect the true value to be inside the confidence interval for the given dataset in 950 cases. The more times we repeat the process, the closer we get to 95% for the proportion of cases where the true value is inside the confidence interval.

Graphical interpretation of a confidence interval

Figure 7.4 shows schematic representations of two 95% confidence intervals for $\mu_A - \mu_B$. The value of $\mu_A - \mu_B$ is represented by the x-axis. For reference, we have also marked the position of zero on the x-axis. The two plots correspond to two different experiments (i.e., two different datasets), so we get two different values of $m_A - m_B$, which we have also marked on the x-axis. The two experiments were done under different conditions. It may be that the true value of $\mu_A - \mu_B$ is genuinely different in the two experiments. In the first experiment (the upper plot), the 95% confidence interval is entirely above zero. So, we can be reasonably confident that the true value of $\mu_A - \mu_B$ is above zero in this first experiment. Contrast this with the second experiment (the lower plot), where the 95% confidence interval straddles the zero value on the x-axis. Despite the value of $m_A - m_B$ being greater than zero in this second experiment, we are not confident enough that the true value of $\mu_A - \mu_B$ is above zero.

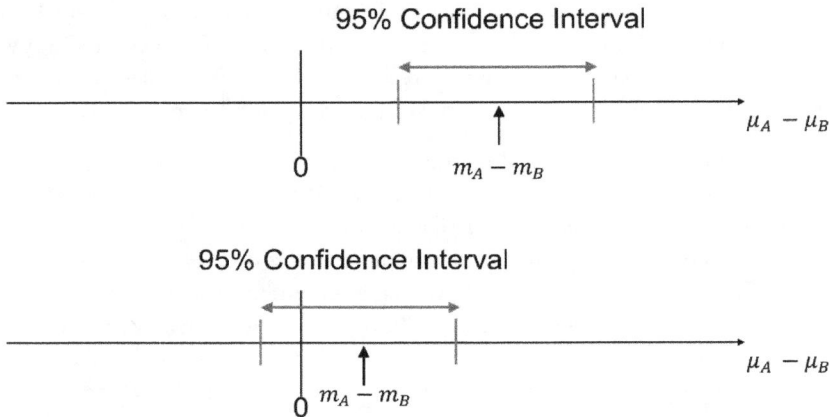

Figure 7.4: Schematic plots of 95% confidence intervals from two experiments

If the preceding discussion sounds similar to how we perform a hypothesis test, that is deliberate. Hypothesis testing and the calculation of confidence intervals are linked. If the 95% confidence interval includes the value represented by the null hypothesis ($\mu_A - \mu_B = 0$ in this case), then we should not be able to reject the null hypothesis at the $\alpha = 5\%$ level. More generally, if a $(1 - \alpha) \times 100\%$ confidence interval includes the value represented by a null hypothesis, we should not be able to reject the null hypothesis at the α level of significance. This means that if we have confidence intervals calculated for us – say, by statistical software – we can quickly eyeball the confidence intervals to determine what should be the outcome of a hypothesis test.

Confidence intervals for any parameter

This last point is highly relevant when we realize that, although the calculation of confidence intervals and hypothesis testing are linked, they are not inseparable. A confidence interval can be calculated for any parameter that we estimate using data, not just for those parameters that appear in hypothesis tests.

Take the μ_A, μ_B parameters, which are the population means of the A and B populations, respectively. In our simple hypothesis test example, we estimated the $\mu_A - \mu_B$ parameter. Let's imagine now that we estimate μ_A and μ_B separately, not as part of some hypothesis test. We could construct 95% confidence intervals for μ_A and μ_B separately. This is shown schematically in *Figure 7.5*. We won't go into details now on how we would calculate confidence intervals for μ_A and μ_B, as that is not important for this qualitative discussion. The value of the parameters and the confidence intervals are now shown on the y-axis for the schematic in *Figure 7.5*. We have used the x-axis to distinguish between the A population and its mean μ_A, and the B population and its mean μ_B.

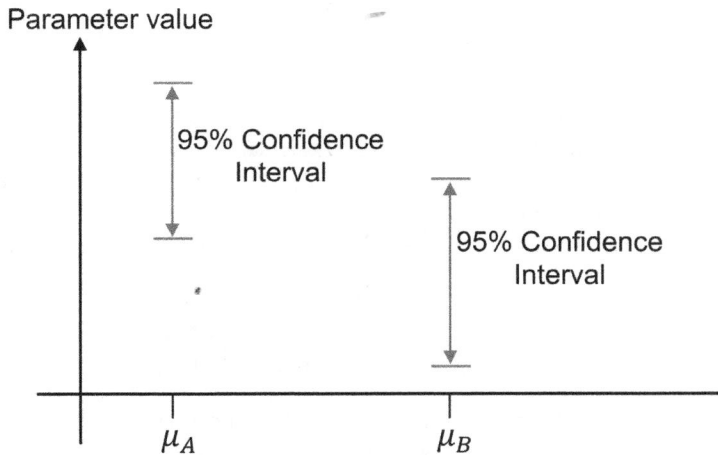

Figure 7.5: A schematic of 95% confidence intervals for separate parameters

The confidence intervals overlap, meaning that we cannot be confident that the μ_A and μ_B parameters have different values. We can reach this conclusion even though we have not performed a hypothesis test for $\mu_A - \mu_B$.

The preceding example was a simple one. Calculating separate estimates for μ_A and μ_B is straightforward, since we can use the sample means m_A and m_B as estimates. Calculation of separate confidence intervals is also straightforward, since we can follow a similar approach to the calculation of the confidence interval for $\mu_A - \mu_B$. This simple example illustrates that we can calculate a confidence interval for any population parameter. However, the more complex the form of the population parameter, the more difficult it is to derive a formula for the confidence interval. As with hypothesis testing, explicit and exact formulae for confidence intervals are typically derived by mathematicians or statisticians, and we as data scientists will then use those formulae. The confidence interval formulae in *Eq. 20, Eq. 21*, and *Eq. 22* are relatively easy to derive because the μ_A and μ_B parameters affect the data distribution in a very simple way – they are the means of distributions from which the data is drawn. We exploit this through a process called **pivoting** to extract an exact formula for the confidence interval. For more complex parameters, we cannot use this pivoting technique to derive an exact confidence interval formula.

As with running t-tests, to calculate a confidence interval in practice, we do not have to calculate *Eq. 22* ourselves. Instead, we can make use of the in-built confidence interval methods in packages such as `statsmodels`. Let's look at a code example.

A confidence interval code example

Let's run a simple confidence interval calculation using classes and methods from the `statsmodels` package. We'll use the data we used in the previous code example, so we'll assume that data is already read in and held as a `pandas` DataFrame, `df_simple_example`. The following code example and more can also be found in the `Code_Examples_Chap7.ipynb` Jupyter notebook in the GitHub repository.

First, we'll use the `statsmodels.stats.weightstats.CompareMeans` class to instantiate a `CompareMeans` object to run the confidence interval calculation. We just pass in our two samples of data, wrapped as `statsmodels.stats.weightstats.DescrStatsW` objects. Since we do not apply any non-uniform weights to the observations, we can pass each `pandas` series to the constructor for the `statsmodels.stats.weightstats.DescrStatsW` class:

```
from statsmodels.stats.weightstats import DescrStatsW, CompareMeans
mean_comparison = CompareMeans(DescrStatsW(df_simple_example['x_A']),
                              DescrStatsW(df_simple_example['x_B']))
```

Now, we'll compute the 95% confidence level for the difference in means, using the `tconfint_diff` method of the `CompareMeans` class. The 95% confidence level is the default:

```
mean_difference_95CI = mean_comparison.tconfint_diff()

mean_difference_95CI
```

This gives the following output:

```
(0.0848686453568156, 0.6211313546431837)
```

We can see that the 95% confidence interval is entirely above zero, consistent with our earlier t-test, which was statistically significant at $\alpha = 0.05$.

With that code example complete, we have covered everything about confidence intervals that we need to, so we'll now summarize what we learned about them.

What we learned

In this section, we learned the following:

- What a confidence interval is and how it quantifies our uncertainty about where the true value of a parameter lies

- How to calculate the confidence interval for the difference between two population means

- What factors affect a confidence interval

- How a confidence interval can be calculated for any population parameter

- How confidence intervals and hypothesis tests are linked and how to interpret a graphical depiction of a confidence interval

Having introduced hypothesis tests and explained how the p-value threshold α controls the false positive rate, we'll now move on to discussing false negatives and how to determine whether a hypothesis test has enough data available to reject the null hypothesis when it should.

Type I and Type II errors, and power

We explained in the first section of this chapter that the significance threshold α is something we set, controlling the rate at which we get false positives. More correctly, the value of α is the probability of falsely rejecting the null hypothesis. It is also called the **Type-I error rate** (i.e., the rate at which we make errors of Type-I). The term "false positive" tends to be used more when we assess accuracy after an event, when we are comparing to some known ground truth (i.e., what the correct decision or classification should have been). Type-I error is a term used more when the null hypothesis is falsely rejected and when discussing a hypothesis test a priori (e.g., when we are designing it). The difference between the terms "false positive" and "Type-I error" is subtle, and for the most part, we ignore it and use the terms interchangeably.

But what about false negatives? What is the hypothesis test equivalent term for a false negative, and is there anything we can do to keep the false negative rate low? The hypothesis test equivalent of the false negative is the **Type-II error**. We make a Type-II error when we don't reject the null hypothesis when we should have done (i.e., when the alternative hypothesis is true, but we accept the null hypothesis). So the Type-II error rate is Prob(Accept null hypothesis | Alternative hypothesis is true).

How can we make it more likely that we will reject the null hypothesis when we should do? Remember that we reject the null hypothesis when the p-value is less than the Type-I error rate threshold α. So, we can make it more likely that we will reject the null hypothesis by, first, increasing α, and second, decreasing the p-value. The first of these options isn't really valid because increasing α would, by definition, increase our Type-I error rate. If we want to decrease the Type-II error rate while keeping the Type-I error rate fixed, then we need to look at what we can do to decrease the p-value. For our simple example of testing whether two means, μ_A, μ_B, are different, we have already looked at what factors determine the p-value. To summarize them again, if we want to decrease the p-value, we need to do the following:

- Have a larger difference in the population means (i.e., increase $|\mu_A - \mu_B|$).

- Decrease the variability in the individual observations, (i.e., decrease the noise variance σ^2)

- Increase the sample size N

The first two of these quantities, $|\mu_A - \mu_B|$ and σ^2, are typically not within our control. So, the only realistic option is to increase N. But by how much? This brings us on to the concept of **power**. We determine a suitable value for N by setting what we consider is an appropriate probability to accept the alternative hypothesis when it is true. The rate at which we accept the alternative hypothesis when it is true is called the power. It is the rate at which we detect true positives. For larger values of N, the power increases. If the power is reasonably high (e.g., > 80%), then we say that the hypothesis test is **sufficiently powered**. It should also be clear that,

$$1 - \text{power} = \text{Type-II error rate}$$

Eq. 23

Since the Type-I error rate is given the symbol α, the Type-II error rate is often given the symbol β, so we have $1 - \text{power} = \beta$. We control the Type-I error rate by specifying a suitable value for α, and we control the Type-II error rate by specifying a suitable value for β or, equivalently, for the power.

Let's look at a concrete example. We'll return to our simple example in the first section of this chapter, where we ran a z-test on the difference between sample means coming from an A population and a B population. Let's say that I want an 80% probability of accepting $\mathcal{H}_1$ if $\mathcal{H}_1$ is indeed true. We'll assume that $N_A = N_B = N$. Now, how big does N need to be to give that 80% probability for a specified value of $|\mu_A - \mu_B|$, σ^2 and α?

First, we'll convert the value of α to its equivalent threshold for the test statistic z. For a two-tailed test, this gives a threshold value of $\sqrt{2}\,\text{erf}^{-1}(1 - \alpha)$. For $\alpha = 0.05$, we already know that this threshold on z is approximately 1.96. So, for a power of 80% with $\alpha = 0.05$, we need $|z| > 1.96$ to occur with a probability of 0.8 when the alternative hypothesis is true. The formula for the test statistic z is shown in *Eq. 15*. Combining these pieces of information gives the criterion,

$$\text{Prob}\left(\left|m_A - m_B\right| > \sigma\sqrt{\tfrac{2}{N}} \times 1.96\right) = 0.8$$

Eq. 24

To calculate the probability on the left-hand side of *Eq. 24*, we essentially repeat the steps that led to *Eq. 11* but with the added fact that $|\mu_A - \mu_B| \neq 0$. Doing so, we get,

$$\text{Prob}\left(|m_A - m_B| > \sigma\sqrt{\tfrac{2}{N}} \times 1.96\right) = 1 - \tfrac{1}{2}\text{erf}\left(\tfrac{1.96}{\sqrt{2}} - \tfrac{(\mu_A - \mu_B)\sqrt{N}}{2\sigma}\right) + \tfrac{1}{2}\text{erf}\left(-\tfrac{1.96}{\sqrt{2}} - \tfrac{(\mu_A - \mu_B)\sqrt{N}}{2\sigma}\right)$$

Eq. 25

To determine the required sample size N we need to get 80% power, we simply set the right-hand side of *Eq. 25* equal to 0.8 and solve the resulting expression for N.

For the t-test given in *Eq. 16* to *Eq. 18*, the calculation of the required value of *N* is slightly more complicated but follows the same principles. Fortunately, we don't usually have to do the detailed calculation ourselves, as most statistical packages will have a power calculation function for each of the main hypothesis tests you are likely to carry out – for example, the `solve_power` function from the `statsmodels.stats.power.TTestIndPower` class. We can illustrate this with a code example.

A t-test power calculation code example

We'll calculate the sample size required to achieve a power of 80% when $|\mu_A - \mu_B|/\sigma = 0.5$ and we set $\alpha = 0.05$. This code example and more can be found in the `Code_Examples_Chap7.ipynb` Jupyter notebook in the GitHub repository:

```
from statsmodels.stats.power import TTestIndPower
print("Sample size required = ", TTestIndPower().solve_power(
    effect_size=0.5, nobs1=None, alpha=0.05, power=0.8,
    ratio=1.0, alternative='two-sided'))
```

This gives us the following output:

```
Sample size required =   63.765610587854034
```

This means that if we have a sample size of approximately $N_A = N_B = 64$, we will have an 80% chance of rejecting the null hypothesis (and accepting $\mathcal{H}_1$) when $|\mu_A - \mu_B|/\sigma = 0.5$ and when we set $\alpha = 0.05$.

Although the preceding code example is very simple, and indeed the concept of power itself is relatively simple, there are still some nuances and subtleties regarding power calculations that you should be aware of. These include the following:

- The power of the t-test increases as the value of $|\mu_A - \mu_B|/\sigma$ increases. This makes intuitive sense. Bigger differences between populations are easier to detect (i.e., to reject the null hypothesis). Similarly, observations with lower noise levels (smaller σ^2) lead to a test with higher power, since the noise obscures the differences between the A and B populations to a lesser extent.

- A smaller value of α requires a larger value of *N* to achieve a specified power. Again, this is intuitive. A smaller value of α means we are applying a more stringent criterion to reject the null hypothesis, and we need more evidence to support that decision.

- The power calculation of the required sample size *N* requires us to know in advance suitable values for $\mu_A - \mu_B$ and σ^2. Where do we get these values from? The answer is from previous datasets, analyses, or experiments we may have run. Sometimes, it is from what seems reasonable values based on expert judgment.

- Ideally, we should always perform a power calculation before running any hypothesis test. In some circumstances, it is mandatory to do so and unethical if we don't. Consider a medical trial where we are going to expose some patients to a stressful or painful procedure as part of the treatment we are testing. If we ran an underpowered trial, we would be exposing patients to pain and stress unnecessarily, as there was no hope that the medical trial could have had a successful outcome (rejecting $\mathcal{H}_0$). We should only run a medical trial if there is a reasonable chance of detecting the kind of difference we expect the treatment to produce, so we should always run a correct and credible power calculation.

- For hypothesis tests that are more complex than a t-test, such as the sort that requires us to estimate the p-value via a computationally intensive sampling method, the calculation of power and estimation of required sample sizes is also computationally intensive. Computationally intensive power calculations are as straightforward as the computationally intensive calculation of p-values. Typically, we can just run several simulations of the hypothesis testing process at different values of N (and any other parameters in the test) to see how often we reject the null hypothesis. Once these "rejection rate curves" have been plotted out, we can read off the required value of N needed to achieve the desired level of power.

With those nuances about power calculations covered, it is time to wrap up this short section and the chapter as a whole. We'll start by recapping what we learned in this section.

What we learned

In this section, we learned about the following:

- Type-I and Type-II errors and how the p-value threshold α sets the Type-I error rate
- How the Type-II error rate is set by power
- How a power calculation is used to determine the sample size needed to achieve a desired level of power
- How to use the inbuilt power calculations in packages such as `statsmodels` to perform power calculations

Summary

The focus of this chapter has been hypothesis testing. The chapter contained only three sections, but those sections contained a wealth of new ideas and concepts. Most of the new concepts centered around p-values – what they are, what they are not, how we calculate them, and how we interpret them. These are concepts that you must get to grips with as a working data scientist. To get to grips with them, we specifically learned about the following,

- How a hypothesis test consists of two hypotheses, the null hypothesis $\mathcal{H}_0$ and the alternative hypothesis $\mathcal{H}_1$, and how we calculate the p-value as the probability of getting the observed test statistic, or larger, if the null hypothesis is true.

- How we use the numerical value of the p-value in comparison to a small threshold value, α, as evidence to reject the null hypothesis and accept the alternative hypothesis

- How the threshold α controls the false positive rate of the hypothesis test

- What the p-value is and what it is not

- What factors affect the p-value

- How all hypothesis tests approximately follow the same five-step process

- How we can use computationally intensive methods to perform any hypothesis test we wish

- How confidence intervals help us quantify the uncertainty associated with a parameter estimate and how to calculate a confidence interval

- What factors affect the width of a confidence interval

- What Type-I and Type-II errors are and how we control the Type-II error rate by specifying the required power of a hypothesis test

- How to compute the sample size required to achieve a specified power from a hypothesis test

In the next chapter, we will cover another self-contained topic that relates to models and how we assess them. We will introduce the idea of model complexity. We will learn how the complexity of a model we build affects its accuracy.

Exercises

The following is a series of exercises. Answers to all the exercises are given in the `Answers_to_Exercises_Chap7.ipynb` Jupyter notebook in the GitHub repository:

1. The `Data/paired_sampled_ttest.csv` file in the GitHub repository contains two columns of data corresponding to observations on paired samples. Use the `scipy.stats.ttest_rel` function from the `SciPy` package to run a two-tailed paired-sample t-test. Is the difference between the sample means statistically significant at the $\alpha = 0.05$ level?

2. The object returned by the `scipy.stats.ttest_rel` function has a method that computes a confidence interval for the difference between the population means. Look at the documentation on `scipy.stats.ttest_rel` to see how to use this confidence interval method, and use it to calculate a 95% confidence interval for the paired sample in question 1.

3. A paired sample test can also be thought of as a one-sample test performed on the differences between the paired observations. The data in question 1 was collected from an experiment where the standardized effect size (the mean difference divided by the standard deviation of the differences) was, a priori, expected to be in the range 0.3–0.65. Now, use the `statsmodels.stats.power.TTestPower.power` function from the `statstmodels` package to calculate the sample size required to achieve a power of 80% for a two-tailed, one-sample t-test, when the standardized effect is at the lower end of the expected range. What is the number of DOF you should use here?

Notes and further reading

1. A copy of the ASA statement on *p*-values and an interesting commentary can be found in this editorial: R.L. Wasserstein and N.A. Lazar, *The ASA's Statement on p-Values: Context, Process, and Purpose*, The American Statistician, 70(2):129–133, 2016: `https://amstat.tandfonline.com/doi/full/10.1080/00031305.2016.1154108#.ZE5wIHbMLb0`.

2. A good online tutorial on non-parametric hypothesis testing with examples in Python is *How to Calculate Nonparametric Statistical Hypothesis Tests in Python* by Jason Brownlee. The article can be found at `https://machinelearningmastery.com/nonparametric-statistical-significance-tests-in-python/`. It is part of Jason Brownlee's excellent *Machine Learning Mastery* website.

3. For a book covering the general area of non-parametric statistics, I recommend *All of Nonparametric Statistics* by Larry Wasserman, Springer, New York, 2010. ISBN: 978-1-4419-2044-7.

4. Some people find the Wasserman book too theoretical for their tastes, so a more applied book that I recommend is *Practical Nonparametric Statistics, 3rd Edition*, by W.J. Conover, John Wiley & Sons, New York, 1998. ISBN: 978-0-471-16068-7. The first edition of the book was written in 1971, so the material is limited in its coverage of non-parametric statistics topics compared to the Wasserman book. However, the focus of the book is very much on learning through application with examples.

Model Complexity

Model complexity may seem like a strange title for a chapter. Why should we care about complexity? One concept you may have already encountered as a data scientist is that of overfitting and how an overfitted model will not make accurate predictions. However, that overfitting stems from using a model whose complexity is greater than that justified by the data. The impact of model complexity on model prediction accuracy is a nuanced one. More specifically, how you decide what is the right level of model complexity can be challenging. To address this challenge requires exploring several new concepts. We will do that exploration in this chapter and do so by covering the following topics:

- *Generalization, overfitting, and the role of model complexity*: Here, we understand how model complexity affects the accuracy of model predictions on unseen data

- *The bias-variance trade-off*: Here, we dig into the mathematical details behind the prediction accuracy ideas we introduced in the preceding section

- *Model complexity measures for model selection*: Here, we introduce commonly used model complexity measures and discuss their strengths and weaknesses

Technical requirements

This chapter will mainly be a visual one. We will introduce the mathematical ideas and concepts through illustrative plots and schematic figures, which we will then unpack and explain at length. Because of this, there are no code examples in this chapter, so no technical requirements.

Generalization, overfitting, and the role of model complexity

What do we mean by a complex model? Very loosely, we think of a more complex model as having more parameters or using more features. This statement is imprecise, but the idea that model complexity broadly follows the number of model parameters/features will be precise enough for the mainly qualitative discussions of this chapter.

A more complex model can fit a training dataset more closely, as it can use the extra features to explain the variation in the response variable/target variable. What are the consequences of this increased flexibility? As a simple example, we'll take a look at *Figure 8.1*, which shows three different models fitted to a small dataset. The black circles in each plot show the training data, while the blue circles show the hold-out sample data points, which, as you can see, represent an extrapolation challenge, since the hold-out data points are all to the right of the training data points. Although using machine learning models in interpolation settings is more common, we've used an extrapolation example here because it is easier to immediately see how well each fitted model predicts the holdout data, and so it is a good experiment to start to introduce some ideas about model complexity. All the data points (training and hold-out) were generated with a quadratic equation (a 2nd-degree polynomial) in a single variable, x. A significant amount of additive noise was also added to the output of the generated data.

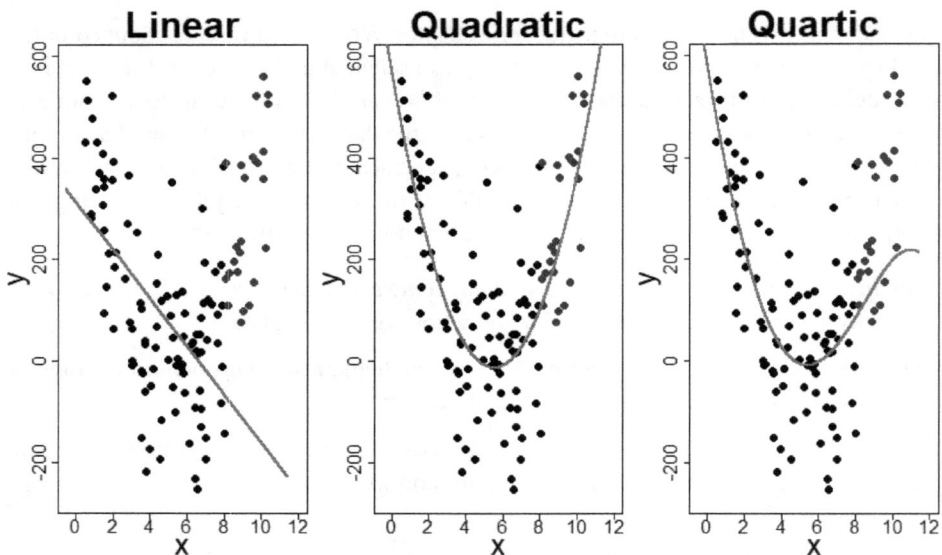

Figure 8.1: Different polynomials fitted to a quadratic dataset

The left-hand plot shows a linear equation in x fitted to the training data. The line shows the predictions from the fitted model across the full range of the feature x. The linear model is obviously incorrect. It contains fewer parameters than the model that generated the data. It clearly doesn't fit the training data well, since it doesn't capture the curvature present in the training data, and there are marked differences between the line (predictions) and the dark circles of the training set. The linear model is also very poor at predicting the hold-out data points. The difference between the line and the lighter circles of the hold-out data points is stark – the trend in the predictions is in a different direction to that of the hold-out data.

The middle plot of *Figure 8.1* shows the predictions (line) from a quadratic model fitted to the training data. The model fits the training data well and makes good predictions on the hold-out set. Any differences between the line and the black or light circles are evenly scattered and look like the result of the significant additive noise present in the data. It is perhaps not surprising that the quadratic model does well. It is of precisely the same form (a 2nd-degree polynomial) as the source of the training and hold-out data.

Or does the quadratic model do better than the linear model simply because it has more parameters and so its shape can "wiggle" more and, therefore, capture more of the variations present in the training data? To test this idea, let's increase the number of parameters in our model even further. The right-hand plot of *Figure 8.1* shows the predictions (red line) from a quartic (a 4th-degree polynomial) model. This model has two more parameters than the quadratic model in the middle plot and three more parameters than the linear model in the left-hand plot. The quartic model fits the training data as well as the quadratic model but predicts poorly on the hold-out data. The trend of the predictions diverges from the trend in the hold-out data. The red line for the quartic model is even beginning to move downwards at the right-hand edge of the plot. If we made predictions for hold-out data at even higher values of the x variable, the quartic model would move in completely the opposite direction to the data.

Let's summarize our findings from that little experiment:

- **Linear model**: One parameter. Fits the training data poorly. Predicts poorly.
- **Quadratic model**: Two parameters. Fit the training data well. Predict well.
- **Quartic model**: Four parameters. Fit the training data well. Predict poorly.

From this simple example, we can see that there appears to be an optimal level of model complexity – a model that is neither too complex nor too simple.

Overfitting

Now, you may wonder whether the poor performance of the quartic model was due to the hold-out data essentially being a test of the extrapolation abilities of the fitted models. Again, we can test this idea with a simple experiment. In *Figure 8.2*, we have plotted the predictions (red line) from a 12th-degree polynomial, fitted to all the data that was present in our previous experiment in *Figure 8.1*. In this case, our model has lots of parameters (13 in total) and has been fitted to all the data (training and hold-out) that we had in *Figure 8.1*.

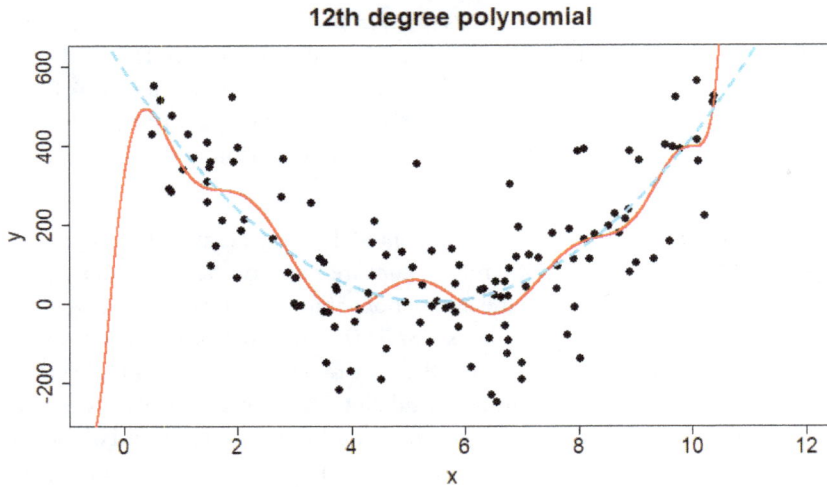

Figure 8.2: A plot of predictions (the red line) from a 12th-degree
polynomial, fitted to the same data in the previous figure

The dashed light-blue line is the true equation that was used to generate the data. The scatter of the individual data points (the black dots) around the dashed line is just the effect of the additive noise we included when generating the data. The dashed line represents the ground truth and is the best we can hope for from any model we fit to the data. A fitted model whose red line coincided with the dashed blue line would be considered "exact," "optimal," or "perfect."

The red line for the 12th-degree polynomial model we have fitted in *Figure 8.2* wiggles around the dashed blue line. This tells us that, first, the fitted model in *Figure 8.2* is not perfect, and second, while the model has broadly fitted the quadratic trend of the true equation, it has also fitted to some of the wiggles in the data that are due to the noise in it. In this case, we say that the fitted model has *overfitted* the data.

Why overfitting is bad

Is overfitting a problem? Yes, it is. Take another look at *Figure 8.2*. What would happen if our fitted model made a prediction for a value of x somewhere in the middle of the training data but where we didn't already have an existing data point? What would happen if we then measured a new data point at x? How close would our prediction be to the new measured data point? We wouldn't expect them to be identical; after all, the new data point, like those in the training set, has noise added to it.

The predicted value of the target variable, y, is given by the position on the red line, corresponding to where the position x is. Our x value may correspond to where those wiggles in the red line are, and so our prediction would follow the pattern present in the noise in the training set. The problem is that for the new data point, the noise, which is random and therefore unpredictable, is highly unlikely to follow the same wiggles that were present in the noise in the training data. The prediction from our overfitted model is likely to be further away from the true value than if we had used a model that didn't follow the wiggles of the noise in the training data. This means that because our model overfitted to the noise in the training set, we will have increased the size of the error the model makes on new unseen data.

Our predictions would have been better on average if our prediction model just followed the more general trend present in the training data, as represented by the dashed blue line. Predictions that follow the blue line would be correct on average. Since fitting to the general trends present in a training set produces a model that predicts well, we say that a good model is one that *generalizes* well.

As we increase the number of parameters in our model, we increase its ability to follow the noise present in its training data. A fitted model with a very high number of parameters would pass almost exactly through the data. This is illustrated schematically in *Figure 8.3*, which shows a highly flexible model (the line) overfitted to a small number of data points (the circles) that broadly follow a quadratic trend.

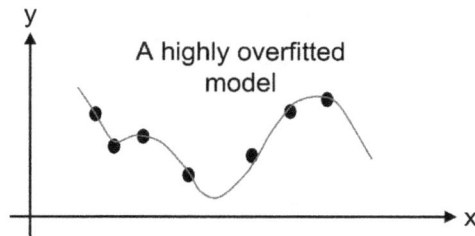

Figure 8.3: A schematic of a highly overfitted model

Although *Figure 8.3* is schematic and the line does not represent a model actually fitted to a dataset, it is still useful to draw out some conclusions:

- The smaller the training dataset is, the easier it is for a complex model to overfit. This highlights that overfitting is dependent on both the model complexity and the training data.

- If we were to use the fitted model in *Figure 8.3* to make a prediction for a data point it has already seen (i.e., one in the training dataset), the model would make a near-perfect prediction. This is because the high complexity of the model has allowed it to effectively "memorize" the training data. No "learning" has occurred. The model is just retrieving data, rather than identifying and learning general trends.

Overfitting increases the variability of predictions

Let's look at another consequence of overfitting. So far, we have discussed what happens to predictive accuracy when we overfit to a single particular dataset. We could get lucky. It could be that we have a highly complex model but, when fitted to our training dataset, we get a model that follows the general trends well and, therefore, generalizes well. What we'll now illustrate is that with increasing model complexity, that is increasingly unlikely to happen. We are unlikely to be lucky.

Take a look at *Figure 8.4*. It shows (dashed) prediction curves from 12th-degree polynomial models, obtained by fitting to different training datasets. The different training datasets were all generated by the same underlying "ground-truth" model that was used to generate the data in *Figure 8.1* and *Figure 8.2*.

12th degree polynomial: Multiple training sets

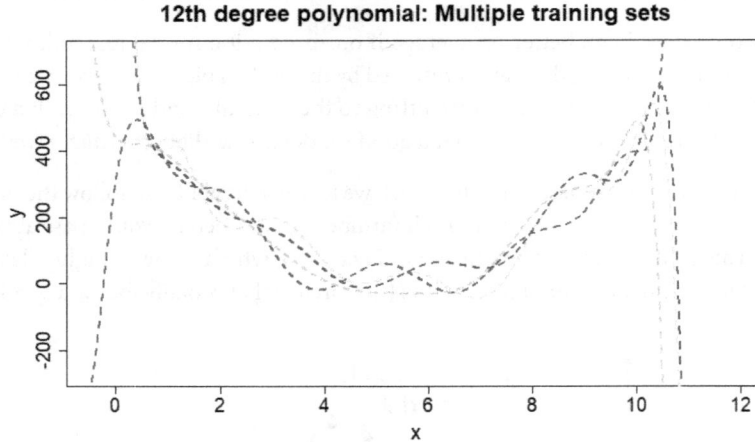

Figure 8.4: The prediction curves of the 12th-degree polynomial
models obtained from different training sets

Since each 12th-degree polynomial model will overfit to the noise present in its corresponding training data, each model displays wiggles (as we can see in *Figure 8.4*). However, each training dataset is slightly different, so each 12th-degree polynomial displays different wiggles. We can see this in the variation of the shapes of the different dashed lines. But the dashed lines show what each model would predict for the given value of x on the x-axis. For a given value of x, there will be significant variation in the prediction at x across the different models fitted to the different training datasets.

As we increase the model complexity (e.g., go from fitting 12th-degree polynomials to fitting 20th-degree polynomials), those fitted models will follow ever more closely the wiggles in the noise in their training dataset. Consequently, the size of the wiggles in each fitted model will get bigger, and the resulting variability in predictions across training sets will increase.

This is a problem. Why so? What *Figure 8.4* illustrates is that at the same value of x, different training sets give different predictions. There is a random element to our prediction, and the variance of that random element increases with increasing model complexity. However, in real life, we will only ever use one training dataset. Are we going to hope that we are lucky and that our particular training data leads to predictions that are close to the true value? A better strategy would be to ensure that we have low sensitivity of the predictions to the choice of training set, by not using an overly complex model.

Underfitting is also a problem

Since overfitting to noise in a training dataset arises because we have too many parameters in our model relative to the amount of data/signal in the training data, you might get the impression that we should use a model with as few parameters as possible. This is not the case. It is possible for a model to be too simple, to contain too few parameters. *Figure 8.5* shows the same dataset as in *Figure 8.2*, but the fitted model represented by the red line in *Figure 8.5* is now just a constant and corresponds to the mean value of the target variable, y, in the training data. Clearly, this fitted model does not follow any wiggles in the data that are due to noise. This model is definitely not overfitted. But, equally, the model is not complex enough to follow the general quadratic trend in the data indicated by the dashed blue line. We say that this model is *underfitted*.

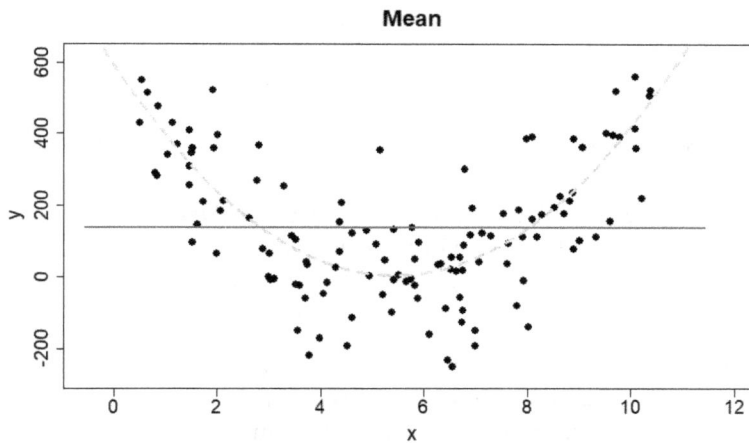

Figure 8.5: A low complexity model consisting of a constant equal to the mean of y

What are the consequences of an underfitted model? It should be qualitatively clear that a low complexity model that underfits a training dataset will make poor predictions on both training data points and on any unseen data points, as it has not learned the general trends that all data points follow. Prediction errors will be large for both training data and any holdout data. As with an overfitted model, an underfitted model generalizes poorly.

In contrast to overly complex overfitted models, low complexity models do not display much variability in their predictions, at a given holdout point x, across different training datasets. *Figure 8.6* shows the prediction curves (dashed lines) for a model that is only a constant (a 0th-degree polynomial) when fitted to the same training datasets used in *Figure 8.4*. For comparison, we have also added the true model line, which is a quadratic and shown by the light blue dashed line.

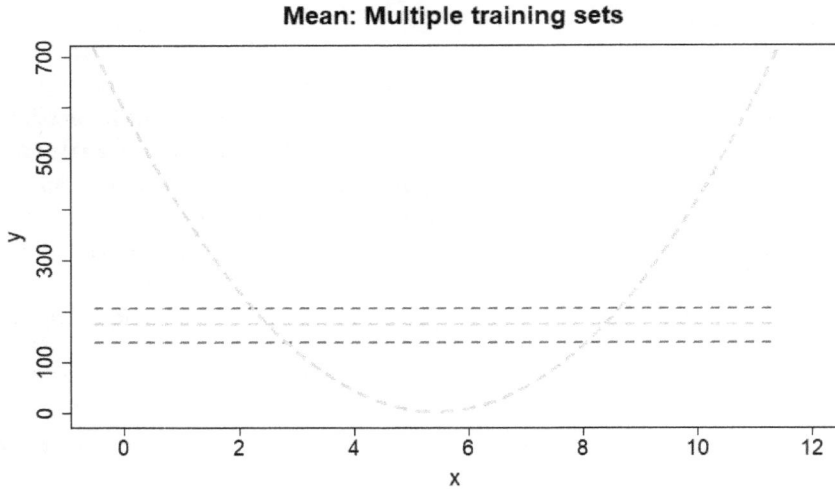

Figure 8.6: The prediction curves of 0th-degree polynomial models obtained from different training sets

There is some variation in the prediction curves of our 0th-degree polynomial, but not much. In fact, the variation is so small that two of the prediction curves lie almost on top of each other – there are, in fact, four horizontal dashed lines plotted in *Figure 8.6*. The variation is considerably smaller than the variation in prediction curves we got from fitting 12th-degree polynomials to the same training datasets.

The benefit of this is that, in contrast to using an overly complex model, it doesn't really matter which training dataset we use; our predictions will come out similar. There is little random variation in predictions between using different training sets of the same size. The downside is that those predictions will be consistently poor. No matter which training dataset we use, predictions from our low complexity model are typically a long way from the true value represented by the curved dashed light blue line. Our low-complexity model displays *bias*.

Clearly, there is a sweet spot in terms of the number of model parameters. At that sweet spot, the model complexity is sufficient to capture the general trends that led to the training data, but it is not sufficient to overfit to the noise in the data. At the sweet spot, the fitted model will generalize well. To identify that sweet spot, we need to make things more quantitative and explain how we define prediction errors and measure generalization. We will do that now.

Measuring prediction error

A data point is a pair of values $(\underline{x}, y)$. The y value is the value of the target variable (or response variable) we observed when we had the corresponding vector, $\underline{x}$, for the feature variables. Remember that for the same $\underline{x}$ value, we could get multiple different values for y because the observation y contains a random component. A dataset is just a set of multiple datapoints. In general, we can denote a dataset as $D = \{(\underline{x}_i, y_i), i = 1, 2, ..., N\}$.

Let's use $g\left(\underline{x} \mid \underline{\theta}, D_{train}\right)$ to denote the model equation of our trained model. The notation $\mid \underline{\theta}, D_{train}$ is used to denote the fact that the trained model will depend on some parameters, $\underline{\theta}$, and the training data, D_{train}. The difference between the model prediction and the observed value of the target variable for the ith datapoint is $y_i - g\left(\underline{x}_i \mid \underline{\theta}, D_{train}\right)$, which we will shorten to just $y_i - g(\underline{x}_i)$. To get a measure of the typical error made by our model, we calculate the **mean squared error** (**MSE**):

$$MSE = \frac{1}{N} \sum_{i=1}^{N} (y_i - g(\underline{x}_i))^2$$

Eq. 1

This is like when we studied least-squares model fitting in *Chapter 4*. We take the square of the individual errors to ensure a positive number so that positive and negative errors don't just cancel out to zero. To get back to a "typical" error, we can finally take the square root of the MSE to get the **root mean squared error** (**RMSE**):

$$RMSE = \sqrt{MSE}$$

Eq. 2

We can calculate the MSE for a given dataset, D. This doesn't necessarily have to be the training dataset. We can evaluate the MSE for the hold-out dataset, $D_{holdout}$, if we want to. Since the accuracy of predictions on the hold-out dataset gives us a feel for how well a model generalizes, we call RMSE $(D_{holdout})$ the generalization error, while we call $RMSE(D_{train})$ the training error. However, you will probably find that MSE and RMSE are used interchangeably – for example, $MSE(D_{holdout})$ can be referred to as the generalization error even though it is obviously an average squared error. From a qualitative perspective, it is inconsequential, and I also tend to use the term "generalization error" when I'm referring to $MSE(D_{holdout})$.

We can now go back and take our previous qualitative conclusions about overfitting and use the MSE to make them more quantitative. We know that as we increase the model complexity by increasing the number of parameters/features in our model, we can fit the training data points evermore closely. This means that for a given training dataset, D_{train}, we expect $MSE(D_{train})$ to decrease monotonically as we increase the model complexity. We also know the generalization ability of our model initially improves as we increase the number of parameters and then deteriorates. This means we expect MSE $(D_{holdout})$ to go through a minimum. These two quantitative statements are summarized schematically in *Figure 8.7*.

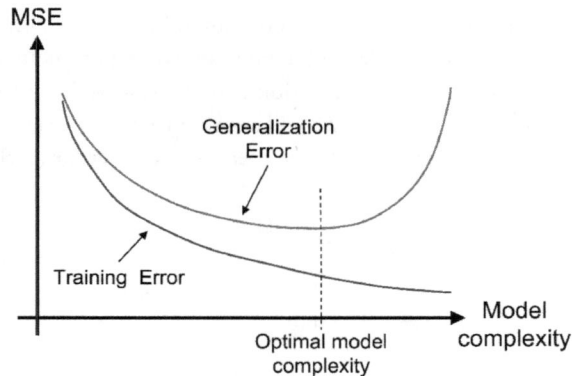

Figure 8.7: The behavior of training and generalization errors with increasing model complexity

In *Figure 8.7*, we have indicated the position where the generalization error is at its minimum. We have marked this as "optimal model complexity." Why does the smallest generalization error identify the optimal model? Why do we want the generalization error to be minimal and not the training error? Well, it is for predictions we want to use the model. We don't need to predict the training data – we already have it. The point where the generalization error is smallest represents the point where the performance of the model is optimal for its intended use. The minimum in the generalization error is the sweet spot we referred to earlier.

Figure 8.7 succinctly represents the key ideas that we have introduced in this section, so this is a good point to summarize what we have learned.

What we learned

In this section, we learned the following:

- Model complexity broadly correlates with the number of features or parameters in a model.

- A model that is too complex will overfit to the noise in training data. It will predict well on the training data points but predict poorly on unseen data.

- A model that is not complex enough will underfit the general trends and patterns present in training data. It will predict poorly on both the training data points and unseen data.

- A model that predicts well on unseen data is said to generalize. It does so by not fitting to the noise in the training data and, instead, by learning the general trends and patterns present.

- Overly complex models have a large random component to their predictions, arising from the choice of training data used. In contrast, low-complexity models will show little variation in their predictions across different training sets but will display bias.

- We can use MSE and RMSE to quantify prediction errors.

- The MSE and RMSE on the training dataset monotonically decrease with increasing model complexity, while the MSE and RMSE on a holdout dataset will display a minimum – first decreasing and then increasing as we increase model complexity.

Having learned the basic ideas behind model complexity and how it affects the generalization abilities of a model, in the next section we will dig deeper into the mathematical detail behind the generalization error curve and introduce a modern twist to its behavior.

The bias-variance trade-off

The generalization error curve in *Figure 8.7* shows a minimum. In the preceding section, we gave a qualitative explanation of why we expected the generalization error to first decrease and then increase with increasing model complexity and why, therefore, this leads to a minimum in the generalization error curve. But to get a quantitative idea of why the generalization error curve displays a minimum and what controls its position, we need to dig into the math behind the curve.

The generalization error curve is made up of two competing contributions, one increasing with model complexity and the other decreasing. It is the competition between these two contributions that leads to the minimum. Those two contributions are, first, the *bias* in a model's prediction at a holdout point, $\underline{x}$, and second, the *variance* in the model's prediction at the holdout point, $\underline{x}$, with the variance arising from the sensitivity of the model's prediction to the precise choice of training data. These two competing contributions are essentially what we highlighted in *Figure 8.4* and *Figure 8.6* in the previous section.

Mathematically, we find that,

$$\text{MSE on holdout data} = \text{Bias}^2 + \text{Variance}$$

Eq. 3

We'll go through the derivation of *Eq. 3* in a moment, as it is instructive to do so and because the derivation of *Eq. 3* is more subtle than we have hinted at. We know from our qualitative discussions of *Figure 8.4* and *Figure 8.6* that with increasing model complexity, first, the bias of predictions decreases, and second, the variance of predictions increases. Schematically, we illustrated this in *Figure 8.8*, which shows the original generalization error curve from *Figure 8.7*, with the bias and variance curves now overlayed.

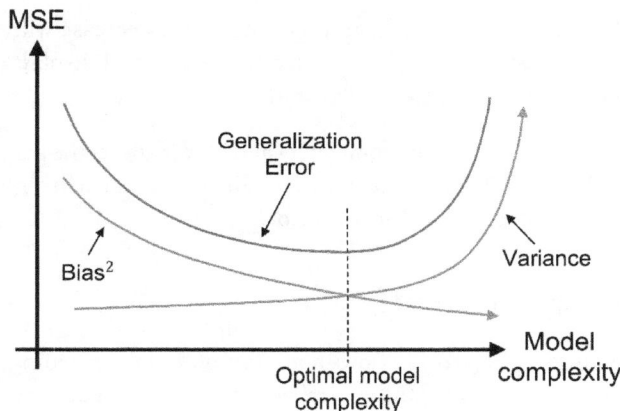

Figure 8.8: A schematic plot of the bias and variance decomposition of the generalization error

What are the consequences of *Eq. 3*? *Figure 8.8* shows that we must always make a trade-off between bias and variance. In fact, the decomposition of the generalization error that *Eq. 3* represents is referred to as the "bias-variance trade-off." At the point of the minimum generalization error, we have managed to optimize this trade-off. But the fact remains that, at any point on the generalization error curve, we have traded bias against variance. For a given training dataset, if we want a model with a smaller bias, we can do so by increasing the model complexity, but we pay the price of increased uncertainty in our model predictions. We must be aware of that; there is no free lunch. Likewise, if we want a model with smaller variance (uncertainty) in its predictions, we can do so by decreasing the model complexity, but we pay the price of increased bias in those predictions.

The only way we can decrease both bias and variance is to increase the size of the training data. This will ensure that any increased model complexity has to be used to explain the extra fine-grained trends and patterns that the extra data reveals, rather than used to overfit to noise in the data.

Proof of the bias-variance trade-off formula

To identify the two contributions to the generalization error curve, we'll calculate the expected generalization error curve. This gives us the typical shape of the generalization error curve, not specific to any particular training dataset. We'll specifically look at the expected MSE of the holdout dataset. This is defined as,

$$\mathbb{E}_{D_{train}} \left(\frac{1}{N_{holdout}} \sum_{(y,\underline{x}) \in D_{holdout}} \left(y - g(\underline{x}| D_{train}) \right)^2 \right)$$

Eq. 4

For simplicity of notation, we have omitted the dependence of $g(\underline{x})$ on the model parameters $\underline{\theta}$ in *Eq. 4*.

To understand *Eq. 4*, we need to understand $\mathbb{E}_{D_{train}}\left(\left(y - g(x|D_{train})\right)^2\right)$ averaged over all the holdout data points (x, y).

If we expand out $\left(y - g(x|D_{train})\right)^2$, we can write this as,

$$\mathbb{E}_{D_{train}}\left(\left(y - g(x|D_{train})\right)^2\right) = \mathbb{E}_{D_{train}}(y^2) - 2\mathbb{E}_{D_{train}}\left(yg(x|D_{train})\right) + \mathbb{E}_{D_{train}}\left(g^2(x|D_{train})\right)$$

Eq. 5

We'll take each of the terms on the right-hand side of *Eq. 5* one at a time. For the first term, we're going to add the assumption that the observed value of the holdout target value, y, is just a noise-corrupted version of a ground-truth value, so we have,

$$y(x) = y_{true}(x) + \varepsilon$$

Eq. 6

We'll assume that ε is a random variable with zero mean and variance σ^2. The $y_{true}(x)$ function is a deterministic function – it has no randomness in it. It is the value we would get for the target variable at a holdout point, x, if there were no added noise. We'll calculate the expectation of $\left(y - g(x|D_{train})\right)^2$ over ε to represent the averaging over the holdout value, y. This means in *Eq. 5*, we replace $\mathbb{E}_{D_{train}}$ with $\mathbb{E}_{D_{train},\varepsilon}$. With that additional change, we get,

$$\mathbb{E}_{D_{train},\varepsilon}(y^2) = \mathbb{E}_{D_{train},\varepsilon}(y_{true}^2) + 2\mathbb{E}_{D_{train},\varepsilon}(\varepsilon y_{true}) + \mathbb{E}_{D_{train},\varepsilon}(\sigma^2) = y_{true}^2 + \sigma^2$$

Eq. 7

In deriving the very right-hand side of *Eq. 7*, we have made use of the fact that the random additive noise ε has mean zero, and that y_{true} is just a fixed number with no randomness.

For the second term in *Eq. 5*, we find,

$$\mathbb{E}_{D_{train},\varepsilon}\left(yg(x|D_{train})\right) = \mathbb{E}_{D_{train},\varepsilon}\left((y_{true} + \varepsilon)g(x|D_{train})\right) = y_{true}\mathbb{E}_{D_{train}}\left(g(x|D_{train})\right)$$

Eq. 8

The final term on the right-hand side of *Eq. 5* is,

$$\mathbb{E}_{D_{train},\varepsilon}\left(g^2(x|D_{train})\right) = \mathbb{E}_{D_{train}}\left(g^2(x|D_{train})\right) = \text{Var}_{D_{train}}\left(g(x|D_{train})\right) + \left(\mathbb{E}_{D_{train}}\left(g(x|D_{train})\right)\right)^2$$

Eq. 9

The very right-hand side of *Eq. 9* follows from the fact that for any random variable z we have, $\text{Var}(z) = \mathbb{E}(z^2) - (\mathbb{E}(z))^2$, and so re-arranging we have $\mathbb{E}(z^2) = \text{Var}(z) + (\mathbb{E}(z))^2$.

Plugging the results from *Eq. 7*, *Eq. 8*, and *Eq. 9* into *Eq. 5*, we get,

$$\mathbb{E}_{D_{train},\varepsilon}\left(\left(y - g\!\left(\underline{x}|D_{train}\right)\right)^2\right) = y_{true}^2 + \sigma^2 - 2y_{true}\,\mathbb{E}_{D_{train}}\left(g\!\left(\underline{x}|D_{train}\right)\right) + \mathrm{Var}_{D_{train}}\left(g\!\left(\underline{x}|D_{train}\right)\right) + \left(\mathbb{E}_{D_{train}}\left(g\!\left(\underline{x}|D_{train}\right)\right)\right)^2$$

Eq. 10

We can rearrange the terms on the right-hand side of *Eq. 10* to give,

$$\mathbb{E}_{D_{train},\varepsilon}\left(\left(y - g\!\left(\underline{x}|D_{train}\right)\right)^2\right) = \left(\mathbb{E}_{D_{train}}\left(g\!\left(\underline{x}|D_{train}\right)\right) - y_{true}\right)^2 + \mathrm{Var}_{D_{train}}\left(g\!\left(\underline{x}|D_{train}\right)\right) + \sigma^2$$

Eq. 11

Now, $\mathbb{E}_{D_{train}}\left(g\!\left(\underline{x}|D_{train}\right)\right) - y_{true}$ is just the expected difference between the prediction of our model at the holdout point, $\underline{x}$, and the true (noise-free) value at that point. This is just the bias. Similarly, the expression $\mathrm{Var}_{D_{train}}\left(g\!\left(\underline{x}|D_{train}\right)\right)$ is just the variance of our model's prediction as we train it on different training datasets. This means we can use *Eq. 11* to write,

MSE on holdout = Bias in prediction^2+ Variance of prediction + Variance of noise

Eq. 12

Eq. 12 is the same as *Eq. 3*, with the addition of the last term, the variance σ^2 of the additive noise. This means that the presence of noise in the data sets a minimum value for the generalization error. No amount of trading off bias against variance will get us below σ^2 for the MSE on the holdout data. This is intuitive.

We have already unpacked the consequences of the bias-variance trade-off equation. This equation neatly summarizes the classical view of model complexity. The schematic plot shown in *Figure 8.7* also neatly summarizes this classical view of model complexity. You may have seen it before, as it is used in many introductions to machine learning. However, for highly parameterized machine learning models, the plot in *Figure 8.7* is not the full story. An interesting twist has arisen in the last few years, emerging from the study of deep learning neural networks. We'll now explain what that twist is.

Double descent – a modern twist on the generalization error diagram

In our previous qualitative discussions, where we fitted various polynomials (from the 0th degree to the 12th degree) to the dataset in *Figure 8.2*, it was probably obvious to you where the sweet spot lay in terms of number of model parameters. The data in *Figure 8.2* was generated using a quadratic equation, so a 2nd-degree polynomial was clearly optimal to use to fit to the training data. In these circumstances, it was self-evident where the generalization error minimum should be because the models we were fitting to the training data were in the same class – polynomials – as the process that generated the data. This will not always be the case.

Imagine that we have a real-world dataset that we model using a neural network. It is highly unlikely that the data was generated using a neural network. However, our qualitative arguments about overfitting that led us to *Figure 8.7* still hold. We still expect to see the training error decrease monotonically as we increase the number of parameters in the neural network, and for the generalization error to display a minimum. For modern machine learning models, what we tend to see is illustrated schematically in *Figure 8.9*.

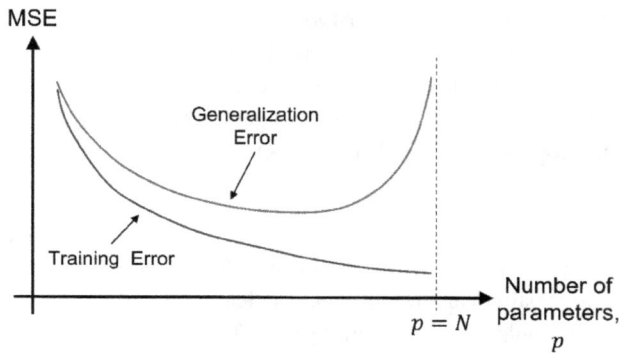

Figure 8.9: A generalization error curve for a modern machine learning model

Schematically, the generalization curve for our neural network model looks like the generalization curve in *Figure 8.7*. The difference is that now the *x*-axis explicitly represents the number of model parameters, *p*. The generalization error curve increases steeply as we increase *p*, with the steep increase occurring as *p* approaches the size of the training dataset, *N*. At first sight, it looks like nothing more interesting is happening and there is no need to explore any further. However, if we continue to increase the number of parameters in our neural network, what we tend to see is illustrated schematically in *Figure 8.10*.

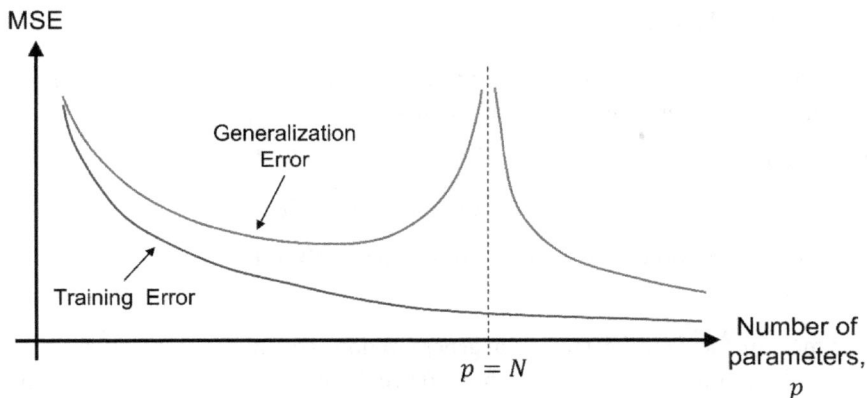

Figure 8.10: A "double descent" generalization error curve for a modern machine learning model

Figure 8.10 shows that as we increase the number of parameters in our neural network beyond N, the generalization error decreases again. This phenomenon is termed *double descent* because the generalization error decreases for a second time as we increase the model complexity. For $p > N$, both the training error and generalization error continue to decrease.

The consequences of double descent

Figure 8.10 tells us that it is possible to have an *over-parameterized* model and still have good generalization performance. This has big implications for large neural network models and, in particular, **Large Language Models (LLMs)**, which can have billions of parameters. *Figure 8.10* shows that this number of parameters in a model is not an obstacle to the model genuinely learning, so double descent gives support to the use and development of deep learning neural networks, LLMs, and other massively parameterized models.

However, that's not to say we fully understand all the details behind double-descent. The phenomenon of double descent is still very much an active area of research. Indeed, at the time of writing (2023), one of the most talked-about papers at the top machine learning conference, NeurIPS, was about double-descent and how you should effectively count parameters in a model – see the *Notes and further reading* section at the end of the chapter for details on this paper. Furthermore, the debate continues in academic circles and on social media about whether LLMs have truly learned from their training data and are displaying understanding, or whether they have simply memorized the training data and stored it in the massive number of model parameters.

Since double descent brings us up to the present day in terms of what is known about generalization error and model predictive accuracy, this seems like a good place to finish this section and recap what we have learned.

What we learned

In this section, we learned the following:

- The generalization error is made up of two contributions, one from the bias in the model predictions and one from the variance of the model predictions
- The bias-variance trade-off and its implications
- The mathematical detail behind the bias-variance trade-off equation
- The phenomenon of double descent in over-parameterized machine learning models and its implications

Having learned more mathematical details about generalization error, in the next section, we will learn about various model complexity metrics and how to use them to select models of the appropriate complexity for a given training set.

Model complexity measures for model selection

Practical model complexity measures tend not to measure model complexity directly. Instead, they measure some sort of trade-off – for example, how much information has been lost by approximating the patterns present in a dataset by using a particular model form, or what evidence a dataset provides for a model form of this level of complexity. These practical metrics don't directly measure model complexity, but they take it into account.

Selecting between classes of models

In the preceding paragraph, we referred to model form. But what do we mean by model form? We mean the mathematical form of the equation that defines a model. So, two models that differ only in their parameter values but otherwise have the same form of mathematical equation have the same model form (e.g., two linear models that use the same features).

A model form represents a whole class of models. Let's go back to our polynomial model example to illustrate. For our polynomial models in *Figure 8.1* and *Figure 8.2*, we had 1st-degree polynomials, 2nd-degree polynomials, 4th-degree polynomials, and 12th-degree polynomials. The model class in this case is just the degree of the polynomial. The 12th-degree model class is the set of all possible 12th-degree polynomials. In this case, the model class is related simply to the number of parameters p in the model because,

$$p = degree\ of\ polynomial + 1$$

Eq. 13

We can denote each model class by M_p, and in this instance, the different model classes M_p represent different levels of model complexity.

The challenge we face is working out which degree of polynomial model to use to model our data. In other words, we need to work out which class of models to use. This process is called **model selection**.

Once we have selected which class of models to use, identifying which model to use within that class is just a case of parameter estimation (e.g., via least-squares fitting) or maximum likelihood estimation. In fact, when comparing two model classes, we often just compare the maximum likelihood models from each class, since this uses the best model in each class as a representative of that class.

We have already met a measure that helps us select between different model forms or classes, the generalization error. We have also seen how it is affected by model complexity. In this section, we will introduce two other commonly used model complexity measures that are used for model selection, the **Akaike Information Criterion (AIC)** and the **Bayesian Information Criterion (BIC)**.

To start, we'll define our model form, g, for which we want to compute the AIC and the BIC. The model form g is a function, $g(\underline{x}|\underline{\theta})$, that takes a feature vector, $\underline{x}$, as input and uses its p model parameters, represented by the vector $\underline{\theta}$, to compute the model output. From now on, we'll use g when we mean either a model or a model form. We also assume we know how to calculate the likelihood, L, of a dataset, D, given the model – that is, we know how to calculate $P(D \mid \underline{\theta}, g)$. With that in place, we can begin to compute the AIC and BIC.

Akaike Information Criterion

The AIC is defined as,

$$\text{AIC} = 2p - 2\log L_{max}$$

Eq. 14

In *Eq. 14*, L_{max} is the maximum likelihood of the model g. The AIC attempts to measure the information lost by the model g when we use it as an approximation of the true process that generated the data. We haven't yet defined what "information" is, and we don't do so until *Chapter 13*. Basically, we can think of the information loss as a measure of how different the general trends and patterns that model g produces are, compared to the general trends and patterns from the true data-generating process.

Obviously, a good candidate model will be close to the true generating process and will have a small information loss, leading to a small AIC value. We can select an optimal model class from a set of candidate model classes by choosing the one that has the smallest AIC.

Since the information loss is not a comparison of the model g to the training data, but a comparison of the model g to the true data-generating process, minimizing the AIC metric automatically guards against overfitting. But wait, I hear you say – doesn't the AIC include the log-likelihood that depends on the training data? Yes, that is true. However, to compute the information loss, we'd have to know the true generating process. Instead, we said that the AIC "attempts" to measure the information loss. The AIC approximates the information loss, but it still retains many of the desirable properties of the information loss, so we can still use it for model selection. Let's look at the AIC formula in *Eq. 14* to see how this model selection is done.

The AIC consists of two contributions, $2p$ and $-2\log L_{max}$. As we increase model complexity by increasing p, we obviously increase the first of these contributions. However, a higher complexity model will be able to fit the training data more closely, so will have a smaller value of $-\log L_{max}$. We can see that the two contributions to the AIC work in competition against each other. Minimizing the AIC will find the optimal trade-off between the two contributions – that is, the optimal balance between p (the model complexity) and $\log L_{max}$ (the measure of the model fit).

Weaknesses of the AIC

The previous sentence makes it sound like the AIC is ideal for what we want to do – find the sweet spot of model complexity. Not quite. There are a few issues we should highlight:

- Firstly, the AIC formula in *Eq. 14* is an asymptotic (i.e., a large sample) result. We said that the AIC approximates the information loss. That approximation is increasingly accurate as, N, the size of the training dataset increases. It is most accurate when N is large. The consequence of this is that for small training datasets, the AIC won't accurately approximate the information loss, and a model selection based on *Eq. 14* may perform poorly. There is a modified or corrected form of the AIC, usually denoted as AICc, that attempts to correct this deficiency. We won't explain the AICc here, other than to say that you can use it for model selection in the same way that you use the AIC – you select the model with the smallest value of AICc.

- Secondly, the use of the AIC follows a different philosophy to how we use the generalization error for model selection. You're probably already familiar with using the MSE on a validation set to perform hyper-parameter optimization for a machine learning model. Also, you've no doubt run cross-validation calculations before. In a machine learning context, we are directly trying to optimize, albeit in an empirical fashion, the thing we care about – the future predictive accuracy of our model. In contrast, when we minimize the AIC, we are optimizing a proxy measure that we believe should be correlated with good predictive accuracy.

The preceding second point is not unique to the AIC. It is also a criticism that we can make of the BIC, which we will explain next.

Why should we use the AIC, then, if it appears to be not as good as the generalization error as a means of selecting the optimal model? The AIC is still a useful metric for many reasons:

- Calculating the MSE on a validation set requires us to have sufficient data to divide into training, validation, and test splits. Similarly, calculating cross-validation measures can be computationally expensive.

- In contrast, the AIC is quick to calculate. If we have fitted our model using maximum likelihood, then we have the AIC essentially with minimal extra computation.

- The theoretical underpinnings of the AIC are easier to understand. Minimizing the generalization error or doing a cross-validation analysis have an intuitive empirical justification to them, but a detailed theoretical analysis of the performance of model selection/hyper-parameter optimization based on validation/cross-validation can be harder to do, so we may have a less-detailed understanding of the weaknesses of such approaches.

Personally, I like to use the AIC in addition to other model selection metrics. The ease of calculation of the AIC is its biggest selling point, but you must always be aware that, first, it is an approximation, and second, it is measures information loss, not predictive accuracy.

Bayesian Information Criterion

The BIC is defined as:

$$BIC = p\log N - 2\log L_{max}$$

Eq. 15

where again L_{max} is the maximum likelihood for our model, g. Like the AIC, the BIC is extremely easy to calculate, particularly if we have already fitted our model g via maximum likelihood estimation.

The formula in *Eq. 15* hides the idea behind the BIC. As you might guess, the BIC is based upon ideas from Bayesian analysis of models. The BIC approximates the *Bayesian evidence* of the model class. Given a dataset, D, the Bayesian evidence for a model class, M, is defined as:

$$Bayesian\ evidence\ =\ P(D \mid M) = \int P(D \mid \underline{\theta}_M, M)\ P(\underline{\theta}_M \mid M)\, d\underline{\theta}_M$$

Eq. 16

In *Eq. 16*, we have used $\underline{\theta}_M$ to denote the model parameters for a model in the class M, and the symbol $P(\underline{\theta}_M \mid M)$ represents the prior distribution on $\underline{\theta}_M$ given the model class M. For our polynomial model class M_{12} this would be the Bayesian prior we put on the 13 coefficients of the 12th-degree polynomial.

In general, the larger the Bayesian evidence of a model class, M, the more evidence the data, D, provides that the model class is the one that generated the data. However, as with the AIC, we run into a technical issue here. Calculating *Eq. 16* is not generally easy. However, when the size of the training dataset, N, is large, we can come up with a general approximation. That approximation is the BIC. In fact:

$$BIC \approx -2\log(\text{Bayesian Evidence})\quad \text{as } N \to \infty$$

Eq. 17

Eq. 17 indicates that maximizing the Bayesian evidence is the same as minimizing the BIC. Consequently, we perform model selection by minimizing the BIC.

As with the AIC, there are several comments and observations we can make about the BIC:

- Like the AIC, the BIC is easy to calculate.

- Again, like the AIC, the BIC is based on a large sample size approximation, so at smaller sample sizes, it may not select the true optimal model class. For smaller sample sizes, it would be better to compute the Bayesian evidence exactly via *Eq. 16*, although this can require considerable mathematical skill or using advanced computationally intensive Monte Carlo techniques.

- The formula for the BIC looks very similar to that for the AIC. The difference between the BIC and AIC is just in the penalty terms; $2p$ for the AIC compared to $p\log N$ for the BIC. So, as N increases, the penalty we pay for increasing the model complexity becomes higher in the BIC than in the AIC, meaning the BIC tends to be a more stringent selection criterion than the AIC.

Having discussed the AIC and BIC at length, it is time to wrap up this section. Let's first summarize what we have learned in this section about practical model complexity measures, and then we'll summarize the chapter overall.

What we learned

In this section, we learned the following:

- The concept of model selection
- The AIC and how we can use it to perform model selection by selecting the model class with the lowest AIC value
- The BIC and how we can use it to perform model selection by selecting the model class with the lowest BIC value
- The strengths and weaknesses of the AIC and BIC

Summary

This chapter has been a shorter one and largely a visual one. The only heavy math came in the proof of the bias-variance decomposition of the generalization error. However, the visual approach has been useful in explaining concepts of overfitting, underfitting, and generalization. At a superficial level, these concepts are intuitive and need very little explanation. You will have probably encountered them before. However, a more thorough understanding of these concepts is crucial if we're not to be misled by them when we're building predictive models. That thorough understanding has required us to learn additional concepts. Across the whole chapter, the concepts we have learned about included the following:

- Model complexity and how we broadly think of this as being related to the number of parameters in a model
- Overfitting to the noise in a dataset and how it increases as we increase the complexity of a model
- Underfitting to the general trends in a dataset and how it increases as we decrease the complexity of a model
- Generalization and how a model that generalizes well is the one that makes accurate predictions on unseen data

- The bias-variance trade-off and how it makes mathematically precise the ideas behind how the minimum in the generalization error arises

- A classical picture of how model complexity affects the generalization error

- That over-parameterized machine learning models such as neural networks have revealed the phenomenon of double descent

- Model selection and how we use model complexity measures to select between different model classes

- The **Akaike Information Criterion (AIC)** as a model selection measure

- The **Bayesian Information Criterion (BIC)** as a model selection measure

Our next chapter is another self-contained topic, function decomposition. In that chapter, we will learn mathematical tools and tricks to build up a function.

Notes and further reading

1. The NeurIPS 2023 conference paper on double descent we referred to in this chapter is *A U-turn on Double Descent: Rethinking Parameter Counting in Statistical Learning*, by A. Curth, A. Jeffares, and M. van der Schaar. A preprint version of the paper can be found on the arXiv archive at `https://arxiv.org/pdf/2310.18988.pdf`.

9

Function Decomposition

The title of this chapter may seem a little odd. Why would we want to decompose a function? The word "decompose" is a bit formal. What we mean is that we're going to break down a function into smaller, easier-to-understand bits. This is very similar to how we decomposed matrices in *Chapter 3*, using eigendecomposition and the **singular value decomposition (SVD)**. The difference is that now, our mathematical object is a function, not a matrix. Function decomposition allows us to see how functions are made, and to see where their properties and characteristics come from. Function decomposition also allows us to do the reverse – that is, build up or compose a function from simple building blocks – and in doing so construct a function with properties and characteristics that are useful to us. This is a beneficial skill to have as a data scientist, where we often want to construct a function with specific characteristics as part of a predictive model. To learn the skill of decomposing functions, we will cover the following topics:

- *Why do we want to decompose a function?*: In this section, we'll learn what function decomposition is and what benefits it gives us.

- *Expanding a function in terms of basis functions*: Here, we'll learn how to calculate the decomposition of a function in terms of simple building block basis functions.

- *Fourier series*: In this section, we'll learn about the most used basis functions, sine and cosine waves, and use them to decompose periodic functions.

- *Fourier transforms*: Here, we'll generalize a Fourier series and use sine and cosine waves to decompose any function.

- *Discrete Fourier transforms*: Finally, we'll use the ideas behind Fourier series and Fourier transforms and use sine and cosine waves to decompose a regularly spaced series of data points.

Technical requirements

All the code examples provided in this chapter (and additional examples) can be found in this book's GitHub repository: `https://github.com/PacktPublishing/15-Math-Concepts-Every-Data-Scientist-Should-Know/tree/main/Chapter09`. To run the Jupyter Notebooks, you will need a full Python installation, including the following packages:

- `pandas` (>=2.0.3)
- `numpy` (>=1.24.3)
- `matplotlib` (>=3.7.2)

Why do we want to decompose a function?

Decomposing a function means breaking it down into its component parts. The reason for doing so is the same as why we wanted to decompose a matrix in *Chapter 3* using the eigendecomposition or using the SVD – to break it down into simpler parts. By simpler, we mean breaking a function down into a sum of functions that are easier to understand, have nicer properties, and behave in ways we understand when we transform them.

What is a decomposition of a function?

We have already answered this question to some degree. Decomposing a function, $f(\underline{x})$, means writing that function as a sum of other functions. In math, that means we write the following:

$$f(\underline{x}) = \sum_{k=1}^{K} \alpha_k h_k(\underline{x})$$

Eq. 1

The number of functions, K, that we decompose, $f(\underline{x})$, into could be finite or infinite. It depends on what we want to do with the decomposition on the right-hand side of Eq. 1.

We also haven't said what the component functions, $h_k(\underline{x})$, are yet. Again, the choice of components, $h_k(\underline{x})$, will depend on how we want to use the decomposition. However, the fact that we have included some coefficients, α_k, in the decomposition hints at the idea that we usually take the component functions, $h_k(\underline{x})$, to be "standardized" in some way. This means they will have properties and behaviors we understand. This is one of the main benefits of decomposing a function. Consider that we have some linear transformation, represented by A, that we want to apply to $f(\underline{x})$. If we use our decomposition in Eq. 1, when we apply A, we get the following:

$$Af(\underline{x}) = \sum_{k=1}^{K} \alpha_k A h_k(\underline{x})$$

Eq. 2

So, if we know and understand what the effect of the transformation, A, has on the component functions, $h_k(x)$, then we understand what effect A has on our function, $f(x)$. Often, we will choose the component functions, $h_k(x)$, because they have a simple behavior when the transformation, A, is applied to them – in other words, the mathematical form of A determines what component functions, $h_k(x)$, we use. We'll learn more about this in the next section. However, to finish this short section, we'll look at a couple of examples where we decompose a function not based on a transformation we want to apply to the function, $f(x)$, but based on the properties of the function, $f(x)$, that we to want understand. This will help reinforce the concept that sometimes, what we decompose depends on what properties we want to analyze or what properties we would like our component functions to have.

Example 1 – decomposing a one-dimensional function into symmetric and anti-symmetric parts

For this example, we'll be looking at one-dimensional functions – that is, functions, $f(x)$, where $x \in \mathbb{R}$, so x can be anywhere on the real number line. Some one-dimensional functions are symmetric, so they have the nice property that $f(-x) = f(x)$. This property can be useful because if we already have code that computes $f(x)$, we don't need any further code to calculate $f(-x)$. Similarly, other mathematical calculations such as integration and differentiation of $f(x)$ can sometimes be simplified using the knowledge that $f(x)$ is symmetric.

Likewise, some one-dimensional functions are anti-symmetric, meaning $f(-x) = -f(x)$. Knowing that a function is anti-symmetric is useful in the same way that knowing a function is symmetric – we can simplify various mathematical and computational calculations.

What happens if a one-dimensional function is neither symmetric nor anti-symmetric? It turns out that any one-dimensional function can be decomposed into (written as a sum of) a symmetric function and an anti-symmetric function. So, we can always write $f(x)$ like so:

$$f(x) = f_{sym}(x) + f_{anti}(x)$$

Eq. 3

The right-hand side of Eq. 3 is a decomposition of $f(x)$ of the form given in Eq. 1. It is a sum of two functions, so $K = 2$, with coefficients, $\alpha_1 = \alpha_2 = 1$. The two functions are a symmetric function, $f_{sym}(x)$, and an anti-symmetric function, $f_{anti}(x)$.

This is great, but how do we work out what $f_{sym}(x)$ and $f_{anti}(x)$ are? Very easily – we calculate $f(x) + f(-x)$. Using the decomposition on the right-hand side of Eq. 3, we find the following:

$$f(x) + f(-x) = f_{sym}(x) + f_{anti}(x) + f_{sym}(-x) + f_{anti}(-x) = 2f_{sym}(x)$$

Eq. 4

The last step in Eq. 4 follows from the fact that $f_{sym}(-x) = f_{sym}(x)$ and $f_{anti}(-x) = -f_{anti}(x)$. From Eq. 4, we have the following:

$$f_{sym}(x) = \tfrac{1}{2}\left[f(x) + f(-x)\right]$$

Eq. 5

A similar calculation reveals the following:

$$f_{anti}(x) = \tfrac{1}{2}\left[f(x) - f(-x)\right]$$

Eq. 6

This was a very simple example, so we'll look at a more realistic one next.

Example 2 – decomposing a time series into its seasonal and non-seasonal components

The left-hand plot in Figure 9.1 shows the daily sales level (number of units sold) of a supermarket product over 3 years. We can think of this time series data as a one-dimensional function. It has a value for each value of the timepoint variable, t.

The sales time series clearly shows a yearly seasonal pattern. This is not unusual for supermarket products. The size of the seasonal variation is large, with around a 400-unit change from low season to high season. Unfortunately, the seasonal variation, which we can't control, is hiding the variation in sales due to price changes, which we can control. The effect of price changes on sales is what we'd like to understand. The daily sales level, with the effect of the seasonality removed, is shown in the right-hand plot of Figure 9.1. Having removed the seasonality, the right-hand plot clearly shows the effect of the price changes that have occurred:

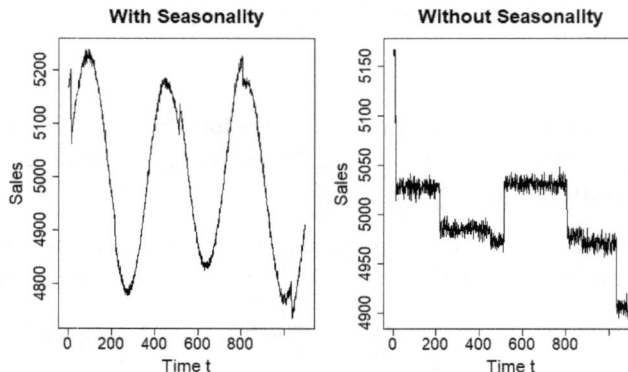

Figure 9.1: Decomposing a sales time series into its seasonal and non-seasonal components

If we can separate the seasonal component from the original data, we can build a model of the remaining part using price as a predictive feature. Then, we can make forecasts of future sales levels at new price points. This means we want to decompose $f(t)$, like so:

$$f(t) = f_{Seasonal}(t) + f_{Non-Seasonal}(t)$$

Eq. 7

Again, Eq. 7 is a simple example of the general form in Eq. 1, with just two component functions with coefficients $\alpha_1 = 1$ and $\alpha_2 = 1$. There are numerous techniques we can use to do this decomposition. For example, the seasonal component in the left-hand plot of Figure 9.1 looks like a sinusoidal pattern, so we could create a new sine wave feature (with a 1-year period) and fit this new feature to the data. We won't go into the details of how to do this. What's more important to understand is that decomposing the time series data into its various components is an example of function decomposition, and that function decomposition serves a useful purpose by allowing us to understand how a function or even data is comprised. This is the main lesson from this short section. Now is a good point at which to summarize that lesson.

What we've learned

In this section, we learned the following:

- Decomposing a function means breaking it down into several component functions that are easier to work with

- When we decompose a function, we write it as a sum of component functions multiplied by a coefficient

- We choose the component functions because they have convenient properties or behavior, such as when we apply a particular transformation to them

Having introduced the basic concept of function decomposition, in the next section, we will move beyond the simple examples we used in this section and show how to decompose using a set of standardized components that have pre-specified properties.

Expanding a function in terms of basis functions

In the previous section, we introduced some key concepts about decomposing functions, namely that we are breaking down or decomposing a function, $f(x)$, into several smaller parts that are easier to interpret and work with. We can also think of those smaller parts as simple building blocks from which we can build up or compose more complicated functions.

Whichever way we choose to look at a function, we need to make the ideas that were introduced in the previous section more concrete. In other words, given a set of simple building block functions, we need to work out how to quantify how much of each of those building block functions are in our function, $f(\underline{x})$. To do so, consider the function shown in Eq. 8:

$$f(\underline{x}) = \sum_{k=1}^{K} \alpha_k h_k(\underline{x})$$

Eq. 8

The coefficient, α_k tells us how much the building block function, $h_k(\underline{x})$, contributes to the function, $f(\underline{x})$. Put another way, α_k tells us how much of $h_k(\underline{x})$ there is in $f(\underline{x})$. We can use this second way of looking at what α_k tells us to work out as a method of calculating α_k. We want to know how much of $h_k(\underline{x})$ there is in $f(\underline{x})$. This is similar to when we were looking at vectors in *Chapter 3*. There, we wanted to know how much of a vector, $\underline{a}$, was in the direction of a particular unit-length basis vector $\underline{v}$. We did this by computing the inner product between $\underline{a}$ and $\underline{v}$. The inner product, $\underline{a}^\top \underline{v}$, tells us the projection of $\underline{a}$ onto $\underline{v}$. This is the amount of $\underline{a}$ that lies along the unit-length vector, $\underline{v}$. The fact that $\underline{v}$ is a unit-length vector is also expressed through an inner product calculation and corresponds to saying $\underline{v}^\top \underline{v} = 1$.

Great, but we're dealing with functions now, not vectors. Is there an equivalent calculation, or inner product between functions? In other words, can we project $f(\underline{x})$ onto $h_k(\underline{x})$? The answer is yes, and we can use the definition of the inner product between vectors to help us define the inner product between functions. Recall that for vectors, the inner product, $\underline{a}^\top \underline{v}$, is calculated as follows:

$$\underline{a}^\top \underline{v} = \sum_i a_i v_i$$

Eq. 9

Here, a_i and v_i are the individual components of vectors $\underline{a}$ and $\underline{v}$, respectively. Since we can loosely think of the N-dimensional vector, $\underline{a} = (a_1, a_2, ..., a_N)$, as a discretized version of a continuous function, it will not surprise you to learn that the inner product between two real functions, $f(\underline{x})$ and $g(\underline{x})$, is calculated as the continuous version of the right-hand side of Eq. 9 – that is, an integral. The inner product between the real functions, f and g, is defined as follows:

$$\langle f, g \rangle = \int f(\underline{x}) g(\underline{x}) d\underline{x}$$

Eq. 10

Note the special symbol, $\langle f, g \rangle$, for the inner product between two functions. Also, note that like the inner product between two vectors, the inner product between functions is a scalar, a number.

Sometimes, an inner product will be defined with respect to a weighting or density function, $\rho(x)$ and we have the following:

$$\langle f, g \rangle = \int f(x) g(x) \rho(x) dx$$

Eq. 11

So far, we have been assuming that the functions, f and g, are real functions. The more general definition of the inner product is as follows:

$$\langle f, g \rangle = \int f(x) g^*(x) dx$$

Eq. 12

Here, $g^*(x)$ is the complex conjugate of $g(x)$. When f and g are real functions, the definition in Eq. 12 is the same as that in Eq. 10. As with vectors, the inner product, $\langle f, g \rangle$, gives us information about the projection of f onto g.

When we were decomposing vectors using unit-length basis vectors, we hit a potential problem if the basis vectors we used, such as v_1, v_2, v_3, and so on, were not orthogonal to each other. Orthogonal basis vectors are like coordinate directions that are at right angles to each other, such as the x, y, or z directions in a conventional 3D plot. Non-orthogonal vectors are like the x-axis and a direction that is at 45 degrees to the x-axis. Orthogonal basis directions or vectors have the advantage that movement along one basis direction does not affect where we are in the other directions – we can operate on one basis vector in isolation and independently from the other basis vectors. This means that the inner product, $a^\top v_1$, told us how much of the vector, a, was made up of the unit-length basis vector, v_1, and likewise, $a^\top v_2$ told us how much of the unit-length basis vector, v_2, was in a.

Is there an equivalent concept of orthogonal functions? There is. Recall that two vectors, a and b, are orthogonal to each other if their inner product, $a^\top b$, is zero. So, we can say two functions, f and g, are orthogonal to each other if $\langle f, g \rangle = 0$. How does knowing this help us? Just like when we decomposed a vector using a set of orthonormal basis vectors, we will choose our functions, $h_1(x), h_2(x), h_3(x)$, and so on, to be orthonormal. This means they satisfy the following properties:

$$\langle h_i, h_j \rangle = 1 \ \text{ if } i = j \quad \text{and} \quad \langle h_i, h_j \rangle = 0 \ \text{ if } i \neq j$$

Eq. 13

We'll postpone our exploration of how to come up with such a set of functions for a moment and look at what we can do with them.

If we take our function defined in Eq. 8 and use basis functions that satisfy the orthonormal conditions in Eq. 13, then when we take the inner product of both the left and right-hand sides of Eq. 8 with $h_i(x)$, we get the following:

$$\langle f, h_i \rangle = \left\langle \sum_{k=1}^{K} \alpha_k h_k, h_i \right\rangle = \sum_{k=1}^{K} \alpha_k \langle h_k, h_i \rangle = \alpha_i$$

Eq. 14

The last step in Eq. 14 follows directly from the properties of the functions in Eq. 13. The result in Eq. 14 tells us that when decomposing our function, $f(x)$, into multiple orthonormal components, $h_1(x), h_2(x), h_3(x)$, and so on, the expansion coefficient, α_i, is given precisely by the inner product, $\langle f, h_i \rangle$.

A set of orthonormal functions that satisfy the relations in Eq. 13 is called an **orthonormal basis** because they provide a basis for decomposing any function into the form in Eq. 8. The functions, $h_i(x)$, are called the basis functions. If the basis functions also allow us to construct any function within a particular space, such as in $\mathbb{R}^d$, then those basis functions are said to be **complete** because they provide us the means to construct all functions in the space.

How do we come up with a set of functions that satisfy the conditions in Eq. 13? The first thing to say is that if we have a set of orthogonal functions, $g_1(x), g_2(x), g_3(x)$, and so on, we can always make them orthonormal. If $g_1(x), g_2(x), g_3(x)$ are orthogonal, it means they satisfy the following conditions:

$$\langle g_i, g_j \rangle \neq 0 \text{ if } i = j \quad \text{and} \quad \langle g_i, g_j \rangle = 0 \text{ if } i \neq j$$

Eq. 15

We can rescale our functions, $g_i(x)$, and define the following:

$$h_i(x) = \frac{g_i(x)}{\sqrt{\langle g_i, g_i \rangle}}$$

Eq. 16

By doing this, we create an orthonormal set of functions from an orthogonal set. It is easy to confirm this: plug the functions, $h_i(x)$, defined by Eq. 16 into Eq. 13. Try it yourself as an exercise.

This means that to construct an orthonormal set of functions, $h_1(x), h_2(x), h_3(x)$, we only need to come up with an orthogonal set of functions, $g_1(x), g_2(x), g_3(x)$. Where do we get those from? Again, we can use the vectors we studied in *Chapter 3* for inspiration. There, we found that the eigenvectors of a square matrix provided us with a set of orthogonal vectors. A matrix represents a linear transformation that is applied to vectors. Could there be an equivalent for functions? In other words, do we have **eigenfunctions** of a linear operator, A, that is applied to functions? Again, as ever, the answer is yes. An eigenfunction, $\phi(x)$, of a linear operator, A, is a function that satisfies the following condition:

$$A\phi(\underline{x}) = \lambda\phi(\underline{x})$$

Eq. 17

The quantity, λ, is the eigenvalue corresponding to the eigenfunction, $\phi(\underline{x})$. It turns out that for many linear operators, A, their eigenfunctions are orthogonal and complete for the space on which A operates. We won't go into the details of which linear operators have orthogonal and complete eigenfunctions as that is beyond the scope of this book. However, if A does have a complete orthogonal set of eigenfunctions, it means that all functions on which A operates can be written in the form of Eq. 8. In the next section, we will encounter a well-known and complete set of basis functions, and we will highlight how they are eigenfunctions of a particular linear operator. For now, let's look at the benefits of using eigenfunctions for our basis functions.

What happens if we apply our linear operator, A, to our decomposition in Eq. 8? We get the following:

$$Af(\underline{x}) = A\sum_{k=1}^{K}\alpha_k h_k(\underline{x}) = \sum_{k=1}^{K}\alpha_k A h_k(\underline{x}) = \sum_{k=1}^{K}\alpha_k \lambda_k h_k(\underline{x})$$

Eq. 18

In Eq. 18, λ_k is the eigenvalue associated with the kth eigenfunction, $h_k(\underline{x})$. Eq. 18 shows us that when we apply A to our function, $f(\underline{x})$, we get back a decomposition of the form of Eq. 8, but just with modified coefficients, $\alpha_k \rightarrow \lambda_k \alpha_k$. This makes it incredibly easy to understand what the impact of applying A to $f(\underline{x})$ will be. It also makes it incredibly easy to calculate the result of applying A to $f(\underline{x})$ as we will already know the expansion coefficients, α_k, and the eigenvalues, λ_k, in advance.

More importantly, Eq. 18 highlights why we have chosen the eigenfunctions of A as our basis for decomposing $f(\underline{x})$. If we know we are going to be doing lots of calculations where we apply a linear operator, A, to various functions, it makes sense to use the eigenfunctions of A as our decomposition basis. If we are going to be doing lots of calculations using a different linear operator, A', we should use the eigenfunctions of A' for our decompositions. This emphasizes the point we made earlier that we often choose our decomposition functions, $h_1(\underline{x})$, $h_2(\underline{x})$, $h_3(\underline{x})$, based on what we want to do with (or to) our function, $f(\underline{x})$.

With that, we've learned the fundamentals of decomposing any function using a set of complete orthonormal basis functions. This provides us with a good place to end this section. So, let's recap what we've learned.

What we've learned

In this section, we learned the following:

- How to calculate the inner product between two functions
- How the inner product between functions can be used to calculate how much of $g(\underline{x})$ is in $f(\underline{x})$

- How to use an orthonormal basis as a convenient set of component functions to decompose a function, $f(\underline{x})$

- The eigenfunctions of a linear operator and how they can naturally provide a complete orthonormal basis for decomposing other functions, $f(\underline{x})$

- How using the eigenfunctions of a linear operator as a decomposition basis makes it easy to understand (and calculate) the effect of that linear operator on a function

Having finished that introduction to using a complete orthonormal basis for function decomposition, over the next three sections, we will learn about the most used complete orthonormal basis, the Fourier functions. We'll start by learning about Fourier series and how to decompose periodic functions.

Fourier series

Fourier series were developed by the French mathematician and scientist Joseph Fourier. Fourier wanted to understand how heat diffuses through materials and doing so required solving what is known as the **Heat equation**. Fourier discovered that solving the Heat equation was made easier if he expressed solutions as a sum of sine and cosine waves. In other words, Fourier decomposed the solution to make solving the equation easier. You will not be surprised to hear, given what we learned in the previous section, that sine and cosine waves are eigenfunctions of one of the operators in the Heat equation.

Since Fourier's original work, using sine and cosine waves to decompose functions has been a standard tool in the toolkit of mathematicians and scientists. As data scientists, we shall learn about their usefulness as well.

Since we are using sine and cosine waves to decompose our function, $f(x)$, it won't surprise you to learn that we use them to decompose periodic functions. An example periodic function is shown in the left-hand plot of Figure 9.2. Since it is periodic, it extends infinitely in both directions along the x-axis, so we have only shown a small snapshot of it in the left-hand plot. The right-hand plot of Figure 9.2 shows one period's worth of the periodic function shown in the left-hand plot. The shape shown in the right-hand plot of Figure 9.2 is repeated to create the shape in the left-hand plot:

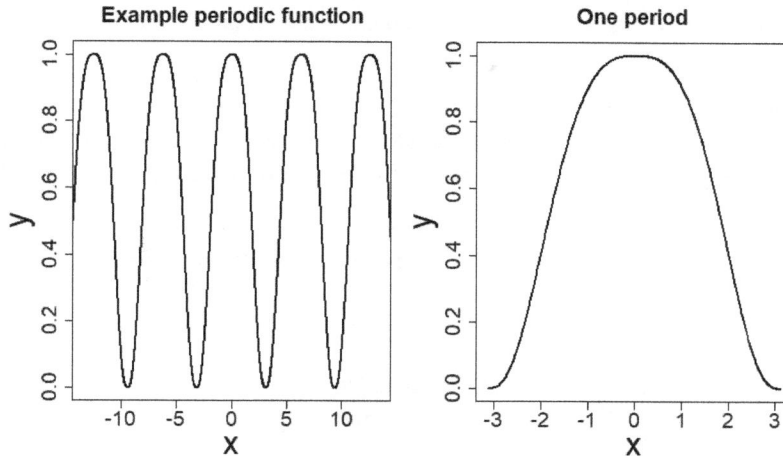

Figure 9.2: Example periodic function

We created the periodic function in Figure 9.2 so that it has a period of 2π. So, we will decompose it using sine and cosine waves that have a period of 2π. These are basis functions of the form $\sin(mx)$ and $\cos(mx)$ with m as an integer, so $m = 0, 1, 2, 3, 4, 5\ldots$ This means we will write our function in Figure 9.2 as follows:

$$f(x) = \sum_{m=0}^{\infty} a_m \cos(mx) + \sum_{m=0}^{\infty} b_m \sin(mx)$$

Eq. 19

You might be wondering why the sums in Eq. 19 don't extend to include negative values of m. This is because $\cos(-mx) = \cos(mx)$, so we can write the following:

$$a_{-m}\cos(-mx) + a_m \cos(mx) = (a_{-m} + a_m)\cos(mx)$$

Eq. 20

This is of the same form as the terms in Eq. 19 but with a modified coefficient. In other words, we only need the cosine terms corresponding to the terms $m = 0, 1, 2, 3, \ldots$ Similarly, since $\sin(-mx) = -\sin(mx)$, we only need the sine terms corresponding to $m = 0, 1, 2, 3, \ldots$

We can simplify the decomposition in Eq. 19 a bit more. Since $\sin(0) = 0$ and $\cos(0) = 1$, we can separate the $m = 0$ terms in Eq. 19 to give us the following:

$$f(x) = \frac{1}{2}a_0 + \sum_{m=1}^{\infty} a_m \cos(mx) + \sum_{m=1}^{\infty} b_m \sin(mx)$$

Eq. 21

The factor of a half in front of the coefficient, a_0, will make sense when we learn how to calculate the values of the coefficients, a_m, b_m.

For now, it is important to emphasize that the coefficients, a_m, b_m, are the equivalent of the coefficients, α_k, in Eq. 8 and that the functions, $\sin(mx)$ and $\cos(mx)$, are the equivalent of the basis functions, $h_k(x)$. It is relatively easy to confirm that our sine and cosine waves are indeed orthogonal since we have the following formula:

$$\int_{-\pi}^{\pi} \sin(mx)\sin(nx)dx = \pi\delta_{mn} \qquad \int_{-\pi}^{\pi} \cos(mx)\cos(nx)\, dx = \pi\delta_{mn}$$

Eq. 22

The δ_{mn} symbol is the **Kronecker delta function** and is 1 if $m = n$ and 0 if $m \neq n$. Sine and cosine waves are also orthogonal to each other, so we get the following:

$$\int_{-\pi}^{\pi} \sin(mx)\cos(nx)\, dx = 0 \quad \text{for all } m, n$$

Eq. 23

This means we can use $\sin(mx)$ and $\cos(mx)$ to decompose any function that is periodic on the interval, $[-\pi, \pi]$. Excellent – it looks like we can do a lot with these sine and cosine waves.

But what if our periodic function is not a repeat of a pattern on $[-\pi, \pi]$, but is a repeat of a function on the interval, $[-L, L]$? Easy – we can just rescale our x variable to create a new variable, $\tilde{x}$, given by the following formula:

$$\tilde{x} = \frac{x}{L}\pi$$

Eq. 24

As a function of $\tilde{x}$, our periodic function is a repeat of a pattern on $[-\pi, \pi]$ and we can use Eq. 21 to decompose it. Overall, this means that once we know the value of the half-period, L, we can decompose any periodic function, $f(x)$, using the following equation:

$$f(x) = \frac{1}{2}a_0 + \sum_{m=1}^{\infty} a_m \cos\left(m\left(\frac{\pi x}{L}\right)\right) + \sum_{m=1}^{\infty} b_m \sin\left(m\left(\frac{\pi x}{L}\right)\right)$$

Eq. 25

The various coefficients, a_m and b_m, are called **Fourier coefficients**, while the overall function on the right-hand side of Eq. 25 is called a **Fourier series** because it expresses the function, $f(x)$, as a series of sine and cosine waves. The value, $m\pi/L$, is called the **wavenumber** because it is a number that describes the main property of the sine or cosine wave: its frequency. Since π/L is just a fixed constant, you will also see the value, m, on its own referred to as the wavenumber as well.

We'll now turn our attention to how we calculate the coefficients, a_m and b_m. We already know how to do this: we can use the expression in Eq. 14 with normalized versions of our orthogonal basis functions, $\sin(mx)$ and $\cos(mx)$. Applying Eq. 14 gives us the following:

$$a_0 = \frac{1}{L}\int_{-L}^{L} f(x)\,dx \qquad a_m = \frac{1}{L}\int_{-L}^{L} f(x)\cos\left(\frac{m\pi x}{L}\right)dx \quad \text{for } m = 1,2,\dots$$

$$b_m = \frac{1}{L}\int_{-L}^{L} f(x)\sin\left(\frac{m\pi x}{L}\right)dx \quad \text{for } m = 1,2,\dots$$

Eq. 26

With the factor of a half in front of a_0 in Eq. 25, the formula for calculating a_0 in Eq. 26 is of a similar form as that for calculating $a_1, a_2, \dots$ This is the reason for including the factor of a half in front of a_0 in the first place.

Let's look at a very simple example of calculating a Fourier series:

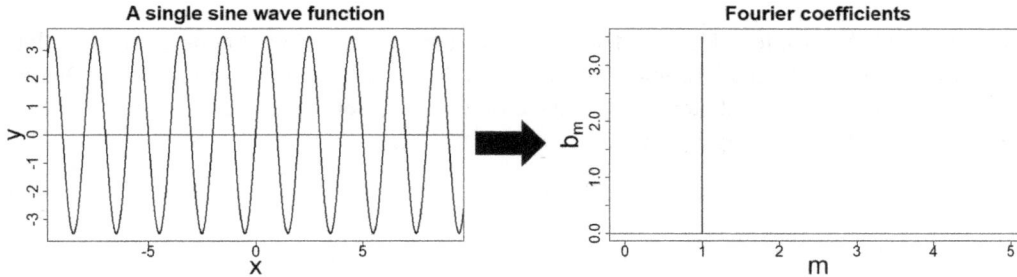

Figure 9.3: A simple Fourier series example

The left-hand plot in Figure 9.3 shows a simple function. It is a single sine wave of period 2 (that is, $L = 1$), and amplitude 3.5. The right-hand plot of Figure 9.3 shows all the Fourier coefficients, b_1, b_2, b_3, that result from the function in the left-hand plot. These are the coefficient values that result when we apply the formulae in Eq. 26 to the function, $f(x)$, shown in the left-hand plot. There is only one non-zero coefficient, which is b_1, and it has a value of 3.5. The Fourier coefficients, a_0, a_1, a_2, are all zero. This is because the function in the left-hand plot of Figure 9.3 can be described by a single sine wave. The function is just $f(x) = 3.5 \times \sin(\pi x)$, so we only need the one coefficient.

As we build up more complex functions, we need more non-zero Fourier coefficients. Let's look at a more complex function:

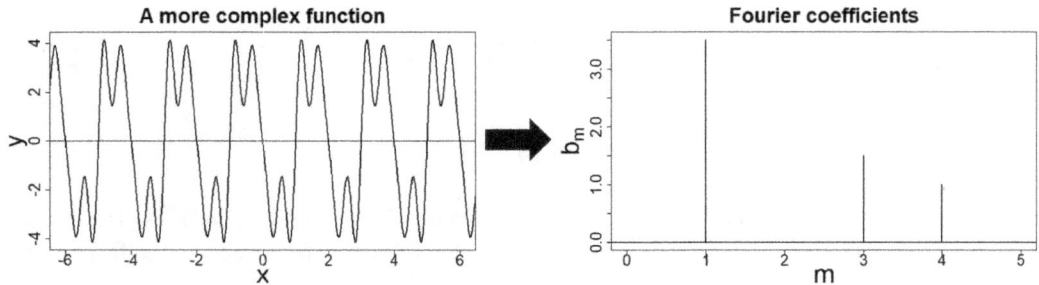

Figure 9.4: A more complex function and its Fourier coefficients

The plot on the left-hand side of Figure 9.4 shows a more complex function. It appears to have a lot more variation (wiggles) than the simple function shown in the left-hand plot of Figure 9.3. But in fact, the function in the left-hand plot of Figure 9.4 is only made up of three separate sine waves. The function is given by the following equation:

$$f(x) = 3.5 \times \sin(\pi(x - 3)) + 1.5 \times \sin(3\pi(x + 7)) + \sin(4\pi x)$$

Eq. 27

The Fourier coefficients for the function in Eq. 27 are shown in the right-hand plot of Figure 9.4. All the coefficients, a_0, a_1, a_2, are zero because the function in Eq. 27 has no cosine waves in it. It is made up of just three sine waves. So, we only have three non-zero coefficients. These are b_1, b_3 and b_4. Their values are plotted in the right-hand side of Figure 9.4. We can also see how they are given by the formula in Eq. 27, which defines our more complex function – we have a sine wave of period 2 and amplitude 3.5, so b_1 = 3.5; we also have a sine wave of period 2/3 and amplitude 1.5, so b_3 = 1.5 ; and finally, we have a sine wave of period 2/4 and amplitude 1.0, so b_4 = 1.0.

We get all that complex variation shown in the left-hand plot of Figure 9.4 from just three sine waves. This highlights how useful Fourier series are. We can understand what looks like quite complex functions very easily by breaking them down into their constituent sine and cosine waves. It also highlights that the more non-zero Fourier coefficients we have, the more complex the variation in the function, and the higher the index, m, on the non-zero Fourier coefficients, the higher the frequency of the variation that the function will display.

Having illustrated the power of Fourier series with those two simple examples, we'll wrap up this section by reviewing what we have learned.

What we've learned

In this section, we learned the following:

- A periodic function can be decomposed into a sum of sine and cosine waves. This is the Fourier series decomposition of the periodic function.

- How to calculate the amplitudes in a Fourier series decomposition of a periodic function.

- How to reconstruct a periodic function from its Fourier series amplitudes.

- Even relatively complex functions can be represented with a simple Fourier series consisting of just a few terms.

Having learned about Fourier series and decomposing periodic functions in terms of sine and cosine waves, in the next section, we'll extend this concept to using sine and cosine waves to decompose any function by learning about Fourier transforms.

Fourier transforms

In the previous section, we learned about Fourier series, and how they allow us to decompose, or equivalently build up, periodic functions in terms of simple building blocks – sine and cosine waves.

But what if our function is not periodic? If a function is not periodic, there is no finite value of L over which the function repeats itself, so effectively, L is infinite. What happens if we make L go to infinity in our Fourier series in Eq. 25? The sums in Eq. 25 become integrals in the limit, $L \to \infty$, and our Fourier series becomes as follows:

$$f(x) = \frac{1}{\sqrt{2\pi}} \int_{-\infty}^{\infty} a(m)\cos(mx)\,dm \; + \frac{1}{\sqrt{2\pi}} \int_{-\infty}^{\infty} b(m)\sin(mx)\,dm$$

Eq. 28

We are still representing our function, $f(x)$, as a superposition of sine and cosine waves, but there are a couple of comments we need to make about Eq. 28:

- Note how the integration variable, m, is playing the equivalent role to the summation index in Eq. 25.

- We have reverted to including sine and cosine waves with values of m between $-\infty$ and ∞, while in Eq. 25, we absorbed the cosine wave amplitudes from $-m$ and m into a single Fourier coefficient, a_m, and likewise for the Fourier coefficient, b_m.

- We have also re-absorbed the Fourier coefficient, a_0, back into the integral, while in Eq. 25, we explicitly separated it from the sum.

- We have what looks like a strange pre-factor of $1/\sqrt{2\pi}$ in front of each integral. We'll learn about why this is so in a moment.

The equivalent formulae to those in Eq. 26 also become integrals when we go to the limit, $L \to \infty$. The new formulae for the coefficients, $a(m)$ and $b(m)$, are as follows:

$$a(m) = \frac{1}{\sqrt{2\pi}} \int_{-\infty}^{\infty} f(x)\cos(mx)dx \qquad b(m) = \frac{1}{\sqrt{2\pi}} \int_{-\infty}^{\infty} f(x)\sin(mx)dx$$

Eq. 29

The formulae in Eq. 29 also have a pre-factor of $1/\sqrt{2\pi}$. They look like symmetric counterparts to the integrals in Eq. 28. This symmetry is the reason for using the pre-factor of $1/\sqrt{2\pi}$.

The previous comments were about how we have re-written the sums in Eq. 25 and Eq. 26 in a way to make them more cosmetically appealing when we go to the limit, $L \to \infty$, and get integrals. Doing so allows us to spot something. Using our knowledge of complex numbers, we know the following:

$$e^{imx} = \cos(mx) + i\sin(mx)$$

Eq. 30

This is Euler's formula. Here, i is the imaginary number that represents the square root of -1. That is $i^2 = -1$. You may know Euler's formula already, but don't worry if you don't – we are only going to use it in its simple form. So, let's define a coefficient:

$$\hat{f}(m) = a(m) - ib(m)$$

Eq. 31

By plugging the expressions for $a(m)$ and $b(m)$ in Eq. 29 into $\hat{f}(m)$ in Eq. 31, we get the following:

$$\hat{f}(m) = \frac{1}{\sqrt{2\pi}} \int_{-\infty}^{\infty} f(x)[\cos(mx) - i\sin(mx)]dx$$
$$= \frac{1}{\sqrt{2\pi}} \int_{-\infty}^{\infty} f(x)\, e^{-imx}dx$$

Eq. 32

From the fact that the sine and cosine functions are orthogonal to each other, we are also able to write the Fourier series in Eq. 28 in the following form:

$$f(x) = \frac{1}{\sqrt{2\pi}} \int_{-\infty}^{\infty} \hat{f}(m)\, e^{imx}\, dm$$

Eq. 33

Eq. 32 and Eq. 33 have a pleasing symmetry to them. They are also the equivalent of Eq. 25 and Eq. 26 but for aperiodic functions. Note that m is also a continuous variable, so our coefficient, $\hat{f}(m)$, is a function and Eq. 32 tells us how to transform the function, $f(x)$, into $\hat{f}(m)$. Because of this, $\hat{f}(m)$ is called the **Fourier transform** of $f(x)$. The function, $\hat{f}(m)$, tells us the amount of sine wave, $\sin(mx)$, and cosine wave, $\cos(mx)$, that are in $f(x)$. Once we know this amplitude function, $\hat{f}(m)$, we can use Eq. 33 to reconstruct $f(x)$. Eq. 33 tells us how to go from $\hat{f}(m)$ back to $f(x)$. Consequently, we refer to Eq. 33 as the **inverse Fourier transform**. The pleasing symmetry between Eq. 32 and Eq. 33 also tells us that $f(x)$ is the Fourier transform of $\hat{f}(-m)$, so we can think of $f(x)$ and $\hat{f}(m)$ as being a pair of related Fourier transforms.

What does the amplitude function, $\hat{f}(m)$, look like for a typical function, $f(x)$? Well, the first thing to point out is that because of its definition in Eq. 31, $\hat{f}(m)$ can have an imaginary part. It doesn't have to. The first function we're going to look at has an entirely real Fourier transform. The function we'll look at is the Gaussian:

$$f(x) = \frac{1}{\sqrt{2\pi}} e^{-\frac{1}{2}x^2}$$

Eq. 34

For the Gaussian function in Eq. 34, its Fourier transform is calculated by applying Eq. 32 to Eq. 34. This gives us the following:

$$\hat{f}(m) = \frac{1}{2\pi} \int_{-\infty}^{\infty} e^{-\frac{1}{2}x^2} e^{-imx}\, dx = \frac{1}{\sqrt{2\pi}} e^{-\frac{1}{2}m^2}$$

Eq. 35

The right-hand side of Eq. 35 is also the form of a Gaussian function. So, a Gaussian transforms into a Gaussian. This is shown in Figure 9.5:

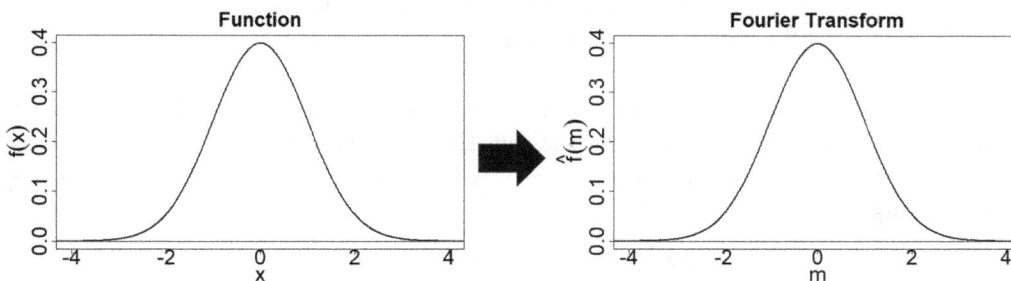

Figure 9.5: Plots of a Gaussian function (left) and its Fourier transform (right)

Let's look at two more Fourier transform pairs. First, let's consider a complex function that is simply a combination of a single sine wave and a single cosine wave:

$$f(x) = \cos(3x) + i\sin(3x)$$

Eq. 36

From the form of Euler's formula in Eq. 30, we can also write $f(x)$ in Eq. 36 as $f(x) = e^{i3x}$, so we can recognize that we only need a single value of m in $\hat{f}(m)$ to reconstruct $f(x)$. This means that the Fourier transform of $f(x)$ in Eq. 36 is a Dirac delta function, $\sqrt{2\pi}\,\delta(m-3)$. Writing the Fourier transform process schematically, we have the following:

$$\cos(3x) + i\sin(3x) \Rightarrow \sqrt{2\pi} \times \delta(m-3)$$

Eq. 37

We'll also illustrate this by plotting the real and imaginary parts of both $f(x)$ and $\hat{f}(m)$ in separate plots. This is shown in Figure 9.6:

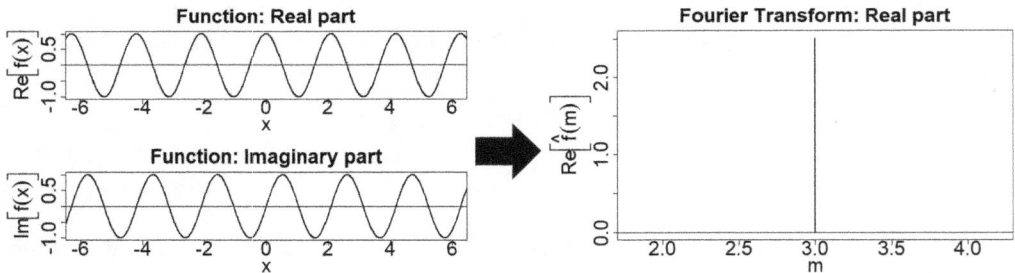

Figure 9.6: A simple function that is a sum of a sine wave and a cosine wave (left) and its Fourier transform (right)

Before we comment on this, let's reflect on the fact that the functions on the left-hand and right-hand side of the arrow in Eq. 37 are a Fourier transform pair. This also tells us that if we transform a Dirac delta function, $f(x) = \delta(x-3)$, we must get a sine wave and cosine wave out. In other words, we have the following:

$$\sqrt{2\pi}\,\delta(x-3) \Rightarrow \cos(3m) + i\sin(3m)$$

Eq. 38

This is also illustrated graphically in Figure 9.7:

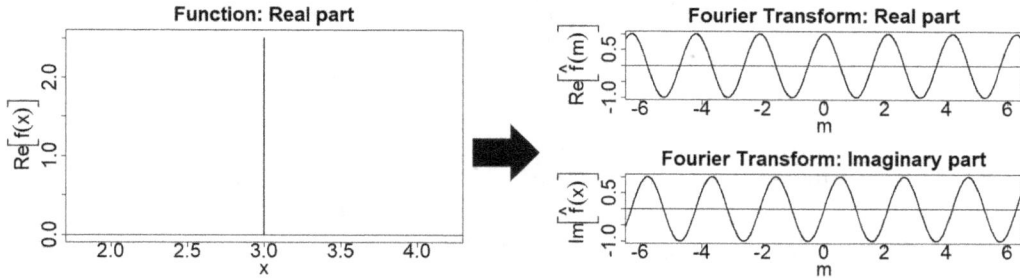

Figure 9.7: A Dirac delta function (left) and its Fourier transform (right)

What this example illustrates is that when we have a function, $f(x)$, that is of a very broad extent, its Fourier transform, $\hat{f}(m)$, is of a very narrow extent. And vice versa – when we have a function, $f(x)$, that is of a very narrow extent, its Fourier transform, $\hat{f}(m)$, is of a very broad extent. The Gaussian example we started with was an exception. It is a function, $f(x)$, whose extent is identical to that of its Fourier transform, $\hat{f}(m)$. The Gaussian function is exceptional in many ways!

A final word on notation. We used $\hat{f}(m)$ to represent the Fourier transform of $f(x)$. You will often see the q symbol used instead of m. Secondly, since calculating the Fourier transform is a process – that is, an operation – you will often see it denoted as some sort of operator, such as using the $\mathscr{F}[f]$ symbol. This says we are applying the process of taking the Fourier transform to the function, f.

The multi-dimensional Fourier transform

So far, when we discussed Fourier series and Fourier transforms, we talked about decompositions of one-dimensional functions. What happens if the function we want to decompose is not one-dimensional? No problem! We can write a d-dimensional function, $f(\underline{x})$, like so:

$$f(\underline{x}) = f(x_1, x_2, \ldots, x_d)$$

Eq. 39

If we fix the values of $x_2, x_3, \ldots, x_d$, we effectively have a one-dimensional function that is a function of x_1. We can calculate the Fourier transform of this one-dimensional function. If we do so, we get the transform, $\hat{f}(m_1)$, which is calculated as follows:

$$\hat{f}(m_1) = \frac{1}{\sqrt{2\pi}} \int_{-\infty}^{\infty} f(x_1, x_2, \ldots, x_d) \, e^{-im_1 x_1} \, dx_1$$

Eq. 40

Once we have calculated the Fourier transform, $\hat{f}(m_1)$, we can use it to reconstruct the function, $f(x_1, x_2, \ldots, x_d)$, using the inverse Fourier transform, as follows:

$$f(x_1, x_2, \ldots, x_d) = \frac{1}{\sqrt{2\pi}} \int_{-\infty}^{\infty} \hat{f}(m_1) e^{im_1 x_1} dm_1$$

Eq. 41

Ok, great, but what progress have we made? If we look at the transform function, $\hat{f}(m_1)$, in Eq. 40 more closely, we'll realize it is also a function of the remaining variables, $x_2, x_3, \ldots, x_d$. If we change the values of any of $x_2, x_3, \ldots, x_d$, we get a different one-dimensional function and hence a different transform, $\hat{f}(m_1)$. This means we can write this transform function as follows:

$$\hat{f}(m_1, x_2, x_3, \ldots, x_d)$$

Eq. 42

We can now consider the function in Eq. 42 as a function of $m_1, x_2, x_3, \ldots, x_d$ and we can repeat our trick of fixing several of the variables. In this case, we'll fix the values of $m_1, x_3, \ldots, x_d$ to give us effectively a one-dimensional function of x_2. We can calculate the Fourier transform of this one-dimensional function to get a transform function, $\hat{f}(m_2)$. Its transform function, $\hat{f}(m_2)$, will be a function of $m_1, m_2, x_3, x_4, \ldots, x_d$ and is calculated as follows:

$$\hat{f}(m_1, m_2, x_3, \ldots, x_d) = \frac{1}{\sqrt{2\pi}} \int_{-\infty}^{\infty} \hat{f}(m_1, x_2, x_3, \ldots, x_d) e^{-im_2 x_2} dx_2$$

Eq. 43

Using Eq. 40, we can re-write the right-hand side of Eq. 43 as follows:

$$\hat{f}(m_1, m_2, x_3, \ldots, x_d) = \frac{1}{2\pi} \int_{-\infty}^{\infty} \int_{-\infty}^{\infty} f(x_1, x_2, x_3, \ldots, x_d) e^{-im_1 x_1} e^{-im_2 x_2} dx_1 dx_2$$

Eq. 44

From the right-hand side of Eq. 44, you can see what we're doing. We're building up a multi-dimensional Fourier transform as a series of one-dimensional Fourier transforms, one variable at a time. If we continue the process, we end up with a transform function, $\hat{f}(m_1, m_2, \ldots, m_d)$, which is calculated as follows:

$$\hat{f}(m_1, m_2, \ldots, m_d) = \frac{1}{(2\pi)^{d/2}} \int_{-\infty}^{\infty} \int_{-\infty}^{\infty} \cdots \int_{-\infty}^{\infty} f(x_1, x_2, \ldots, x_d) e^{-im_1 x_1} e^{-im_2 x_2} \ldots e^{-im_d x_d} dx_1 dx_2 \ldots dx_d$$

Eq. 45

Let's say we collect all the angular wavenumbers, $m_1, m_2, \ldots, m_d$, into a vector, $\underline{m} = (m_1, m_2, \ldots, m_d)$, called the (angular) wave vector, and recognize that we can use the inner product notation to write the following:

$$m_1 x_1 + m_2 x_2 + \ldots + m_d x_d = \underline{m}^\top \underline{x}$$

Eq. 46

In this case, we can write the Fourier transform function, $\hat{f}(m_1, m_2, \ldots, m_d)$, as $\hat{f}(\underline{m})$ and calculate it as follows:

$$\hat{f}(\underline{m}) = \frac{1}{(2\pi)^{d/2}} \int f(\underline{x}) \, e^{-i\underline{m}^T \underline{x}} \, d\underline{x}$$

Eq. 47

The integral in Eq. 47 is over the entire domain of the function, $f(\underline{x})$, which is $\mathbb{R}^d$. From the transform function, $\hat{f}(\underline{m})$, we can reconstruct the original function, $f(\underline{x})$, via the inverse multi-dimensional Fourier transform, which is calculated using the following formula:

$$f(\underline{x}) = \frac{1}{(2\pi)^{d/2}} \int \hat{f}(\underline{m}) \, e^{i\underline{m}^T \underline{x}} \, d\underline{m}$$

Eq. 48

Again, the integral in Eq. 48 is over $\mathbb{R}^d$. Eq. 47 and Eq. 48 form a multi-dimensional Fourier transform pair. Unsurprisingly, they are very similar in structure to the one-dimensional Fourier transform pair of Eq. 32 and Eq. 33. Also, Eq. 47 and Eq. 48 show that we can decompose multi-dimensional functions into multi-dimensional sine and cosine waves. You will also not be surprised to learn that there is a multi-dimensional equivalent of the Fourier series – that is, we can calculate multi-dimensional coefficients, $a(\underline{m})$ and $b(\underline{m})$, that allow us to express functions that are periodic in multiple dimensions as sums of multi-dimensional sine and cosine waves. For brevity, we won't go further into multi-dimensional Fourier series here, but the generalization of Eq. 26 to the multi-dimensional situation is straightforward.

This subsection on multi-dimensional Fourier transforms brings this section to a close. Despite this being a short section, we have learned a lot, so let's summarize our findings.

What we've learned

In this section, we have learned about the following:

- Almost any function can be decomposed into a continuous superposition of sine and cosine waves. This is the Fourier transform of a function.

- How to calculate the amplitudes in the Fourier transform of a function.

- How to reconstruct a function from its Fourier transform. This is the inverse Fourier transform.

- The multi-dimensional Fourier transform and how we can view it as a series of one-dimensional Fourier transforms, one applied to each dimension individually.

Having learned about Fourier series and then Fourier transforms, both of which allow us to decompose functions, in the next section, we'll come back to one of the main aspects of data science – data. We'll learn about the **discrete Fourier transform** (DFT), which can be used to decompose discrete sequences of data points in terms of sine and cosine waves.

The discrete Fourier transform

Having learned about Fourier series and Fourier transforms, it seems we have strayed a bit from the path of data science. Where are the data aspects of all of this? Fourier series and Fourier transforms are about decomposing functions, not data. Yes, it is useful to be able to decompose and reconstruct a function, but what happens if we don't have the exact mathematical equation of our function and only have data points taken from the function? Is there a Fourier-like decomposition that works with data, not mathematical expressions? This is where the DFT comes in.

Often, we only have observations from a function, say the value of the function, $f(x)$, taken at regularly spaced intervals. In this case, we would have the N observations taken at x values, $x_0, x_1, \ldots, x_{N-1}$, with corresponding function values, $f_0, f_1, \ldots, f_{N-1}$. Without loss of generality, we can assume the spacing between the x values is 1 as we can simply rescale the X-axis if not. This means our x values are $0, 1, 2, \ldots, N - 1$. We'll use f_n to denote the function value observed at $x = x_n = n$. The values of n are $n = 0, 1, 2, \ldots, N - 1$.

We can use one of our usual tricks and represent this set of observations as a single function using a set of Dirac delta functions. Let's call that function $\tilde{f}(x)$. So, $\tilde{f}(x)$ is given by the following formula:

$$\tilde{f}(x) = \sum_{n=0}^{N-1} f_n \, \delta(x - n)$$

Eq. 49

Now that we have a single function that represents our data, we could just apply the Fourier transform formula to calculate $\mathscr{F}\left[\tilde{f}\right](m)$. However, there is a subtlety to deal with. We have only N data points, so we only need, at most, N Fourier amplitudes to represent the data series. It turns out that when we take the Fourier transform of the function in Eq. 49, we only need to consider values of the wavenumber, m, that are of the form $m = \frac{2\pi k}{N}$, $k = 0, 1, \ldots, N - 1$. The number, k, is just a re-scaled wavenumber. Having defined k in this way, we define the DFT as follows:

$$F_k = \sum_{n=0}^{N-1} f_n e^{-i\frac{2\pi}{N}kn}$$

Eq. 50

The value, F_k, in Eq. 50 is an amplitude. It converts the entire data series, $f_0, f_1, \ldots, f_{N-1}$, into a single number, F_k. We compute this amplitude, F_k, for each value of $k = 0, 1, \ldots, N - 1$. Just like the Fourier transform, the DFT amplitude, F_k, tells us how much of a particular sine and cosine wave is present in our original series. This is the insight we are looking to extract from our data. And just like the Fourier transform, there is an analogous formula to Eq. 50 that tells us how we can reconstruct the original data series, $f_0, f_1, \ldots, f_{N-1}$, once we know all the amplitudes, $F_0, F_1, \ldots, F_{N-1}$. This is the inverse DFT formula, and it is given by the following equation:

$$f_n = \frac{1}{N} \sum_{k=0}^{N-1} F_k e^{i\frac{2\pi}{N}kn}$$

Eq. 51

Eq. 50 and Eq. 51 are the DFT equivalent of the Fourier transform pair of equations, Eq. 32 and Eq. 33. Like the Fourier transform equations, the DFT equations have a pleasing symmetry to them. And like the Fourier transform equations, the DFT equations define a pair – in this case, a pair of amplitude sets, $\{f_n\}_{n=0}^{N-1}$ and $\{F_k\}_{k=0}^{N-1}$. In contrast, the Fourier transform equations define a pair of functions, $f(x)$ and $\tilde{f}(m)$.

One final point to illustrate about the DFT before we move on to a code example is what happens when we use Eq. 50 but with wavenumber $N - k$. Let's find out. If we replace k with $N - k$ in Eq. 50, we get the following:

$$F_{N-k} = \sum_{n=0}^{N-1} f_n e^{-i\frac{2\pi}{N}(N-k)n} = \sum_{n=0}^{N-1} f_n e^{-i2\pi n} e^{i\frac{2\pi}{N}kn} = \sum_{n=0}^{N-1} f_n e^{i\frac{2\pi}{N}kn}$$

Eq. 52

In deriving Eq. 52, we have made use of the fact that $e^{-i2\pi n} = 1$ for any integer, n. The very right-hand side of Eq. 52 tells us that for any real sequence, f_n, its DFT has the property $F_{N-k} = F_k^*$. More generally, the DFT amplitude at wavenumber $N - k$ is related to the DFT amplitude at wavenumber k. Consequently, even if our sequence, f_n, is a set of observations from a single sine wave, its DFT will have two amplitudes that have a non-zero value! Let's see this in action by looking at a code example.

DFT code example

We'll use the data plotted on the left-hand side of Figure 9.4. The data we will use can be found in the `discrete_fourier_transform_data.csv` file in the `Data` folder of this book's GitHub repository. The data consists of $N = 200$ values sampled from Eq. 27 between $x = -10$ to $x = 9.9$ at intervals of $\Delta x = 0.1$. The following code example can be found in the `Code_Examples_Chap9.ipynb` Jupyter Notebook in this book's GitHub repository. First, we'll read in the data:

```
import numpy as np
import pandas as pd
import matplotlib.pyplot as plt
df_dft = pd.read_csv('../Data/discrete_fourier_transform_data.csv')
```

Next, we'll look at the data:

```
df_dft.head()
```

This gives us the following output:

```
        X           Y
0    -10.0    1.010775e-14
1     -9.9   -1.344028e+00
2     -9.8   -2.896048e+00
3     -9.7   -3.882870e+00
4     -9.6   -3.398076e+00
```

Here, we have two columns – one for the *X* values and one for the function values that are held in the *Y* values. Now, let's plot the data:

```
# Plot of the data to which we will apply the DFT
plt.plot(df_dft.x, df_dft.y, 'black')
plt.title('Sampled data', fontsize=24)
plt.xlabel('x', fontsize=20)
plt.ylabel('y', fontsize=20)
plt.xticks(fontsize=12)
plt.yticks(fontsize=12)
plt.show()
```

Here's the plot:

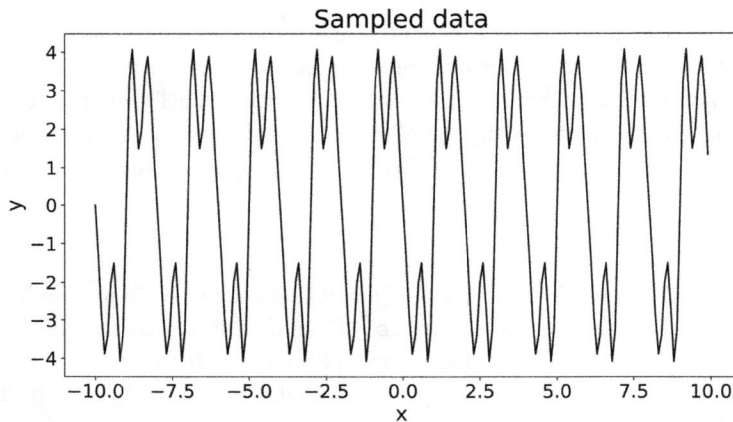

Figure 9.8: Plot of data sampled from Eq. 27

To calculate the DFT, we must pass a NumPy array into the NumPy FFT function, which we can call using `numpy.fft.fft`:

```
dft_y = np.fft.fft(df_dft['y'].to_numpy())
```

We'll calculate the absolute values (modulus) of the DFT values to get a real-valued measure of how much of each wavenumber is present in our data series. Then, we'll plot those amplitudes. Because there are only three waves present in our sample data, we expect to see only six spikes in the amplitude plot, symmetrically distributed around $N/2 = 100$:

```
# Take modulus of amplitudes
dft_amplitudes = np.absolute(dft_y)
# Plot the modulus of the amplitudes
plt.stem(dft_amplitudes)
```

```
plt.title('DFT amplitudes', fontsize=24)
plt.xlabel('Wave number index', fontsize=20)
plt.ylabel('Modulus of DFT amplitude', fontsize=20)
plt.xticks(fontsize=12)
plt.yticks(fontsize=12)
plt.show()
```

Here's the plot:

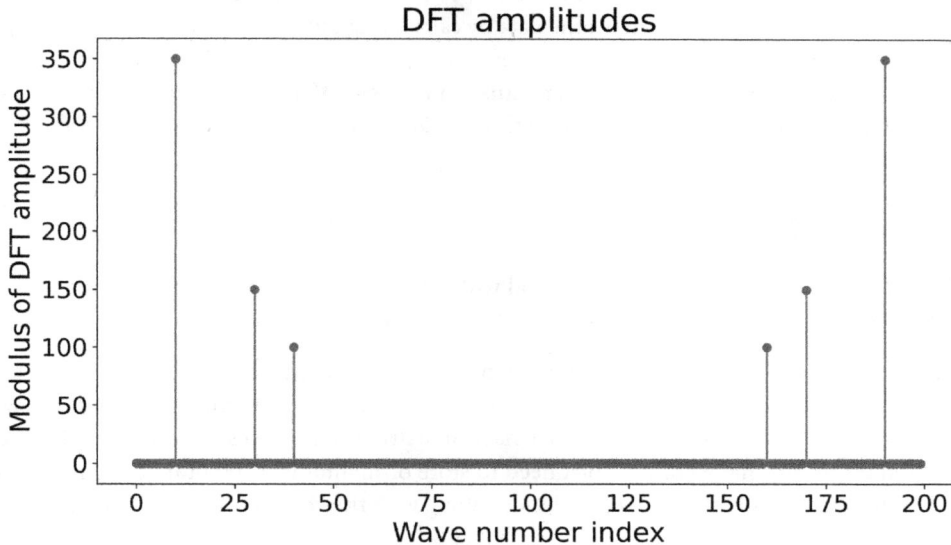

Figure 9.9: Plot of the absolute (modulus) value of DFT amplitudes

There are indeed six spikes in the DFT amplitude plot in Figure 9.9, symmetrically placed around 100 on the x-axis. Consequently, we'll only need to look at the first three spikes, which are at index values of $k = 10, 30, 40$. Since the data series spans the interval from $x = -10$ to $x = 9.9$ – that is, a gap of 19.9 in $N = 200$ observations – the spikes correspond to the period of lengths given by the following formula:

$$L = 2 \times \frac{N-1}{k \times 19.9} = 2 \times \frac{199}{k \times 19.9}$$

Eq. 53

Thus, the first three spikes correspond to the following periods:

$$2 \times \frac{199}{199} = 2 \quad \text{and} \quad 2 \times \frac{199}{30 \times 19.9} = \frac{2}{3} \quad \text{and} \quad 2 \times \frac{199}{40 \times 19.9} = \frac{2}{4}$$

Eq. 54

These periods agree with our starting equation for y in Eq. 27. Also, notice that the relative ratios of the amplitudes of the first three spikes in Figure 9.9 agree with the relative ratios of the amplitudes of the three sine waves in Eq. 27, namely 3.5:1.5:1.0.

You might be wondering, if we only needed the first three spikes in Figure 9.9 to understand what was going on with the DFT in the code example, then why did we have six spikes in the plot? Why not just plot three spikes only? Part of the reason is that in our code example, we were working with a real data series, so from Eq. 52, we know that the DFT amplitudes obey the relationship $F_{N-k} = F_k^*$. We can easily get the amplitude for any value of $k > N/2$ just from the amplitude at k. This means that for real data series, we can reconstruct the data by calculating only the amplitudes, F_k, for integer values of k from $k = 0$ up to $k = N/2$ if N is even, and up to $k = \frac{N}{2} - 1$ if N is odd. Doing this is known as calculating the **real discrete Fourier transform** or **real DFT**. Again, many packages, such as NumPy, will provide functions for calculating the real DFT of a real data series and for computing the inverse real DFT.

Uses of the DFT

The DFT looks like it could be fun to play around with, but is it useful to us as data scientists? Why should I go to all that effort to compute the DFT of a data series?

The preceding code example illustrates one of the main practical benefits of the DFT and one that is of particular use to us as data scientists. In the code example, we represented the whole of a 200 data point series with just three amplitudes. We have an extremely compressed – that is, efficient – representation of the original data. We don't need to store or manipulate 200 numbers. Instead, we only need to store and manipulate three amplitudes along with the corresponding wavenumbers, k.

Okay, I hear you say. But the reason we can do this is that the original data series was generated from three sine waves. It's hardly surprising that we can represent it with three sine waves. Real data is unlikely to be generated from just a small number of sine waves. That is true, and the amplitude plot of the DFT of a real data series will have many more non-zero amplitudes.

However, the DFT approach to representing a data series still gives us a means of representing that data series efficiently. The high wavenumber components of a DFT correspond to high-frequency sine and cosine waves. Usually, the variation in a data series that corresponds to a genuine signal is slowly varying, coming from slowly varying external influences. Extremely high-frequency components in a data series are usually considered to be a sign of noise. This is not a mathematical proof, just a general rule of thumb.

This means we can use the DFT of a data series to identify and remove the noise in the data series. By setting the amplitude of the high-frequency components to zero, we remove their contribution to the data series. It also means we have reduced the number of non-zero amplitudes. By keeping only the low wavenumber amplitudes, we have an efficient, but slightly lossy, representation of the original data. In **digital signal processing** (**DSP**), this is called a **low-pass filter** because only the low frequencies have been allowed to pass unaltered. Part of the art and science of DSP is choosing the threshold on the wavenumber at which we filter out the amplitudes.

At the start of this chapter, we decomposed functions because we wanted to operate on them efficiently, such as with some linear operator, A. Similarly, here, we have decomposed a data series using the DFT because it gave us a useful representation in which the noise components were clear and could then be filtered out efficiently. If we think of the DFT as a way of efficiently representing any sequence of values or coefficients, not necessarily just a data series, then unsurprisingly, other numerical operations can be performed more efficiently using the DFT. These include the following:

- Multiplication of high-order polynomials

- Multiplication of large integers

- Image processing using the multi-dimensional DFT

- Approximate solution of partial differential equations

What is the difference between the DFT, Fourier series, and the Fourier transform?

A final word on the DFT to wrap up this section. Here, we will emphasize the differences between Fourier series, the Fourier transform, and the DFT.

When looking at the DFT code example, you might have wondered why we sampled data points from the range $x = -10$ to $x = 9.9$. One reason was that with a sampling interval of $\Delta x = 0.1$, we get a total of $N = 200$ data points, which is a convenient number. If we had sampled between $x = -10$ to $x = 10$, we would have ended up with $N = 201$ data points. We would have also ended up with a different amplitude plot to the one in Figure 9.9 – we would have a lot more than just six wavenumbers with non-zero amplitudes. The new amplitude plot would look similar to that in Figure 9.9 but the differences would be noticeable. If you don't believe me, give it a go – I have set it as an exercise at the end of this chapter. At this point, you may be wondering how this can be. How can going from $N = 200$ to $N = 201$ sampled data points lead to a noticeable difference in the DFT? After all, it is the same function we are taking the DFT of. And that is the key difference between Fourier series and Fourier transforms on the one hand, and the DFT on the other. Fourier series and Fourier transforms are used to decompose functions, while the DFT is used to decompose data series. We can summarize this as follows:

- A Fourier series is used to decompose a periodic function

- A Fourier transform is used to decompose a non-periodic function

- A DFT is used to decompose a data series consisting of regularly spaced data points

The reason why we see a difference in the DFT amplitude plot when we go from a sample of $N = 200$ data points to a sample of $N = 201$ data points is that they are different data series. Yes, they are sampled from the sample underlying function – the one in Eq. 27 – but they are two different data series.

This discussion on the differences between Fourier series, Fourier transforms, and the DFT brings us neatly to the end of this section, as well as this chapter. Next, we'll summarize what we've learned about the DFT before summarizing this chapter.

What we've learned

In this section, we learned the following:

- A regularly spaced series of data points can be represented as a sum of sine and cosine waves. This is the DFT of the data series.

- How to calculate the amplitudes in a DFT of a data series.

- How to reconstruct a data series from its DFT. This is the inverse DFT.

- The differences between Fourier series, Fourier transforms, and the DFT.

Summary

This chapter has been another chapter on a specific mathematical technique. There have been a lot of formulae, particularly when it comes to learning about the various forms of Fourier decompositions. However, the formulae are not the key point here. Formulae can always be looked up. It is more important to understand and remember the concepts. Of these, the most important concepts we have covered are as follows:

- Why we decompose a function into a set of simpler building block functions.

- How a function can be decomposed using a set of basis functions, and how we can use the inner product between functions to calculate the coefficients or amplitudes in a decomposition.

- How to use an orthonormal basis as a convenient set of component functions to decompose a function, $f(x)$.

- The eigenfunctions of a linear operator and how they can naturally provide a complete orthonormal basis for decomposing other functions, $f(x)$.

- That a periodic function can be decomposed into a sum of sine and cosine waves, as a Fourier series.

- Almost any function can be decomposed into a continuous superposition of sine and cosine waves. This is the Fourier transform of a function.

- The multi-dimensional Fourier transform and how we can view it as a series of one-dimensional Fourier transforms, one applied to each dimension individually.

- A regularly spaced series of data points can be represented as a sum of sine and cosine waves. This is the DFT of the data series.

- How to calculate the amplitudes in a DFT of a data series.

- The differences between Fourier series, Fourier transforms, and the DFT.

Our next chapter is another self-contained topic and is on networks. There, we will learn what networks are and about data from networks.

Exercises

The following is an exercise to test your understanding of the material covered in this chapter. The answer to this exercise can be found in the `Answers_to_Exercises_Chap9.ipynb` Jupyter Notebook in this book's GitHub repository:

- Re-run the code example at the end of the section on the DFT. Now, repeat this exercise but use the data in the `discrete_fourier_transform_amended_data.csv` file in the `Data` folder of this book's GitHub repository. Look at the differences between the two data series and the differences in the amplitudes from their respective DFTs. What do you notice?

10

Network Analysis

This chapter is about networks and datasets represented by networks. Networks link things together. Since many things in real-world data science are linked to each other, you will encounter networks and network data a lot as a data scientist. Therefore, as a data scientist, you must learn something about networks and how to analyze them. To learn about networks, we will cover the following topics:

- *Graphs and network data*: In this section, we'll learn why network data is important for data science and what a graph is

- *Basic characteristics of graphs*: Here, we'll learn the essential concepts and terminology relating to graphs, and in particular about adjacency matrices

- *Different types of graphs*: In this section, we'll learn about some of the main classes of graphs you will encounter as a data scientist and the behavior and properties of those different classes of graphs

- *Community detection and decomposing graphs*: Finally, we'll learn about breaking a graph down into its important sub-graphs

Technical requirements

All the code examples provided in this chapter can be found in this book's GitHub repository: https://github.com/PacktPublishing/15-Math-Concepts-Every-Data-Scientist-Should-Know/tree/main/Chapter10. To run the Jupyter Notebooks provided, you will need a full Python installation, including the following packages:

- numpy (>=1.24.3)

- matplotlib (>=3.7.2)

- NetworkX (>=3.1.0)

Graphs and network data

In the introduction, we mentioned that much of the real-world data you will encounter as a data scientist is network data. However, not all real-world data is network data. So, how do we recognize when we are dealing with network data, and perhaps more importantly, how do we recognize when the network aspect of the data is relevant to how we analyze the data?

Network data is about relationships

In the introduction, we explained that we need to learn about network data because the things that produce the data are linked to each other. This tells us that network data is about relationships. Or rather, network data arises when we have relationships between many of the data-generating entities we are studying. This also gives us a useful rule-of-thumb for when we should take the network aspect of the data into account in our analysis:

- If the relationships between the entities we are studying are strong, then we can't ignore the network aspect of the data

- Conversely, if the relationships between the entities we are studying are all weak, then we can probably ignore the network aspect of the data

It is important to realize that when relationships are weak, we still have a network and network data. But by ignoring the network structure in our analysis, we are making an approximation. Analyzing the data on this basis will be easier and will still yield valuable insights, so the approximation of ignoring the network aspect of the data is a useful one. However, we would get more accurate analysis and insights if we did consider the network aspect of the data, but this would be at the expense of having to use or implement more complex analyses and algorithms.

So, how do we decide whether the relationships are weak enough to ignore? This is a skill that's acquired through experience. To help you with this, we'll look at a couple of real-world examples where the presence of relationships is central to the data to be analyzed. We'll touch on these examples throughout this chapter.

Example 1 – substituting goods in a supermarket

We'll start with a commerce example – goods sold in a supermarket. When a shopper buys food in a supermarket, they have many choices. Let's consider the example of a shopper wanting to buy a frozen pizza. In many supermarkets, the shopper will have the choice between a couple of different brands and several different flavors. *Figure 10.1* shows a hypothetical but realistic example of the set of brands and flavors that a shopper might choose between when buying a frozen pizza:

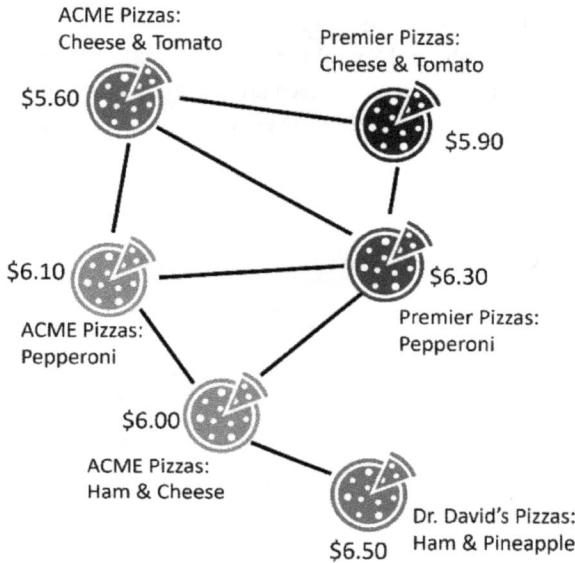

Figure 10.1: Frozen pizza brands and flavors

When choosing a pizza, shoppers are happy to swap between certain brands and certain flavors if the price is right. The solid line between two pizzas in Figure 10.1 indicates that a shopper considers those pizzas interchangeable or substitutes for each other. This means that there is a relationship between those two pizzas, that of substitutability.

From the supermarket's perspective, it is crucial to understand which pairs of pizzas are substitutes for each other. The more substitutes a shopper considers there to be, the more options they have if one of the pizzas is priced too high, and so the more price-sensitive the shopper is when choosing between the substitutes. This means that pizzas with a lot of substitutes must be priced very competitively. In contrast, the ham and pineapple pizza in *Figure 10.1* has only one substitute – the ham and cheese pizza – and so doesn't have to be priced as competitively as the pepperoni and cheese and tomato pizzas. From the supermarket's perspective, knowing this substitutability network structure determines their pricing strategy and ultimately their profitability.

Example 2 – international trade

If we want to understand economics at an international level, we need to understand the imports and exports from each country as trade between countries contributes significantly to the balance sheet of each country. This means we cannot ignore trading relationships between countries when analyzing the GDP of each country.

Figure 10.2 shows the cash value (in US dollars) of the total exports from five different European countries (UK, Germany, France, Italy, and Spain) for 2022. The export figures were obtained from `https://comtradeplus.un.org/`. Each arc represents the total export from one country to the other. The color of the arc matches the exporting country. The thickness of the arc is proportional to the total export value:

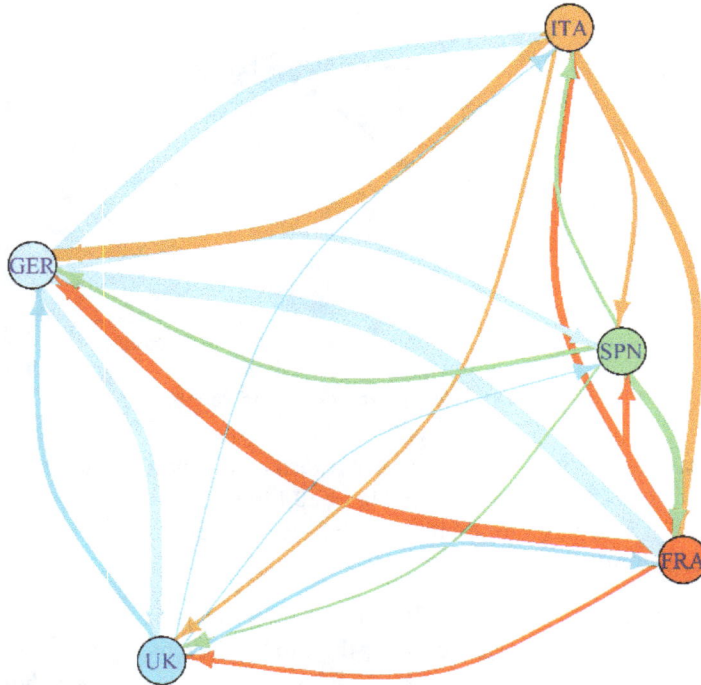

Figure 10.2: The network of exports between Germany (GER), France (FRA), Italy (ITA), Spain (SPN), and the United Kingdom (UK) in 2022

The network structure of international trade is clear from this visualization. The visualization also makes it immediately clear which countries export the most, and which have the strongest reciprocal trading relationships.

Now that we have introduced two examples of networks in real-world data science, let's look at what they have in common and the terminology we can use to describe those common features. This brings us to the mathematical concept of a **graph**.

What is a graph?

In both the real-world examples provided, we visualized the network as a 2D object, consisting of a set of entities with connections or arcs between them. This is because this is a very natural and intuitive way of visualizing a network. Because of this, these 2D objects are considered mathematical objects in their own right and have their own branch of mathematics that studies them. In mathematics, a 2D object such as that shown in *Figure 10.1* or *Figure 10.2* is called a **graph**. The mathematical study of graphs is called **graph theory**.

Consequently, the terms **network** and **graph** are used interchangeably. Likewise, network analysis and graph theory are used interchangeably, although network analysis tends to be preferentially used when we are dealing with real-world networks or data generated by real-world networks, while graph theory tends to be preferentially used when we are dealing with the more abstract analysis of graph structures and their properties.

What is central to graph theory is that a graph consists of a set of entities called **nodes** and a set of connections between nodes called **edges**. In graph theory, a node tends to be called a **vertex**, so the terminology nodes and vertices are interchangeable. Consequently, a graph, G, consists of a set of vertices, V, and a set of edges, E. In math, we denote this as $G = (V, E)$, meaning the graph, G, has vertices, V, and edges, E. This is illustrated schematically by the generic graph in Figure 10.3:

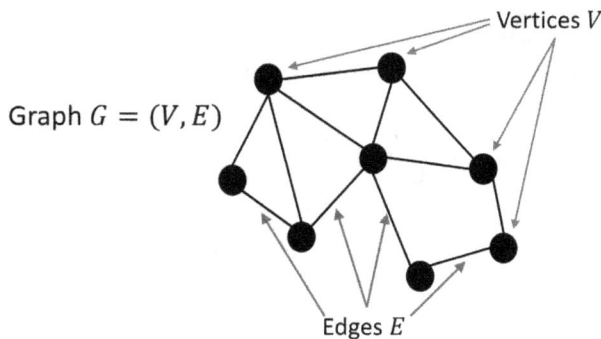

Figure 10.3: Schematic of a generic graph, G.

In our first real-world example, the nodes or vertices are the different pizzas, and the edges are the solid lines indicating a substitutability relationship between two types of pizza. In our second real-world example, the nodes or vertices were the European countries, and the edges were the trading relationships between them.

In a graph, both the nodes and edges can have attributes – that is, additional data or values associated with them. Typically, in a data science context, it is these attributes that we are interested in, and we study how the network structure or **topology** affects them. In our first real-world example, it was the price of the pizzas that we were interested in, and this was an attribute of the pizzas – that is, a node attribute. In our second real-world example, it was the total exports from one country to another that we were interested in, and this was an edge attribute. In general, an edge attribute is termed an **edge weight**, as it is typically used to quantify the strength or weight of the relationship between the entities represented by the two nodes connected by the edge.

From our real-world examples, we can see that although graphs consist of just two simple things, nodes and edges, graphs are far from being simple mathematical objects. There are a lot of different quantities and characteristics of a graph, G, that we can calculate. In the next section, we will begin to outline some of the more commonly calculated graph characteristics and introduce some of the more commonly used concepts and terminology in graph theory. However, this is also a good place to end this introductory section on network data and what networks are, so we'll recap what we have covered.

What we've learned

In this section, we learned the following:

- Networks imply relationships and we use networks to represent data where relationships are important. This is usually where those relationships have a significant effect on the data associated with the entities between which the relationships exist.

- A network is also a graph. A graph consists of nodes (vertices) and edges. The nodes and edges can have attributes or weights associated with them.

Having introduced the basic idea of what a network is and how one can be used to represent relationship data, let's learn how to mathematically characterize the properties of a graph.

Basic characteristics of graphs

A graph, G, has nodes (vertices), V, and edges, E, but to uncover and analyze the rich structure present in a network, we'll need to introduce additional concepts and terminology beyond just nodes and edges. Let's start with those edges, which come in two flavors.

Undirected and directed edges

A key difference between our two real-world examples was that in our pizza example, the presence of an edge denoted a substitutability relationship that applied in both directions. An ACME Pizzas cheese and tomato pizza is considered by shoppers to be a substitute for a Premier Pizzas cheese and tomato pizza, and vice versa. In contrast, the arcs in our trade network in Figure 10.2 have a direction associated with them, indicated by the arrow at the end of each arc. The exports from the UK to Germany do not match the exports from Germany to the UK. In Figure 10.2, we represented this by having two directed edges between the UK and Germany nodes, with different edge weights. Schematically, we used the arc thickness to represent the edge weight, so you can see the asymmetry in exports between the UK and Germany in Figure 10.2.

More generally, we refer to an edge as being **directed** if it has a specific direction associated with it. If an edge does not have a specific direction associated with it, we say it is an **undirected** edge. Edges that represent flows of some quantity are directed, while an edge that represents a symmetric relationship is undirected. Figure 10.4 shows the different types of edges we can get between two nodes:

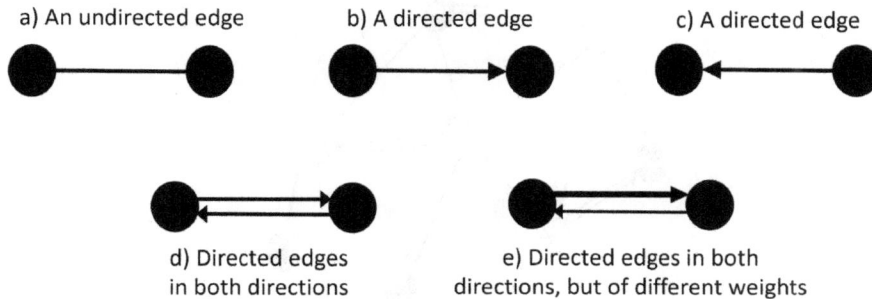

a) An undirected edge   b) A directed edge   c) A directed edge

d) Directed edges
in both directions

e) Directed edges in both
directions, but of different weights

Figure 10.4: A schematic of the different types of edges

By comparing examples *a* and *d* in Figure 10.4, we can see that two directed edges of equal weight between two nodes are not equivalent to an undirected edge between those two nodes. If we want to represent two flows of equal strength between the two nodes, we should use directed edges, while if we want to denote the presence of some symmetric logical property between the nodes, we should use an undirected edge.

Having introduced the idea of a directed edge, we can now introduce the idea that a node can have a relationship with itself. We can do this by using a directed edge of the form shown in Figure 10.5:

Figure 10.5: A schematic of the directed edge between a node and itself

A directed edge between a node and itself is ideal when we want to represent a dynamic flow of some quantity out from the node, but some of that quantity may be retained by the node. For example, let's say we were modeling flows over time of internet users between different pages of a website. From one time point to the next, a user may navigate to a new page within a website or remain on the current page. Here, nodes would represent pages within the website and directed edges would represent the flows. The fact that in a single timestep, a proportion of users don't move pages – that is, they effectively flow from the page to itself – means we would represent that using an edge of the form in Figure 10.5.

Once we have the concept of a directed edge, we naturally have the concept of a **directed network** or **directed graph**. This is a graph consisting of directed edges. An example directed graph is shown by the schematic in Figure 10.6. Likewise, a graph made up of undirected edges is called an **undirected graph** or an **undirected network**:

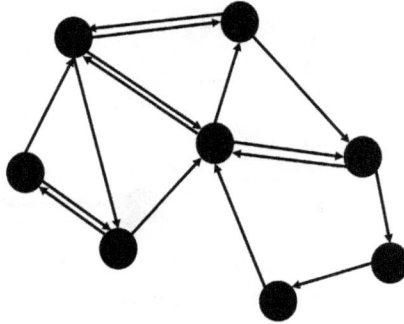

Figure 10.6: A simple directed graph

Now that we have introduced the different types of edges, let's look at how to mathematically encode a graph and how that encoding can change according to the different edge types in the graph.

The adjacency matrix

So far, we have been focusing on the visual representation of a graph. This is good, but what if we want to do some mathematics on a graph? How can we represent a graph mathematically? We introduced the notation that a graph is $G = (V, E)$, but that is a bit abstract. What if want to work with actual numbers? How do we encode the presence of a relationship or edge mathematically? We do this using the idea of an **adjacency matrix**. For a graph, G, with N nodes, its adjacency matrix, $\underline{A}$, is an $N \times N$ matrix whose elements are defined as follows:

$$A_{ij} = \begin{cases} 1 \text{ if an edge exists between node } i \text{ and node } j \\ 0 \text{ otherwise} \end{cases}$$

Eq. 1

The adjacency matrix encodes with 1s and 0s, whether two nodes are connected or not. If two nodes are connected, we say they are neighbors or adjacent to each other, hence the name adjacency matrix. Another way of looking at this is that $A_{ij} = 1$ if we can get from node i to node j in one step, and if we can they are nearest neighbors.

Properties of the adjacency matrix

Figure 10.7 shows a simple undirected graph where we have numbered the nodes, along with their corresponding adjacency matrix:

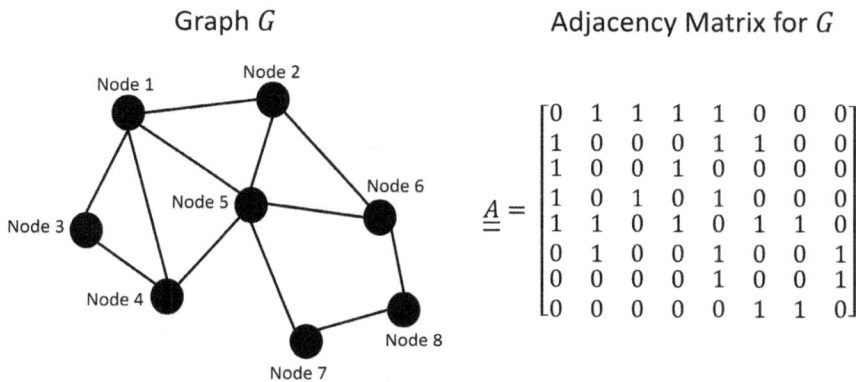

$$\underline{\underline{A}} = \begin{bmatrix} 0 & 1 & 1 & 1 & 1 & 0 & 0 & 0 \\ 1 & 0 & 0 & 0 & 1 & 1 & 0 & 0 \\ 1 & 0 & 0 & 1 & 0 & 0 & 0 & 0 \\ 1 & 0 & 1 & 0 & 1 & 0 & 0 & 0 \\ 1 & 1 & 0 & 1 & 0 & 1 & 1 & 0 \\ 0 & 1 & 0 & 0 & 1 & 0 & 0 & 1 \\ 0 & 0 & 0 & 0 & 1 & 0 & 0 & 1 \\ 0 & 0 & 0 & 0 & 0 & 1 & 1 & 0 \end{bmatrix}$$

Figure 10.7: An example undirected graph and its corresponding adjacency matrix

Because the adjacency matrix encodes the connections of a graph, it has some very nice properties and allows us to do some neat calculations. We'll look at two of those properties now:

- **Symmetry**: For an undirected graph such as that in Figure 10.7, the adjacency matrix is symmetric. We can easily see this from the fact that if $A_{ij} = 1$, then there is an edge between nodes i and j, so node i is also a neighbor of node j, which implies that $A_{ji} = 1$. A similar argument shows that if $A_{ij} = 0$, then i and j are not neighbors, so $A_{ji} = 0$. So, overall, for an undirected graph, we always have $A_{ij} = A_{ji}$, meaning the matrix is symmetric.

- **Next-nearest-neighbors**: The adjacency matrix allows us to identify next-nearest-neighbors and next-next-nearest-neighbors and so on. Consider the product of matrix elements, $A_{ik}A_{kj}$. This number is 1 only if $A_{ik} = 1$ and $A_{kj} = 1$, so only if there is a connection between node i and node k and also a connection between node k and node j. So, $A_{ik}A_{kj} = 1$ if we can get from node i to node j in two steps via node k, and 0 otherwise. Now, let's consider all possible intermediate nodes, k. If we compute $\sum_k A_{ik}A_{kj}$, it will simply count a 1 every time there is a two-step path between node i and node j. So, $\sum_k A_{ik}A_{kj}$ gives the number of possible two-step routes between i and j. But wait a minute, $\sum_k A_{ik}A_{kj}$ looks like something we've seen before! It is just matrix multiplication. In *Chapter 3*, I mentioned that matrices would crop up everywhere in data science. Here, $\underline{\underline{A}}^2 = \underline{\underline{A}} \times \underline{\underline{A}}$ is a matrix whose matrix elements tell us the number of two-step paths between each pair of nodes in the network. If the matrix element, $(\underline{\underline{A}}^2)_{ij}$, is non-zero, it means that nodes i and j are connected by a two-step path and so are next-nearest-neighbors. Unsurprisingly, we can continue this logic and we find that the matrix, $\underline{\underline{A}}^3$, has matrix elements that count the number of three-step paths between each pair of nodes and can be used to easily identify next-next-nearest-neighbors. More generally, the matrix, $\underline{\underline{A}}^n$, counts the number of n-step paths between each pair of nodes.

The adjacency matrix for a directed graph

We have used undirected graphs to introduce the idea of the adjacency matrix, but can the concept be extended to directed graphs? The answer is yes. Again, the definition is simple:

$$A_{ij} = \begin{cases} 1 \text{ if a directed edge exists going from node } i \text{ to node } j \\ 0 \text{ otherwise} \end{cases}$$

Eq. 2

Figure 10.8 shows an example directed graph and its corresponding adjacency matrix:

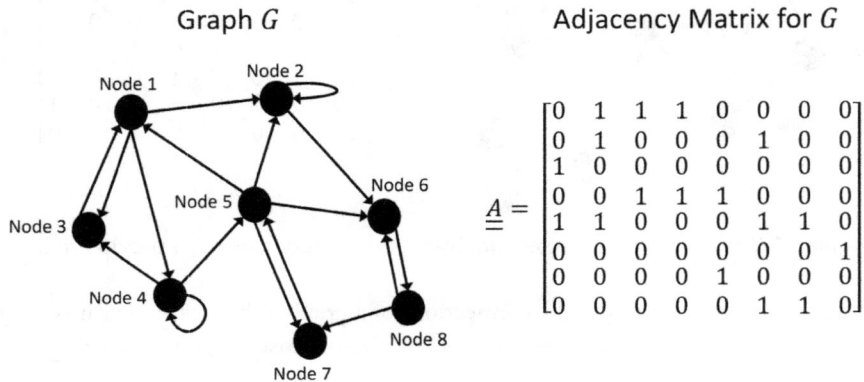

Graph G Adjacency Matrix for G

$$\underline{\underline{A}} = \begin{bmatrix} 0 & 1 & 1 & 1 & 0 & 0 & 0 & 0 \\ 0 & 1 & 0 & 0 & 0 & 1 & 0 & 0 \\ 1 & 0 & 0 & 0 & 0 & 0 & 0 & 0 \\ 0 & 0 & 1 & 1 & 1 & 0 & 0 & 0 \\ 1 & 1 & 0 & 0 & 0 & 1 & 1 & 0 \\ 0 & 0 & 0 & 0 & 0 & 0 & 0 & 1 \\ 0 & 0 & 0 & 0 & 1 & 0 & 0 & 0 \\ 0 & 0 & 0 & 0 & 0 & 1 & 1 & 0 \end{bmatrix}$$

Figure 10.8: An example directed graph and its corresponding adjacency matrix

From Figure 10.8, we can see that a directed graph does not necessarily have a symmetric adjacency matrix. It will only be symmetric if, between a pair of nodes, there are edges in both directions or no edges at all.

We can also see that we now have the possibility of non-zero matrix elements on the diagonal of the adjacency matrix. These represent edges from a node to itself. In the example in Figure 10.8, we have included an edge from node 2 to itself and likewise an edge from node 4 to itself.

As with the undirected case, we can use the adjacency matrix of a directed graph to identify next-nearest-neighbors and so on. The matrix elements of the matrix, $\underline{\underline{A}}^n$, still correspond to counts of the number of n-step paths between each pair of nodes; it's just that when following an n-step path, we are only allowed to move in the direction given by the edge. Consequently, as with the undirected case, a non-zero value in the matrix element, $(\underline{\underline{A}}^n)_{ij}$, means that there is at least one n-step path between nodes i and j.

Adjacency matrices for weighted directed graphs

Again, it is natural to ask if we can take the edge weights into account when defining the adjacency matrix. Since the adjacency matrix captures the neighbor structure of the graph, can we not include the edge weight as a measure of "neighborliness?" Again, the answer is yes by extending the definition of the matrix elements to the following:

$$A_{ij} = \begin{cases} \text{Weight } w_{ij} \text{ if a directed edge exists going from node } i \text{ to node } j \\ 0 \text{ otherwise} \end{cases}$$

Eq. 3

Since we only have a non-zero edge weight, w_{ij}, if there is an edge going from node i to node j, the matrix of edge weights, $\underline{w}$, is the adjacency matrix in this general definition. So, in general, $\underline{A} = \underline{w}$. This means we can also think of the absence of an edge as simply an edge with zero weight.

Figure 10.9 shows an example of a weighted directed graph and its corresponding adjacency matrix:

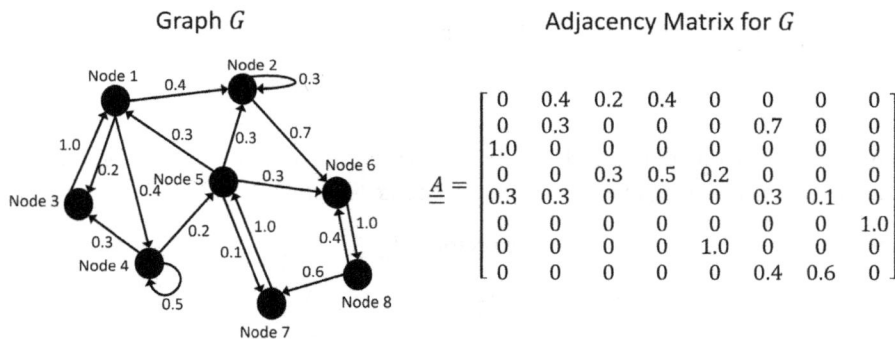

Figure 10.9: An example weighted directed graph and its corresponding adjacency matrix

The most obvious difference between Figure 10.9 and the previous examples in Figure 10.7 and Figure 10.8 is that the matrix elements are no longer just 1s and 0s. This means that the matrix, $\underline{A}^n$, no longer counts the number of n-step paths between nodes. In fact, if the edge weights are allowed to be negative, the interpretation of $\underline{A}^n$ is not always clear. However, if the edge weights are restricted to being positive or zero, then a non-zero value for the matrix element, $(\underline{A}^n)_{ij}$, still indicates that there is at least one n-step path between nodes i and j.

The different forms of the adjacency matrix that we have illustrated here show how useful it is since it encodes almost everything about a graph. Consequently, almost all calculations involving graphs make use of the adjacency matrix. We'll look at more characteristics of graphs and in some cases show how those characteristics can be calculated from the adjacency matrix.

In-degree and out-degree

We can also see from Figure 10.8 that taking edge direction into account when defining the adjacency matrix can change the matrix markedly. Other than the addition of an edge from node 2 to itself and likewise for node 4, the topology of the graph in Figure 10.8 is the same as that for the undirected graph in Figure 10.7. However, some nodes in Figure 10.8 have few edges. Take node 6, for example. The directed graph in Figure 10.8 only has one edge, while the undirected graph in Figure 10.7, which has the same topology, shows that node 6 has three edges. This naturally introduces the idea that we should distinguish between counts of edges that are incoming to a node and counts of edges that are outgoing from a node.

The number of edges coming into a node is known as the node's **in-degree**, while the number of edges leaving a node is known as the node's **out-degree**. How do we calculate the in-degree and out-degree? Using the adjacency matrix, of course. If we have an adjacency matrix of the form shown in Figure 10.8, consisting of just 1s and 0s, then the in-degree for the ith node is just the sum of the matrix elements in the ith column of the adjacency matrix. In math form, this looks as follows:

$$\text{In-degree for node } i = \sum_j A_{ji}$$

Eq. 4

This also means that if we put all the in-degree values for all the nodes, $i = 1, 2, \ldots, N$, into a vector, we can calculate that vector using matrix multiplication like like so:

$$\text{In-degree vector} = \mathbf{1}_N^{\mathsf{T}} \underline{A}$$

Eq. 5

In Eq. 5, $\mathbf{1}_N^{\mathsf{T}}$ means an N-element row-vector consisting of all 1s. A similar bit of logic shows that the out-degree of node i is just the sum of all the matrix elements in the ith row of the adjacency matrix. In math terms, we have the following:

$$\text{Out-degree for node } i = \sum_j A_{ij}$$

Eq. 6

So, the vector of out-degree values can calculated as a matrix multiplication via the following equation:

$$\text{Out-degree vector} = \underline{A}\, \mathbf{1}_N$$

Eq. 7

Centrality

The in-degree and out-degree are node-level characteristics. Adding the in-degree and out-degree values together gives us the total number of edges connected to a node – its total degree. A node with a high total number of edges is connected to a lot of other nodes and so is likely to be at the heart of the network. This means that the total degree value of a node gives a measure of how important or how central it is to the network. We say that the total degree value is a **centrality measure**.

As you can imagine, there are many ways in which you can judge the importance of a node, so there are many node centrality measures. The centrality measure based on the number of edges connected to a node is called **degree centrality**. The following are some other node centrality measures to consider:

- **Betweenness centrality**: This attempts to measure to what extent a node lies between key parts of the network, and so acts as a node through which paths must pass when going between any other two nodes in the network.

- **Closeness centrality**: As the name suggests, this measures how close the node is to all the other nodes, based on path distances along the edges of the network. A node that has a short average distance from other nodes is at the heart of the network.

- **Eigenvector centrality**: A measure of node importance based on the likelihood of ending up at a node while performing a long random walk on the network. This measures how popular the node is, as voted for by the other nodes.

Different node centrality measures capture different aspects of node importance. A node can be very important by one centrality measure, and not so important by another measure. We need to be aware of the nuances of different centrality measures when we use them.

But why do we care about node centrality measures? Why do we care about knowing whether a node is important or not? In a real-world setting, an important node is typically an influential node, and therefore one we want to identify by ranking on a particular node centrality measure. More interestingly, graphs with certain types of highly influential nodes can have some surprising behaviors. We'll learn about these graphs and their behaviors in the next section, but for now, we'll summarize what we have learned in this section about the basic characteristics of graphs.

What we've learned

In this section, we learned the following:

- How edges in a graph can be undirected or directed, meaning a graph can be an undirected graph or a directed graph

- How we can have a directed edge from a node to itself

- How edges can have a weight associated with them

- Adjacency matrices and how they mathematically encode the network structure

- How an adjacency matrix can be used to identify and count n-step paths between each pair of nodes in a network

- The in-degree and out-degree values of a node and how they can be calculated from the adjacency matrix

- How the total degree value of a node (the sum of the node's in-degree and out-degree) is a centrality measure that measures the importance of a node

- How there are many other different node centrality measures

Having learned the basic concepts and terminology that describe a graph, next, we'll look at different types or families of graphs and focus on some of their specific but very interesting properties.

Different types of graphs

There are many different graphs you may encounter as a data scientist. Many of these graphs can be grouped into different classes. In this section, we will outline some of the most important classes of graphs you will encounter. The list of classes we'll cover here is not intended to be exhaustive. It will introduce you to the concepts and terminology associated with the most common classes of graphs you will encounter.

Fully connected graphs

One of the differences between our two real-world examples is that in our trade network example, each node (country) is connected to every other node. We say that the trade network is **fully connected**. In contrast, in our pizza network, every pizza is not connected to every other pizza.

The left-hand graph in Figure 10.10 shows a graph with four nodes. Each of the nodes is connected to every one of the other three nodes. It is fully connected. In contrast, the graph on the right-hand side of Figure 10.10 also contains the same four nodes but is not fully connected:

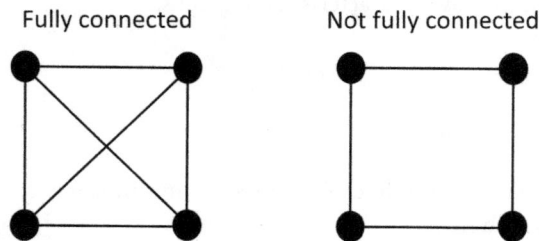

Figure 10.10: A graph that is fully connected and one that is not

Disconnected graphs

Our real-world pizza example graph is not fully connected. Some edges that could be possible are absent. What would happen if we removed even more edges from our pizza network? Removing edges from a network will eventually cause it to break into separate distinct and un-connected graphs. Figure 10.11 shows such an example:

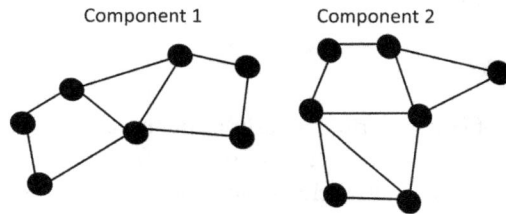

Component 1   Component 2

Figure 10.11: Disconnected components in a graph

We refer to the separate disconnected parts of the graph as **components**. In Figure 10.11, there are two components. One can have graphs that have many disconnected components. Each component is a graph in its own right, and any analysis we apply to a graph can be applied to each separate component of a graph. Consequently, from now on, we shall only discuss how to analyze single-component graphs.

Directed acyclic graphs

Directed graphs are great for representing flows. They are also great for representing how one quantity (node) depends on another, or how one quantity (node) influences another. The presence of a directed edge from node i to node j in a directed graph could be used to indicate that the quantity represented by node i influences the quantity represented by node j. Or equivalently, node j has a dependency on node i. Directed graphs are often used to represent patterns of dependencies. An important requirement of any such dependency graph is that there are no **cycles**. A cycle is a set of steps (hops) that take us from a node back to itself. A cycle would mean we have a chain of dependencies from a node back to itself – the quantity would have a dependency on itself, something that cannot be resolved.

The graph on the left-hand side of Figure 10.12 shows a directed graph with a cycle. The highlighted cycle is a path between three nodes:

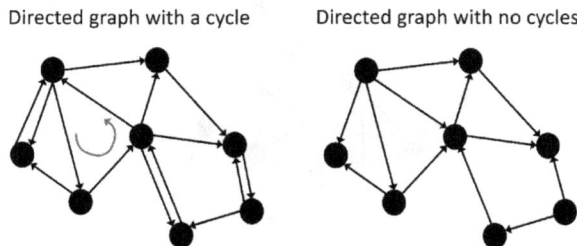

Directed graph with a cycle   Directed graph with no cycles

Figure 10.12: Directed graphs with and without cycles

There are many cycles in the left-hand graph in Figure 10.12 – there are several pairs of nodes with directed edges in both directions between them. Also, any edge from a node to itself automatically represents a cycle.

For dependency graphs, we want graphs that do not have any cycles. Since such graphs have no cycles or are **acyclic**, they are referred to as **directed acyclic graphs** (**DAGs**). The right-hand graph of Figure 10.12 shows an example of a DAG.

Small-world networks

You may have already heard of small-world networks as many real-world networks are said to be small-world networks. Small-world networks are not so much a class of networks but are networks that display the small-world effect or small-world phenomenon. The small-world phenomenon is also known colloquially as the "six degrees of separation" effect. It refers to the observation that in many networks, the shortest path between any two nodes is not very long – typically only six steps or "hops," meaning that we can get from any node on the network to any other node on the network in typically six hops or less. In a real-world setting, you may have heard of the example that any Hollywood actor can be connected to actor Kevin Bacon via a chain of six movies or less. This is the "six degrees of Kevin Bacon" phenomenon.

The small-world phenomenon is widespread in real-world networks. Theoretical models of how networks evolve and grow suggest that the small-world phenomenon should be expected in most networks. The phrase "small-world network" itself stems from the real-world phenomenon we often experience in social-network settings, where we meet someone new from another part of the world and it turns out they already know one of our best friends or one of the friends of our friends. This strikes us as an unlikely occurrence and we exclaim, "What a small world!"

At the heart of the small-world phenomenon is the presence of nodes in the network that have a high total degree. As we said when introducing centrality measures, total degree is one way of identifying the important nodes. A node with a high total degree is connected to a lot of other nodes. If you want to find a short path between two nodes, going via a node that is connected to almost every other node in the network would seem like a smart idea. High-degree nodes allow us to efficiently move between different parts of the network:

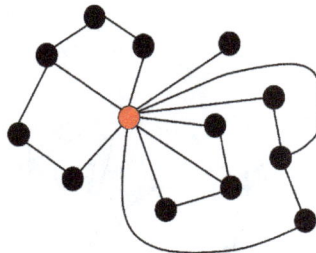

Figure 10.13: A high-degree node that connects all parts of the network

Figure 10.13 shows an example of a high-degree node, colored in red, that acts as a connection between any other two nodes. By passing through the red node, we can go from one side of the network to the other in three steps or less.

Let's put a bit of math detail on the small-world phenomenon. The presence of high-degree nodes means the typical shortest-path distance between any two nodes is always small, no matter how big the network gets. If we denote the shortest-path distance between node i and node j as D_{ij}, then the mean shortest-path distance is given by the following equation:

$$\overline{D} = \frac{1}{N^2} \sum_{i=1}^{N} \sum_{j=1}^{N} D_{ij}$$

Eq. 8

In Eq. 8, N is the number of nodes in the network. A mathematical statement of the small-world phenomenon is that $\overline{D}$ grows very slowly as N is increased. Theoretical analysis of small-world networks shows that $\overline{D}$ typically grows as log$N$. This means that if we double the size of a small-world network by doubling N, we only increase the typical shortest-path distance between nodes by an amount, log2 – that is, a small increase. A consequence of this is that typical distances between nodes on even very large real-world networks, such as the World Wide Web or popular social networks, are not large (between 10-20 hops).

Another common feature of small-world networks is that some of the highly connected nodes also exhibit **long-range** connections. This means they are not just connected to lots of nodes in the same part of the network, but they are connected to lots of different parts of the network. In Figure 10.13, the red node is connected to nodes on the very right-hand side of the network via those curved edges. The red node is connected to almost every other node in the network. In social network settings, such nodes are the people who seem to know everyone. In our Hollywood actor example, Kevin Bacon is a node who is connected to many other Hollywood actors.

Nodes with many long-range connections not only have a large total degree, but they also have a total degree that is significantly above what we might expect given the values from most of the other nodes. This raises the question of what sort of distribution of node degree values we see in different types of networks. In a small-world network, we expect to see an extended right-hand tail to the distribution, meaning that there is a small, but non-negligible, probability of getting a node with a very large total degree value. Beyond this, what kind of degree distribution should we expect for a graph? Does it have a natural shape or scale? Some of the most interesting graphs that graph theorists have studied recently are those whose degree distributions have no scale to them at all. These are called **scale-free** graphs, and we'll introduce them next.

Scale-free networks

A scale-free network is a network whose statistical properties have no natural scale to them. This means that if the node degree value is x, then the probability distribution of x can't be written as some function, $f(x/\lambda)$. If it could, it would mean that the value of λ would provide a natural scale against which to measure the values of x. The only possible distributions that satisfy the requirement of not having a scale are **power-law distributions**. This means the probability of getting a degree value of x is proportional to $x^{-\alpha}$. The parameter, $\alpha > 0$, is the exponent of the power-law distribution. The degree value is a discrete quantity, so we can write the probability distribution as follows:

$$\text{Prob}(\text{Total Degree} = x) = \begin{cases} 0 & \text{if } x < x_{min} \\ \dfrac{x^{-\alpha}}{\zeta(x_{min}, \alpha)} \end{cases}$$

Eq. 9

The value of x_{min} is the minimum value that x can take. The function, $\zeta(x_{min}, \alpha)$, in Eq. 9 is known as the **Hurwitz zeta function**. It ensures that the probability distribution is properly normalized – that is, the sum of all the probabilities is 1. This means that the function, $\zeta(x_{min}, \alpha)$, is given by the following equation:

$$\zeta(x_{min}, \alpha) = \sum_{n=0}^{\infty} (n + x_{min})^{-\alpha}$$

Eq. 10

The power-law distribution in Eq. 9 is a slowly decreasing function. The slow decrease in probability as x increases means that there is a small but non-negligible probability of getting a very large value of x. In simple terms, it means that in any large scale-free network, there will be a few nodes that are connected to a large proportion of the network.

What does a power-law distribution look like? Pretty boring in fact. If we look at Eq. 9 in a bit more detail and take the logarithm of it for $x \geq x_{min}$, then we have the following:

$$\log \text{Prob}(\text{Total Degree} = x) = -\alpha \log x - \log \zeta(x_{min}, \alpha)$$

Eq. 11

Eq. 11 is just the equation of a straight line with $\log x$ being the variable on the x-axis. This means that on a logarithmic scale, our power-law probabilities decrease linearly. Figure 10.14 shows an example of a power-law distribution for $\alpha = 2.5$ and $x_{min} = 1$. The line shows the probabilities calculated according to Eq. 9. Note that the red line is linear. That is because we have plotted both the x-axis and y-axis in Figure 10.14 on a logarithmic scale to emphasize the power-law decay of the probabilities:

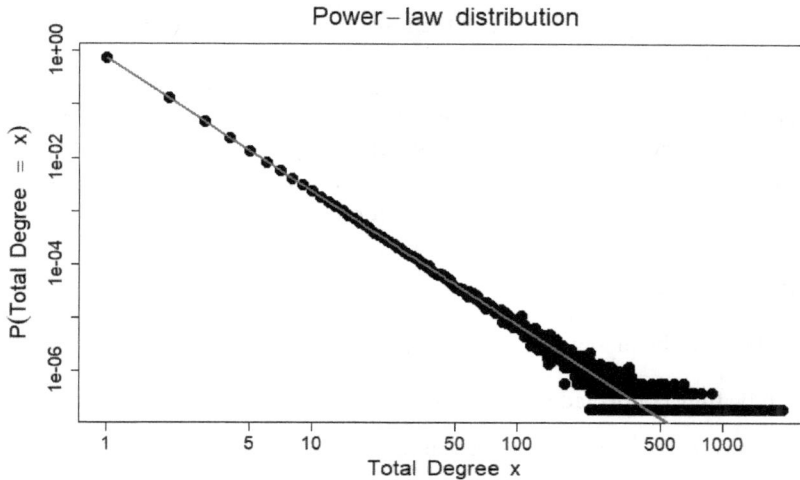

Figure 10.14: A power-law distribution of node degree values

The black dots in Figure 10.14 represent sample proportions of the different node degree values when I generated a sample of 5,000,000 degree values from the probability distribution represented by the red line. As you would expect, the sample proportions follow the line closely. We can see from those black dots that there is a small but not insignificant fraction of total degree values greater than 1,000. This would represent nodes that were connected to more than 1,000 other nodes.

From the example in Figure 10.14, it should be apparent that scale-free networks exhibit the small-world phenomenon, so scale-free networks are also small-world networks. Perhaps more interesting is that for scale-free networks, the mean shortest path distance between nodes, $\overline{D}$, only grows as $\log\log N$ as the size of the network, N, increases. Now, $\log\log N$ is an even more slowly increasing function of N than $\log N$, meaning that in scale-free networks, the typical distance between nodes hardly changes, even as we get to very big networks. We can think of scale-free networks as being super-small-world networks.

In reality, any finite-sized network always has a scale associated with it – the size of the network. That means a real-world network cannot be truly scale-free. Does this mean scale-free networks are useless to us? No, far from it. Many large real-world networks, while not **exactly** scale-free, are extremely well approximated by scale-free networks, so studying and understanding the properties of scale-free networks helps us understand the properties of those real-world networks. For example, the node degree values of real-world networks can display linear behavior like that in Figure 10.14 or very close to it. Examples of real-world networks whose node degree distributions are modeled well by a power-law distribution include the following:

- The World Wide Web, where both the node in-degree and out-degree values of websites follow power-law distributions closely

- Scientific paper citation networks, where the in-degree values appear to follow a power-law distribution

That concludes our introduction to different classes of graphs. We have covered only a small number of classes in this section. There are many more. However, you will have gained a flavor of and feel for the rich variety of structures, behaviors, and uses that you can see across different classes of graphs. Let's remind ourselves what we have covered in this section.

What we've learned

In this section, we learned the following:

- A graph can be fully connected if every node is connected to every other node.
- A graph can consist of several disconnected components, each of which is a graph in its own right.
- A directed graph can be used to represent a set of dependencies, in which case it is desirable not to have any cycles in the directed graph. Such graphs are called DAGs.
- Many real-world graphs exhibit the small-world phenomenon, where the typical distance between any two nodes is small and only a weakly growing function of the network size.
- Scale-free networks have node degree distributions that are scale-free and follow a power-law. Scale-free networks exhibit the small-world phenomenon.

Having learned about some different types of graphs, including graph types that occur frequently in real-world settings, in the next section, we'll do some calculations with graph data. We will look at community detection, where we take a graph and try to identify the main sub-graphs within it that represent distinct communities of nodes.

Community detection and decomposing graphs

Community detection is a common data science task and a useful technique to have in your data science toolkit, but let's start by describing what we mean by a community.

What is a community?

In many real-world networks, nodes are used to represent people. Consequently, when we have a collection of highly connected nodes, forming almost a fully connected separate graph, we can think of this as a **community** of interacting people.

We can extend this idea to situations where the nodes do not represent people. For example, our trade network example at the beginning of this chapter was fully connected, but if it wasn't, there might be groups of countries that preferentially trade with each other and don't trade with other countries. We would have separate trading blocks or trading communities. Similarly, in our pizza example, we have groups of pizzas that are more similar to each other and hence interchangeable. This means we have communities of pizzas.

Why is knowing about communities useful to us? A community represents a group of similar entities. In data science, it is often useful to be able to group things – to analyze them together, to model their response to a feature using a single parameter. This means we often want to take a graph, as represented by an adjacency matrix, and identify the community structure present in the graph. This is the problem of **community detection**. We are taking a graph and breaking it down into a set of sub-graphs. In other words, community detection is about decomposing a graph.

How to do community detection

We have said that we can think of a community as a highly connected set of related nodes. This means that when we have isolated communities, detecting them is easy. Figure 10.15 shows an example of two distinct communities of nodes. The different communities are obvious. The different communities correspond to separate graphs. The communities are disconnected from each other. If Figure 10.15 looks familiar, that is because it is the same as Figure 10.11, which we used to illustrate a graph consisting of disconnected components:

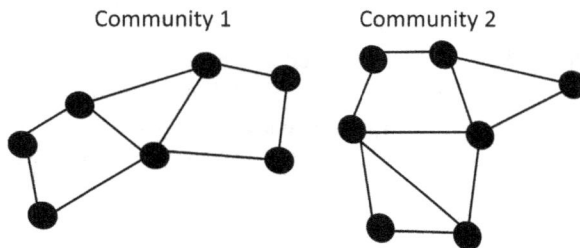

Figure 10.15: Two distinct communities of nodes

Mathematically, detecting communities is easy when they correspond to disconnected components of a graph. However, consider the graph in Figure 10.16. It is a single graph, but how many communities are there?

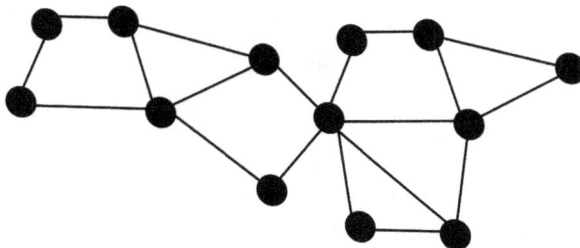

Figure 10.16: A single graph with two communities

To most people, the graph in Figure 10.16 would also appear to consist of two communities. The communities are clear. There is one community on the left-hand part of the graph in Figure 10.16 and one on the right-hand side. To any human observer, the identification of the communities is obvious and corresponds to that shown in Figure 10.17:

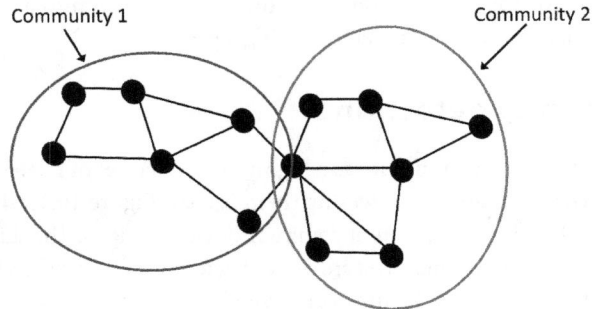

Figure 10.17: Identifying two communities in a single graph

The two communities in Figure 10.17 are connected and have a node in common. As humans, we are OK with the idea that communities are connected, with nodes that act as links between them. Mathematically, we need algorithms that are comfortable with the ambiguity of not having perfectly disconnected components and can identify the parts of the graph that are most community-like. We will briefly describe some of these algorithms next.

Community detection algorithms

Community detection algorithms attempt to identify the parts of a graph that look most like a distinct community. As you can imagine, they do this by attempting to find a subset of nodes that are tightly connected and weakly connected to nodes outside of the subset. Essentially, community detection corresponds to a process of using the information encoded in a graph's adjacency matrix to assign each node to a subset (the community).

Once each node has been assigned to a community, we can assess the quality of the assignment by measuring how inter-connected nodes within the same community are compared to nodes from separate communities. If we have a quantitative metric that measures this quality – that is, measures the within-community connectedness compared to the between-community connectedness – then we can iterate the overall process to maximize this community quality metric. This is the essence of a community detection algorithm.

As you can imagine, there are different metrics we can use to measure the quality of a specific node-to-community assignment and different optimization approaches we can use to maximize the chosen assignment quality metric. This means there are many different types and flavors of community detection algorithms.

The most commonly used type of community detection algorithm is **modularity maximization**. As the name suggests, this algorithm maximizes the total modularity score of the community assignment. The modularity score of a community is the difference between the actual number of edges between nodes within the community and the expected number of edges within the community if the edges of the graph were placed at random. So, we can think of the modularity score as a measure of the observed within-community excess number of edges. If our community assignment genuinely reflected some true underlying community structure, then we would expect this excess number of edges to be high as nodes within a genuine community have a high probability of being connected. So, a high modularity score is indicative of a good community assignment. Modularity maximization can be applied to split a graph into just two communities or more than two communities. There are also different algorithms to perform the maximization, including the following:

- Greedy algorithms, which perform the maximization iteratively

- The spectral method, which uses the eigen-decomposition (see *Chapter 3*) of the modularity matrix (which is derived from the adjacency matrix) to find an optimal community assignment.

- The Louvain algorithm, which performs the modularity maximization agglomeratively. It starts by joining single nodes into pairs, then joining other pairs or single nodes together, and so on, each time measuring the modularity score until no improvement in the modularity score is obtained.

There are also other classes of community detection algorithms, including the following:

- **Model-based approaches**: These fit a probabilistic model (see *Chapter 5*) to the graph structure, with the model assuming some community structure. This allows us to compute the likelihood of the observed network structure given the assumed community structure. The likelihood can then be maximized with respect to the community structure to obtain a community structure that is highly compatible with the given network adjacency matrix.

- **Betweenness-based approaches**: These attempt to uncover the community structure present in a graph by breaking the graph apart into disconnected components. By iteratively identifying and then removing edges that have a high "betweenness" score – that is, those edges that connect a high number of different parts of the network – the idea is that the graph will quickly separate (fall apart) into disconnected communities after the removal of these high "betweenness" edges. Perhaps the most well-known and commonly used betweenness community detection algorithm is that of Girvan and Newman – see the *Notes and further reading* section at the end of this chapter for details on this algorithm.

Having introduced community detection algorithms, let's try some out with a code example.

Community detection code example

The following code example and more can be found in the `Code_Examples_Chap10.ipynb` Jupyter Notebook in this book's GitHub repository. It uses the example of the graph in Figure 10.16. We'll use the `NetworkX` Python package to do the community detection. The graph in Figure 10.16 is represented as a series of nodes and edges and stored in the `Data/example_network_adjlist.txt` file in this book's GitHub repository. The data in the file is in a format that the `NetworkX` package can understand. Let's start by reading in the graph:

```
import network as nx
G = nx.read_adjlist("../Data/example_network_adjlist.txt",
                    nodetype=int)
```

The `nodetype=int` option tells `NetworkX` that the nodes are represented (labeled) by integers. Now that the graph is stored in G, we can visualize it:

```
nx.draw(G, with_labels=True, node_color='lightblue',
        pos=nx.spring_layout(G, seed=293))
```

This results in the following graph. We have overridden the default node color to make the node labels more visible:

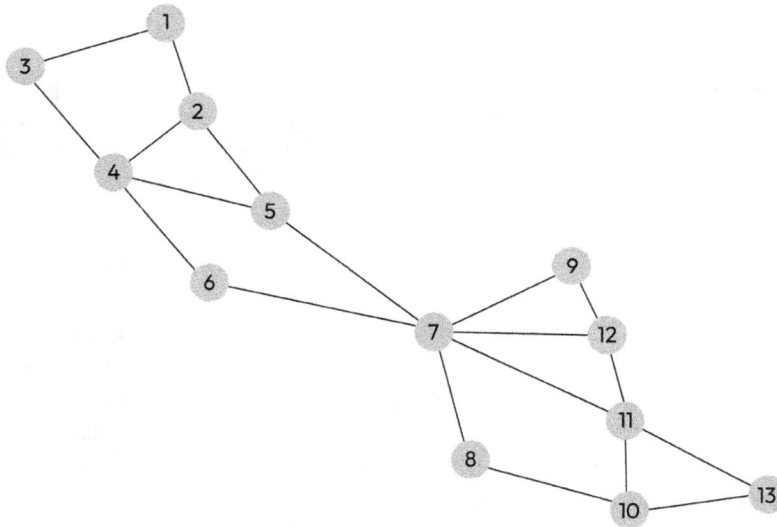

Figure 10.18: Our starting labeled graph

Now, we'll apply a modularity maximization algorithm to the graph, G. In this case, we'll use the greedy modularity maximization algorithm that the `NetworkX` package supplies:

```
communities = nx.community.greedy_modularity_communities(G)
```

We've stored the result in the `communities` object. Let's look at it:

```
communities
```

This gives us the following output:

```
Out[5]:
[frozenset({7, 8, 9, 10, 11, 12, 13}), frozenset({1, 2, 3, 4, 5, 6})]
```

Here, we can see that the `communities` object is two lists of integers, corresponding to the node IDs that the greedy modularity maximization algorithm has placed into two communities. In other words, the algorithm has identified two communities, one consisting of nodes [1,2,3,4,5,6], and the other corresponding to nodes [7,8,9,10,11,12,13]. We'll add some colors to the nodes according to which community they are in:

```
community_colors = ['lightblue', 'orange']
color_dict = {}
for i in range(len(communities)):
    x = list(communities[i])
    for j in range(len(x)):
        color_dict[x[j]] = community_colors[i]

color_map = [color_dict[node] for node in G.nodes]
```

Note that we first created a dictionary that mapped the node label to a community color and then created a list that used that dictionary to map the node index to a community color. This is because the order in which the nodes are indexed does not necessarily match the node labels, even if those labels are integer values. So, the node with a label of "6" is not necessarily stored as the 6th node. To check the order in which the nodes are indexed, you can use `G.nodes`:

```
G.nodes
```

This gives us the following output:

```
Out[7]:
NodeView((1, 2, 3, 4, 5, 7, 6, 8, 9, 11, 12, 10, 13))
```

Here, we can see that node 7 is indexed in the 6th position, and node 6 is indexed in the 7th position. With that little nuance sorted out and a color map created, we can now redraw the graph with each node colored according to which community it is in:

```
nx.draw(G, with_labels=True, node_color=color_map,
        pos=nx.spring_layout(G, seed=293))
```

This gives us the following graph:

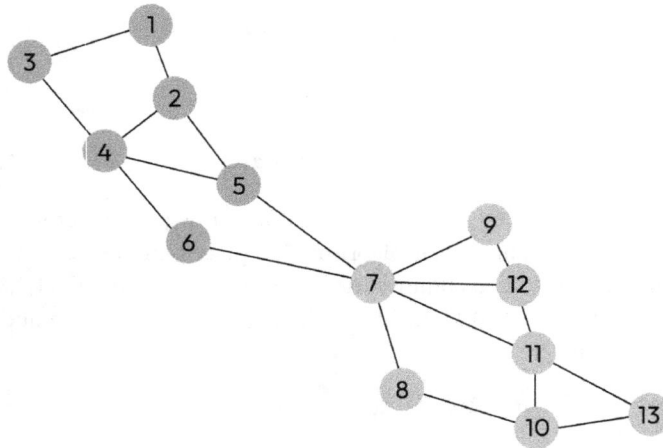

Figure 10.19: Our graph with two communities identified and nodes
colored according to which community they belong to

From Figure 10.19, we can see that the community detection algorithm has uncovered what we would regard as the obvious communities that are present in the starting graph in Figure 10.18. While it is a simple example, it does illustrate how easy it is to run community detection algorithms. It also illustrates the usefulness of those algorithms. It is usually relatively easy to construct a pairwise measure or weight of how closely related two nodes or objects are. This means we can easily construct a weighted adjacency matrix for a set of interacting objects, such as a retailer's products. Community detection algorithms then allow us to easily uncover the natural groupings of those objects, with minimal extra coding.

Having run a simple code example to illustrate how community detection is performed, we'll wrap up this section and this chapter.

What we've learned

In this section, we learned the following:

- A community represents a collection of similar or related nodes
- Community detection attempts to break a graph into distinct sub-graphs or communities
- How community detection algorithms work by optimizing a metric that measures the quality of the assignment of nodes to communities
- Modularity maximization community detection algorithms

- Other approaches to community detection, such as model-based approaches and betweenness-centrality-based approaches

Summary

This chapter has been about a specific type of data – network data. We have learned that networks are used to represent relationships. Since relationships are prevalent in many real-world scenarios, it is essential that, as a data scientist, you are familiar with the main concepts and terminology relating to networks. Of these, the most important concepts we have covered in this chapter are as follows:

- What a network represents and that a network is a graph

- A graph, G, consists of a set of nodes (vertices), V, and a set of edges, E, between those nodes

- The edges of a graph can be undirected or directed and can have weights associated with them

- The structure of a graph is encoded in the adjacency matrix

- The in-degree and out-degree of a node can be calculated from the adjacency matrix and tell us the number of edges coming into a node and leaving a node, respectively

- The sum of the node in-degree and out-degree values gives us the total degree value for a node and is a node centrality measure, which is a measure of the importance of the node

- A graph can be fully connected if every node is connected to every other node

- If a directed graph doesn't have any cycles, then it is called a DAG, and DAGs are used to represent dependency structures between entities

- Many graphs exhibit the small-world phenomenon, whereby the typical distance between two nodes on the graph is small and a weakly growing function of the network size

- Scale-free graphs have node degree distributions that are power-laws

- Community detection algorithms attempt to break down a graph into its natural sub-graphs of closely connected subsets of nodes

- Key community detection algorithms such as modularity maximization

The next chapter, like this chapter on network data, is about another specific type of data and the systems that generate it. We'll be looking at dynamical systems, the data they generate, and the underlying equations that control the evolution of those dynamical systems.

Exercises

The following is a series of exercises. Answers to all these exercises can be found in the `Answers_to_Exercises_Chap10.ipynb` Jupyter Notebook in this book's GitHub repository:

1. The Zachary Karate Club is a well-known network in the field of network science, so much so that a copy of the network is stored in the `NetworkX` package and can be accessed via the `karate_club_graph()` function. Use this function to create the karate club graph and then use the `community.greedy_modularity_communities` function to identify the communities within the graph. You can assume that there are two communities, so you should look at how to use the `cutoff` and `best_n` parameters of the `community.greedy_modularity_communities` function to ensure that only two communities are found. Which nodes do you think are at the center of each of the two communities found?

2. Use the `scale_free_graph` function of the `NetworkX` package to create a scale-free graph with 10,000 nodes. Having generated the scale-free graph, use the `degree` function of the `NetworkX` package to calculate the degree value of each node. Sort the node degree values in descending order and plot them against their rank on a log-log plot. You should see a linear relationship on the log-log plot. You'll need to pay attention to the object type returned by the `degree` function of the `NetworkX` package. You may find it useful to convert it into a dictionary by wrapping it inside a call to `dict()`.

3. Look at the `NetworkX` package documentation for the `watts_strogatz_graph` function. This uses the Strogatz and Watts algorithm to generate simulated small-world networks. Use this function to generate a series of graphs with the number of nodes N =10,20,40,80,160,320,640,1280. For each graph, set k=5 and p=0.3. These are the initial degree values and edge re-wiring probability of the Strogatz and Watts algorithm, respectively. For each graph, use the `shortest_path_length` function of the `NetworkX` package to compute the shortest path distance between each of the N^2 pair of nodes in the graph. For each value of N, calculate the average shortest path distance, $\overline{D}$, as defined by Eq. 8, and plot $\overline{D}$ against log$N$. You should see a linear relationship between $\overline{D}$ and log$N$. See if you can spot a short-cut where you don't have to compute all N^2 shortest path distances to calculate $\overline{D}$.

Notes and further reading

1. For details on the Girvan and Newman algorithm, see the research paper by M. Girvan and M.E.J. Newman, *Community structure in social and biological networks*, Proceedings of the National Academy of Sciences USA, 99:7821-7826, 2002. A preprint version of the paper can be found in the arXiv archive at `https://arxiv.org/pdf/cond-mat/0112110.pdf`.

Part 3:
Selected Advanced Concepts

In this part, we will introduce a selection of advanced math concepts. As with Part 2, each concept is a standalone topic. But, in contrast to Part 2, we're now introducing topics at the cutting edge of data science and data science research. There is still a high probability you will encounter these concepts in your data science work, especially the longer you work in data science. Because of the advanced nature of the topics, each chapter is only designed to give you a basic grounding in that topic. But by the end of Part 3, you will understand the core ideas of each of these topics and be able to use that understanding to guide your own studies.

This section contains the following chapters:

- *Chapter 11, Dynamical Systems*

- *Chapter 12, Kernel Methods*

- *Chapter 13, Information Theory*

- *Chapter 14, Bayesian Non-Parametric Methods*

- *Chapter 15, Random Matrices*

11
Dynamical Systems

In this chapter, we introduce the concept of modeling a dynamic system – a system that changes over time according to some mathematical law. Why is that useful? Many systems we encounter in real-world data science are dynamical systems. They output data, and to understand that data we need to understand the underlying system. If we have a mathematical understanding of the underlying dynamical system, we can also use data from that system to make inferences and predictions about how the system will behave in the future – one of the key goals of doing useful data science.

This chapter will take a deliberate data science perspective on dynamical systems. Many traditional applied math books on dynamical systems will explain concepts such as **phase plane diagrams**, **fixed points**, **basins of attraction**, **bifurcation points**, and **deterministic chaos**. While highly relevant for dynamical systems and extremely interesting, we don't have space to cover such topics in this chapter. Instead, we want to focus on those dynamical systems that are commonly used in data science modeling and what their main behaviors are. That is, we focus on dynamical systems as a practical data science tool. That means we will focus almost exclusively on simple discrete dynamical systems. To understand dynamical systems from a data science perspective, we will cover the following topics:

- *What is a dynamical system and what is an evolution equation?*: In this section, we introduce the very basic concepts behind dynamical systems, why we study them, and what controls them

- *First-order discrete Markov processes*: In this section, we introduce the simplest discrete-time dynamical system and show how it can be used to model real-world dynamical processes

- *Higher-order discrete Markov processes*: In this section, we extend the principles behind first-order discrete Markov processes, and where we show how these higher-order models can be related back to first-order discrete Markov processes

- *Hidden Markov models (HMMs)*: In this section, we extend discrete Markov processes to situations where not every state transition results in an observable output

Technical requirements

All code examples given in this chapter can be found in the GitHub repository, `https://github.com/PacktPublishing/15-Math-Concepts-Every-Data-Scientist-Should-Know/tree/main/Chapter11`. To run the Jupyter notebooks, you will need a full Python installation, including the following package:

- `numpy` (>=1.24.3)

What is a dynamical system and what is an evolution equation?

The term *dynamical system* implies that something is changing over time. That is what a dynamical system is: it is a system that changes over time. Typically, we will measure something about that system, giving us a series of data values from different timepoints.

Wait – isn't this just time-series data that we met in *Chapter 6*? No – there is a difference here. When looking at time-series data, we have a series of data values, but we typically don't consider the underlying system that generated that data. We only consider the relationships between data values at different timepoints, such as any auto-regressive structure present. In contrast, when studying dynamical systems, it is the underlying system we are interested in modeling.

The most important thing dynamical systems have is **state**. The state of a dynamical system is a variable or attribute, or a collection of variables or attributes, that determines how the dynamical system behaves. The state of a dynamical system changes or evolves over time. Most commonly, it is the state of a dynamical system that we observe or measure, although not always. In the last section of this chapter, we will meet a class of dynamical systems whose state is hidden or **latent**.

Time can be discrete or continuous

We have said that when studying dynamical systems, a state changes over time. The time variable can be continuous or discrete. What determines whether the time variable is continuous or discrete depends on two factors:

- Whether the system naturally changes according to a discrete timestep or evolves continuously
- Whether we can observe and measure the dynamical system continuously or only at regular intervals

Some dynamical systems naturally evolve continuously; for example, a classical particle in a force field – such as a particle with mass in a gravitational field or a particle with charge in an electric field.

Other dynamical systems naturally evolve according to discrete timesteps. Or we may have a continuous driven system, but the main driving factor determining the evolution of the dynamical system undergoes its biggest changes according to the tick of some regular clock.

How we measure time can be up to us. If we have the capability to record measurements from the dynamical system continuously or near continuously so that it makes no difference, then we can take the time variable to be continuous.

If we can only measure at discrete timepoints – for example, measuring weekly sales volumes of a business – then the underlying system may evolve continuously but our model of it will use a discrete-time variable. Because of this, in commercial data science, it is more common to model dynamical systems using a discrete-time variable, and we will focus exclusively on discrete-time dynamical systems in the remaining sections of this chapter. In contrast, in data science studies of scientific data, the underlying systems tend to evolve continuously according to some well-known and already discovered scientific laws, such as Newton's laws of motion that determine how our classical particle moves in a gravitational field.

Time does not have to mean chronological time

So far, I'm guessing that when we have been discussing time you assumed that we meant chronological time; that is, time that measures how far through the day, week, month, year we are. However, when we discuss dynamical systems, there is no reason why we must restrict ourselves to chronological time. For discrete-time systems, we have said the time variable represents the regular tick of a clock. This means we could interpret any one-dimensional variable that increases in steps as being a "clock." And so, any system that has a state that changes as we move in steps along some one-dimensional direction can be considered as being a discrete-time dynamical system. Take the following DNA sequence:

```
ATGGTGCATCTGACTCCTGAGGAGAAGACTGCTGTC
```

Our DNA holds the instructions for making all the proteins in our body. Those instructions are encoded as a linear sequence of letters taken from a 4-letter alphabet (consisting of A, C, G, and T). The preceding sequence is part of the DNA sequence of the gene that codes for the human hemoglobin delta sub-unit protein. The linear nature of DNA sequences means we can think of a DNA sequence as being generated by a discrete-time dynamical system. The state is the letter at each point in the sequence, and that state changes as we step along the sequence, so position along the sequence is our discrete-time variable.

But what determines the next letter in our DNA sequence? More generally, for any dynamical system, what determines the evolution of state? This introduces the concept of an evolution equation that controls how the state evolves.

Evolution equations

An evolution equation is an equation from which we can determine how the state of a dynamical system evolves over the time variable. Since an evolution equation tells us how state evolves over time, the equation must involve the time variable either explicitly or implicitly.

For a concrete example, take our classical particle moving in a gravitational field. The evolution of the particle's position $\underline{x}(t)$ is determined by Newton's second law of motion; the particle's mass multiplied by its acceleration is equal to the force acting on the particle. The force, in this case, is given by the particle's mass multiplied by the negative gradient of the gravitational potential field $\Phi(\underline{x})$. This gives us the following equation:

$$m\frac{d^2\underline{x}}{dt^2} = -m \; \nabla \; \Phi(\underline{x})$$

Eq. 1

There are several observations we can make about the evolution equation in *Eq. 1*:

- The evolution equation does not directly tell us what the position of the particle, $\underline{x}(t)$, will be at a later time t'. To find that, we must solve the differential equation in *Eq. 1*. This is what we meant when we said earlier that an evolution equation is an equation from which we can determine how the state evolves.

- The evolution equation is a local equation. It tells us how the state evolves over a small time period, infinitesimally small in this case since we have a differential equation. To determine the evolution of the state over longer time periods, we must effectively apply the evolution equation iteratively; that is, multiple times. This is overwhelmingly the most common feature of evolution equations for dynamical systems.

- Finally, the evolution equation in *Eq. 1* is deterministic. If we know the starting position of our particle and we can solve *Eq. 1*, we know our particle's position exactly at any later timepoint. This means our dynamical system is deterministic in this example.

As we'll find out in the next section, not all dynamical systems are deterministic. Some are stochastic, and the evolution equation for these systems tells us how the state probability distribution evolves. But for now, we'll recap what we have learned about dynamical systems so far, before moving on to the next section.

What we learned

In this section, we have learned the following:

- A dynamical system has a state, and that state changes over time.

- For some dynamical systems the time variable is continuous, while for other dynamical systems the time variable is discrete.

- The time variable of a dynamical system does not necessarily have to represent chronological time. Instead, it can represent any one-dimensional variable along which the state of a system changes.

- An evolution equation determines how a dynamical system evolves.

- If the dynamical system is deterministic, then the evolution equation tells us precisely how the state evolves, while if the dynamical system is stochastic, the evolution equation tells us precisely how the state probability distribution evolves.

Having introduced the basic ideas of what a dynamical system is, we're going to look at some very specific but widely used dynamical systems. As we have said, we will focus on discrete-time dynamical systems. In the next section, we will look at one of the simplest discrete-time dynamical systems: the first-order discrete-time Markov process.

First-order discrete Markov processes

A discrete first-order Markov process is one of the simplest dynamical systems we can study. Our time variable is discrete, and we have a finite number of states between which we can move from one timepoint to the next.

Since the possible states are discrete, we can label them using integer values 1,2,3… and so on. We also use i and j to represent generic states.

The state at time t is denoted by the variable X_t, and so a trajectory of observations from timepoint $t = 0$ to timepoint $t = T$ would be represented by the sequence $X_0, X_1, X_2, …, X_T$. For example, if we have five possible states $i = 1, 2, …, 5$ and we set $T = 10$, then we may have the specific trajectory $X_0 = 3$, $X_1 = 4, X_2 = 4, X_3 = 3, X_4 = 4, X_5 = 3, X_6 = 5, X_7 = 1, X_8 = 2, X_9 = 2, X_{10} = 4$. We can shorten the representation of the trajectory and say the observed trajectory is the sequence [3,4,4,3,4,3,5,1,2,2,4].

By taking the possible states that the system can be in to be a finite set, we simplify the mathematics considerably. However, since we can make that finite set as large as we want, we have considerable flexibility in the real-world systems to which we can apply first-order discrete Markov processes.

Variations of first-order Markov processes

We have used the words *discrete* and *process* to describe our Markov models. But be aware that *Markov process* is a very general term that applies to a variety of different model types. You will also see the term *Markov chain* used to describe the discrete-time discrete state-space models we study here. That is because we have a sequence or **chain** of states.

We can, of course, have different variants of a Markov process. The possible variants are the following:

1. Discrete-time and discrete state space – what we study in the rest of this chapter
2. Continuous time and discrete state space
3. Discrete-time and continuous state space
4. Continuous time and continuous state space

All of these are Markov processes, but the term *Markov chain* tends to be reserved for variants *1-3*, where either the time variable or the space is discrete. We will study discrete-time and discrete state-space Markov processes (Markov chains) in the rest of this chapter; for shorthand, we will refer to them just as discrete Markov processes.

A Markov process is a probabilistic model

What determines the value of X_t? The interesting aspect of a first-order discrete Markov process is that the trajectory is stochastic, not deterministic. In our previous example, where we had five possible states, if we are in state 4 at $t = 3$ (that is, $X_3 = 4$), then the state we observe at the next timepoint $t = 4$ could be any of the five states 1, 2, 3, 4, or 5. In our example, the state X_4 turned out to be 3. This makes a first-order discrete Markov process a probabilistic model, so we'll need to introduce some probabilities.

In general, we consider X_t to be a random variable. At timepoint t, the probability distribution from which X_t is drawn we'll denote as π_t, meaning that $X_t \sim \pi_t$ in the notation from *Chapter 2*. If we have a total of N possible states, then π_t is a vector of N probabilities. It gives us the proportions of observations we would expect to see in each state at timepoint t. We can write this vector out in long hand as an array:

$$\pi_t = \left[\pi_t(1), \pi_t(2), \ldots, \pi_t(N) \right]$$

Eq. 2

On the right-hand side of *Eq. 2*, we have denoted the probabilities as $\pi_t(i)$ to indicate that the probability of each state i can change over time. This is because in a first-order Markov process, the probability of being in state j at the next timepoint $t+1$ changes according to which state we are currently in. In more formal notation, we write this as the following:

$$\text{Prob}\left(X_{t+1} = j \mid X_t = i\right) = p_{ij}$$

Eq. 3

This means that we have a probability matrix $\underline{P}$, with matrix elements p_{ij}, with the probability p_{ij} being the probability of being in state j at timepoint $t + 1$ given we were in state i at timepoint t. Notice that the probabilities p_{ij} do not depend upon time. They are static.

Because which state i we have at timepoint t determines the probabilities of which state we get at the next timepoint $t + 1$, it means the state probability distribution π_t influences the state probability distribution π_{t+1}. In other words, the state probability distribution changes over time. It evolves. In a moment, we'll derive the equation that shows precisely how π_t evolves over time, but first, we'll look at the probability matrix $\underline{P}$ in more detail.

The transition probability matrix

We can think of *Eq. 3* as saying that the probability of going from state i to state j is p_{ij}. That is, *Eq. 3* describes the probabilities of transitioning from state i to state j. Consequently, we refer to the probabilities p_{ij} as **transition probabilities**, and the matrix $\underline{P}$ as a **transition matrix**. You will also see it referred to as a **transition probability matrix**, a **transition rate matrix**, or just a plain **rate matrix** since it describes the rate (frequency) with which we transition from one state to another.

Properties of the transition probability matrix

Since the transition matrix elements p_{ij} are probabilities, they must satisfy the rules of probabilities and probability distributions. Firstly, this means the following:

$$0 \leq p_{ij} \leq 1 \quad \text{for all } i \text{ and } j$$

Eq. 4

Secondly, the total probability of going from state i to any state must be 1 because we must end up in some state after having been in state i. In math, this means the following:

$$\sum_{j=1}^{N} p_{ij} = 1 \quad \text{for all } i$$

Eq. 5

In matrix algebra, we can write this as follows:

$$\underline{P}\,\underline{1}_N$$

Eq. 6

Here, $\underline{1}_N$ is a column vector with N elements that are all 1. To make this more concrete, let's look at an example.

Epidemic modeling with a first-order discrete Markov process

When modeling the progression of a disease outbreak or an epidemic, one of the standard models used by epidemiologists and mathematical modelers is the SIR model. The acronym **SIR** refers to the three states in the model; **S** = susceptible to infection, **I** = infected, **R** = recovered from infection. In one timestep, a person can move between these states. For example, a person who is susceptible to infection may become infected, while a person who has been infected may recover from their infection. For a person who had been infected and recovered, we would expect them to have built up some immunity to future infections. It is not guaranteed that a recovered person won't become susceptible in the future, but we would expect the probability of transitioning from a recovered state to a susceptible state to be small, certainly smaller than the probability of transitioning from a susceptible state to an infected state. Similarly, a person who is infected can't go straight back to being susceptible – they would have to recover first. Likewise, a person can't jump straight from being in a susceptible state to a recovered state – they must become infected first. We'll map our S, I, and R states to the integers 1,2, and 3, respectively. So the susceptible state is equivalent to a state value of 1; the infected state is equivalent to a state value of 2; and the recovered state is equivalent to a state value of 3. Given this mapping and the aforementioned considerations, the transition matrix may look like this:

$$\underline{P}_{SIR} = \begin{bmatrix} 0.9 & 0.1 & 0.0 \\ 0.0 & 0.8 & 0.2 \\ 0.05 & 0.0 & 0.95 \end{bmatrix}$$

Eq. 7

The transition matrix in *Eq. 7* is a simplification. In real epidemic modeling, the transition probabilities would be dependent on the proportions of the population in each state. For example, the more people who are infected, then the higher the probability that someone susceptible will become infected because there is a higher probability of that susceptible person encountering an infected person. This means that in real-world epidemic modeling the transition probabilities are dynamic themselves. However, for simplicity of explanation, we will take the SIR transition probabilities to be static and we will use the matrix in *Eq. 7*.

The transition probability matrix is a network

The matrix in *Eq. 7* looks like the weighted adjacency matrices we studied in the previous chapter on networks. Does this mean we can think of our states and the transition rate matrix as a network and use network diagrams to represent it? The answer is yes. In fact, *Figure 11.1* shows a network diagram representation of the transition matrix in *Eq. 7*:

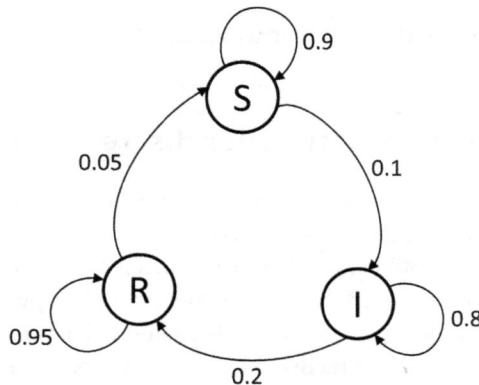

Figure 11.1: Network representation of the SIR transition rate matrix in Eq. 7

If we look at both *Figure 11.1* and the transition rate matrix in *Eq. 7*, we notice several things:

- The network in *Figure 11.1* has arcs from each node to itself. This is because there is a nonzero probability of staying in each state from one timepoint to the next.

- The transition rate matrix is not symmetric. This makes sense from the perspective of our SIR model but is a very general occurrence. It would be unusual to study a problem where the transition rate matrix was symmetric.

- There are matrix elements in the transition rate matrix that are zero. This means that the transition occurs with zero probability. It is effectively forbidden. For example, the transition from a susceptible state to a recovered state has zero probability because you can't recover if you haven't been infected, and so you must transition from susceptible to infected first before you can get to the recovered state.

Using the transition matrix to generate state trajectories

The transition rate matrix in *Eq. 7* tells us how to evolve the state from one timepoint to the next. That means we can use *Eq. 7* to generate an example sequence of states for an individual person. We can start the sequence in the susceptible state. This is the $t = 0$ state. From there, the first row in the transition rate matrix in *Eq. 7* tells us the probability distribution from which the next state is drawn. This is the $t = 1$ state. Once we know the $t = 1$ state, we can select the corresponding row in the rate matrix and use that to sample the $t = 2$ state. How do we draw states from a probability distribution? We learned about this in *Chapter 2*. We can use the numpy.random.choice function to do this for us. In fact, we'll now look at a code example doing just that.

Code example – using the transition rate matrix to evolve the state

The code example given next, and more, can be found in the Code_Examples_Chap11.ipynb Jupyter notebook in the GitHub repository. We're going to start from the susceptible state, and sample the next state using the probabilities from the first row of the rate matrix in *Eq. 7*. We'll repeat a further nine times, each time using the probability distribution from the row of the rate matrix corresponding to the current state. At the end, we will have a sequence of 11 states. This will be an example trajectory from our SIR first-order Markov process and could represent the trajectory of an individual person:

```python
import numpy as np
# Set the seed for the numpy random number generator.
# This ensures that we get reproducible results
np.random.seed(17335)

### Define our states
# Our states are "Susceptible" = index 0,
# "Infected" = index 1
# "Recovered" = index 2
```

```
state_map = {0:'S', 1:'I', 2:'R'}

### Define our rate matrix
rate_matrix = np.array([
    [0.9, 0.1, 0.0],[0.0, 0.8, 0.2],[0.05, 0.0, 0.95]])

### We'll start in the susceptible state and then sample
### our next state and repeat for 10 iterations
n_iter=10

# Set initial state and its label
current_state = 0
states_sequence = [state_map[current_state]]

# Do 10 iterations
for i in range(n_iter):
    # Get transition probabilities when starting from
    # our current state
    next_state_probs = rate_matrix[current_state,:]

    # Use numpy function to sample an integer from [0,1,2]
    # with the specified transition probabilities
    next_state = np.random.choice(3, p=next_state_probs)

    # Update our current state to this new state
    # and get its label
    current_state = next_state
    states_sequence.extend(state_map[current_state])
```

Let's look at the generated sequence:

```
print(states_sequence)
```

This gives the following output:

```
['S', 'S', 'S', 'S', 'S', 'I', 'R', 'R', 'R', 'R', 'R']
```

We can see that the individual here remained in the susceptible state for five timepoints, then became infected, but quickly recovered.

The preceding code example shows how to generate an example trajectory from a first-order discrete Markov process, but one person's trajectory from an SIR model does not give us a complete picture of the epidemic. What is happening across the whole population of individuals? How are the proportions of people in the S, I and R states changing over time? We'll answer that question in the next sub-section.

Evolution of the state probability distribution

We can calculate the distribution of states at the next timepoint just using the rules of probability (see *Chapter 2*). We have the following:

$$\text{Prob}(X_t = j) = \sum_{i=1}^{N} \text{Prob}(X_{t-1} = i)\text{Prob}(i \rightarrow j)$$

Eq. 8

In *Eq. 8*, the notation $\text{Prob}(i \rightarrow j)$ is the probability of transitioning from state i to state j and so is just our transition matrix probability p_{ij}. *Eq. 8* says that the probability of ending up in state j at timepoint t is the sum of the probabilities of all the ways we can get to j; that is, a sum over all the starting states i at timepoint $t - 1$. Recall our shorthand notation $\pi_t(i)$ for $\text{Prob}(X_t = i)$ and $\underline{\pi}_t$ for the vector of probabilities $[\pi_t(1), \pi_t(2), ..., \pi_t(N)]$. In terms of $\underline{\pi}_t$ and our transition rate matrix $\underline{P}$, we can write *Eq. 8* as follows:

$$\underline{\pi}_t = \underline{\pi}_{t-1}\underline{P}$$

Eq. 9

Eq. 9 tells us how the probability distribution over states evolves over time. It tells us in the language of linear algebra that we learned in *Chapter 3*. *Eq. 9* is our evolution equation for first-order discrete Markov processes.

We can iterate *Eq. 9*. The probability distribution $\underline{\pi}_{t-1}$ can also be written, using *Eq. 9*, as $\underline{\pi}_{t-1} = \underline{\pi}_{t-2}\underline{P}$. Plugging this back into *Eq. 9*, we get the following:

$$\underline{\pi}_t = \underline{\pi}_{t-2}\underline{P}\,\underline{P} = \underline{\pi}_{t-2}\,\underline{P}^2$$

Eq. 10

Continuing this iteration, we get the following:

$$\underline{\pi}_t = \underline{\pi}_0\underline{P}^t$$

Eq. 11

Eq. 11 tells us what our probability distribution will be at any later timepoint t, given our starting distribution $\underline{\pi}_0$ and our rate matrix $\underline{P}$. This is true even if we start at $t = 0$ in a definite state. If at $t = 0$ we are in a susceptible state in our SIR example, then we just set $\underline{\pi}_0 = [1,0,0]$. This reflects the fact that at $t = 0$, before any epidemic has begun, 100% of people are in a susceptible state, no one is infected, and no one has yet recovered from an infection. Note that due to the stochastic nature of the state evolution, even if we start off with a 100% probability of being in the susceptible state, at a later timepoint t we will have a probability distribution that is spread out, and the probabilities of being infected or recovered will be nonzero.

What happens over the long term? How much spreading out of the state probability distribution happens? What happens if we make t big? What happens if we take $t \to \infty$? We'll find out in the next sub-section.

Stationary distributions and limiting distributions

Let's increase t in *Eq. 11*. We will assume (for now) that as we increase t, the state probability distribution settles down and gets closer and closer to some particular set of values. This is the **limiting distribution**. We'll use the notation π_∞ to represent this limiting distribution. By definition of taking $t \to \infty$, the limiting distribution π_∞ is mathematically defined from *Eq. 11* as follows:

$$\pi_\infty = \lim_{t \to \infty} \pi_0 \underline{P}^t$$

Eq. 12

You'll notice that we have written π_∞ on the left-hand side of *Eq. 12* as though it doesn't depend upon the starting distribution π_0. That is because it doesn't. No matter what starting distribution π_0 we start with, if we iterate the evolution equation in *Eq. 9* for an infinite amount of time, we will always end up with the distribution π_∞. This is not quite true, but will be for many real-world situations, and will certainly be true for all the examples we present here – see point *1* in the *Notes and further reading* section at the end of the chapter for more details.

With that caveat in place, can we unpack what π_∞ represents more intuitively? Mathematically, π_∞ is what we get if we iterate the evolution equation in *Eq. 9* for an infinite amount of time. But this also means that π_∞ is close to the distribution we would get if we iterated *Eq. 9* for a **long but finite** amount of time. In our SIR example, this means that π_∞ would be close to the distribution we would get at the end of a long epidemic. It is as though our system is approaching a stable stationary equilibrium. So, how do we identify this stationary equilibrium? This brings us to another concept and helps us address the question of whether we always have a limiting distribution.

The new concept we need is that of a **stationary distribution**. As the name suggests, a stationary distribution is a state probability distribution that doesn't change in time. Because it doesn't change in time, we'll use the notation π to represent it. In comparison to π_t, we have dropped the subscript t because by definition, the stationary distribution does not depend upon time. Because π doesn't change with time, it means that when we apply the evolution equation in *Eq. 9*, we must get π back. In other words, the stationary state distribution satisfies the following equation:

$$\pi = \pi \underline{P}$$

Eq. 13

This equation looks familiar. It is saying that when we operate on the vector π with the matrix $\underline{P}$, we get the vector π. It is an eigenvector equation that we met in *Chapter 3*. Formally, the stationary distribution π is a left eigenvector of the matrix $\underline{P}$ corresponding to an eigenvalue of 1. It is a left eigenvector because we are multiplying by $\underline{P}$ with π to the left of $\underline{P}$.

Another way of saying *Eq. 13* is that we can only get a stationary distribution if the transition matrix $\underline{P}$ has a left eigenvector with eigenvalue 1. This gives a mathematical way of identifying stationary distributions.

We can also take the transpose of *Eq. 13* and get the following:

$$\underline{P}^\mathsf{T} \pi^\mathsf{T} = \pi^\mathsf{T}$$

Eq. 14

This means that the stationary distribution, or rather its transpose π^T, is also a right eigenvector of $\underline{P}^\mathsf{T}$ corresponding to an eigenvalue of 1.

But what is so special about stationary distributions? Well, let's take our limiting distribution π_∞ and apply our evolution equation *Eq. 9* to π_∞ one more time. What does $\pi_\infty \underline{P}$ give us? Because π_∞ is what we got when we applied $\underline{P}$ to π_0 an infinite number of times, applying $\underline{P}$ to π_∞ is the same as applying $\underline{P}$ to π_0 infinity+1 number of times. But infinity+1 is still infinite, so it can't have changed anything. This means applying $\underline{P}$ to π_∞ must give us back π_∞. In math, we have the following:

$$\pi_\infty = \pi_\infty \underline{P}$$

Eq. 15

Comparing *Eq. 15* to *Eq. 13* immediately tells us that our limiting distribution π_∞, if it exists, is a stationary distribution. This also means that stationary distributions are candidates for a limiting distribution. And we already know how to identify stationary distributions from the left eigenvectors of $\underline{P}$. The gotcha here is that a limiting distribution is a stationary distribution, but not all stationary distributions are limiting distributions. That is why we said the stationary distributions of $\underline{P}$ only give us candidates for the limiting distribution. It is possible for a transition matrix $\underline{P}$ to have stable distributions but no limiting distributions.

With that gotcha understood, let's look at a code example. We'll revisit our SIR example and calculate the left eigenvectors of $\underline{P}_{SIR}$ in *Eq. 7*, and we'll iterate the evolution equation *Eq. 9* for a large number of times.

Code example – stationary and limiting distributions

The code example given next, and more, can be found in the `Code_Examples_Chap11.ipynb` Jupyter notebook in the GitHub repository. Any stable distribution of the first-order Markov process is given by a right eigenvector of the transpose of the transition rate matrix corresponding to eigenvalue 1. We can use the `numpy.linalg.eig` function, which computes the right eigenvectors and eigenvalues of a square matrix. We'll assume we have already imported the `numpy` package as `np` and still have the rate matrix from our SIR previous code example in memory. First, we compute the right eigenvectors of the rate matrix:

```
# Compute the right-eigenvectors and eigenvalues of our rate matrix
rate_eigen = np.linalg.eig(np.transpose(rate_matrix))
```

We can then look at the eigenvalues:

```
# The eigenvalues are held in the first element of the tuple
rate_eigen[0]
```

This gives the following output:

```
array([0.825+0.06614378j, 0.825-0.06614378j, 1.0+0.j])
```

The last eigenvalue is 1 (its imaginary part is zero). Therefore, we can use the last eigenvector as the stable distribution. It will be normalized to unit length, but we need its elements to sum to 1. But remember from *Chapter 3* that any multiple of an eigenvector is still an eigenvector with the same eigenvalue, so we can just re-scale it to sum to 1:

```
# Extract the last eigenvector (we can drop the zero imaginary part)
pi_stable = np.real(rate_eigen[1][:,2])

# Rescale the vector so its elements sum to 1
pi_stable /=np.sum(pi_stable)
```

Let's look at the stable distribution:

```
# Print the stable distribution
pi_stable
```

This gives the following output:

```
array([0.28571429, 0.14285714, 0.57142857])
```

From this, we can see that for the stable distribution, 28.6% of people are in the susceptible state, 14.3% are in the infected state, and 57.1% are in the recovered state.

So, we have a stable distribution, but do we have a limiting distribution, and if so, is it the same as the stable distribution? To approximate any limiting distribution, we'll start from a definite initial state; that is, a distribution that has only one nonzero value. We'll then apply the transition rate matrix a large number (100) of times to get an approximation of the limiting distribution:

```
# Set the initial state distribution. We'll
# set it to a distribution representing 100% of people
# being in the susceptible state
current_distribution = np.array([1.0, 0.0, 0.0])

n_iter = 100
for i in range(n_iter):
    # Get the state distribution at the next timepoint by
    # multiplying by the transition rate matrix
    next_distribution = np.matmul(current_distribution, rate_matrix)
    current_distribution = next_distribution

current_distribution
```

This gives the following output for the approximation of the limiting distribution:

```
array([0.28571429, 0.14285715, 0.57142856])
```

By comparing our approximation of the limiting distribution to the stable distribution we obtained previously, we can see they are numerically identical (up to the precision of the calculation). Our limiting distribution is the same as our stable distribution. If our epidemic were to run for a long time without any further interventions, we would have 28.6% susceptible, 14.3% infected, and 57.1% recovered.

Having done that simple but instructive code example, we'll now look at an interesting characteristic of first-order Markov processes: the fact that they are memoryless.

First-order discrete Markov processes are memoryless

Wait! What? What do we mean by saying our first-order Markov process is memoryless? The transition probabilities depend upon the current state. The next state is strongly influenced by what has come before. Surely that is memory?

Not quite. What we're talking about here is **conditional independence**. Once we know (or specify) the current state, we have all the information we need to produce the next state (by sampling from the appropriate row of the transition matrix). The state preceding the current state has no influence on the next state once we have specified the current state. Given the current state, the next state is independent of the preceding state. That is what we mean by conditional independence. This property is also termed the **Markov property** and is what we mean when we say a system is **Markovian**.

In practical terms, what does this mean? In a first-order model, specifying the current state means we completely forget about the preceding state. That is what we mean by memoryless. It is a strength of Markov models as it simplifies the mathematics. But it is also a weakness of Markov models, as it means beyond the current state the state history has no effect on the next state. We'll address this weakness in part in the next section, but we won't be able to eliminate this weakness.

To finish this section on first-order discrete Markov processes, we'll return to the fact that they are probabilistic models. If you remember from *Chapter 5*, a probabilistic model allows us to calculate things such as the likelihood of the data given the model and its parameters. We can do lots of useful things with the likelihood once we have a formula for it. So, what is the formula for the likelihood of an observed state sequence given a first-order discrete Markov model?

Likelihood of the state sequence

Imagine we have a state sequence for timepoints $t = 0, 1, \ldots, T$. This means we have a set of random variables $X_0, X_1, \ldots, X_T$ for which we got the following observed state values: $i_0, i_1, \ldots, i_T$. What is the likelihood of that sequence of observed state values? Remember from *Chapter 5* that the likelihood is the probability of the data given the model parameters. In this case, the model parameters are the transition probabilities and the initial state distribution. Can we write the probability of the data in terms of these model parameters? To see that we can, we'll first write the probability of the data as follows:

$$\text{Prob}(X_0 = i_0, X_1 = i_1, \ldots, X_T = i_T) = \text{Prob}(X_0 = i_0) \times \text{Prob}(X_1 = i_1 | X_0 = i_0) \times \text{Prob}(X_2 = i_2 | X_1 = i_1) \times \ldots \times \text{Prob}(X_T = i_T | X_{T-1} = i_{T-1})$$

Eq. 16

The right-hand side of *Eq. 16* follows from the conditional independence property of the first-order Markov process. Once we know the state at time $t - 1$, the probability distribution of the next state at time t is entirely known and given by the appropriate row of the transition matrix. This means we can write the right-hand side of *Eq. 16* as follows:

$$\pi_0(i_0) \times p_{i_0 i_1} \times p_{i_1 i_2} \times p_{i_2 i_3} \times \ldots \times p_{i_{T-1} i_T}$$

Eq. 17

In *Eq. 17* π_0 is the state probability distribution at $t = 0$. *Eq. 17* has expressed the likelihood in terms of the model parameters, but we can simplify the expression in *Eq. 17* further. Because many of the transitions we observe in the state sequence will be of the same type (for example, a $2 \rightarrow 3$ or an I→R in our SIR epidemic example), we can combine them in *Eq. 17*. We then get the following:

$$\text{Likelihood}(i_0, i_1, i_2, \ldots, i_T) = \pi_0(i_0) \prod_{i=1}^{N} \prod_{j=1}^{N} p_{ij}^{N_{ij}}$$

Eq. 18

Here, N_{ij} is the number of times we observe a transition $i \to j$ in the state sequence $i_0, i_1, i_2, \ldots, i_T$. The expression in *Eq. 18* is very compact and easy to calculate, as all we need to do is count the number of observed transitions of each type.

Having obtained a simple expression for the likelihood of an observed state sequence, can we use it to do useful things, such as estimating the model parameters via **maximum likelihood estimation (MLE)**? The answer is yes, and so next, we'll look at how to use Eq. 18 to estimate the transition probabilities p_{ij} via maximum likelihood.

MLE of the transition probabilities

The log-likelihood of the observed state sequence is given by taking the logarithm of *Eq. 18*. We get the following:

$$\text{log-likelihood} = \log \pi_0(i_0) + \sum_{i=1}^{N} \sum_{j=1}^{N} N_{ij} \log p_{ij}$$

Eq. 19

We want to maximize the log-likelihood in *Eq. 19* with respect to each p_{ij} and subject to the constraints, $\sum_j p_{ij} = 1$ for all i. We do this maximization using the differential calculus recapped in *Chapter 1* and using Lagrange multipliers to impose the constraints. Differentiating *Eq. 19* with respect to p_{ij} and setting the derivative to zero, we get the following solution:

$$\hat{p}_{ij} = \lambda_i N_{ij}$$

Eq. 20

In *Eq. 20*, λ_i is the Lagrange multiplier for the constraint $\sum_j p_{ij} = 1$. The hat symbol on top of $\hat{p}_{ij}$ is there to remind us that this is an **estimate** of p_{ij} and not the true value of p_{ij} itself. The expression in *Eq. 20* is very simple, and we can easily adjust λ_i to match the constraint, to finally get the following:

$$\hat{p}_{ij} = \frac{N_{ij}}{\sum_{j'=1}^{N} N_{ij'}}$$

Eq. 21

The expression in *Eq. 21* is very simple and very pleasing. It says that the maximum likelihood estimates of the transition probabilities are just the observed proportions of each transition type.

We have covered a lot in this section on first-order discrete Markov processes. That is because they are extremely useful for building simple models of dynamical processes. They are a workhorse model – used everywhere. Let's recap what we have learned about them.

What we learned

In this section, we have learned the following:

- How first-order discrete Markov processes model how a system evolves from one state to another over time
- That the time variable in a first-order discrete Markov process is discrete; that is, it increases in steps $t \rightarrow t + 1$
- That a first-order discrete Markov process is a probabilistic model
- That the probability distribution from which the next state is drawn depends only on the current state and is given by the corresponding row of the transition probability matrix
- That the transition matrix can be represented as a network (a graph)
- How to generate trajectories from a first-order discrete Markov process using the rate matrix
- About stable distributions of first-order discrete Markov processes and how they are given by left eigenvectors of the rate matrix corresponding to an eigenvalue of 1
- About limiting distributions of first-order discrete Markov processes
- That we don't always have a limiting distribution, but if we do, it will correspond to a stable distribution of the Markov process
- That first-order discrete Markov processes are memoryless
- How to calculate the likelihood of an observed state sequence and how to use the likelihood formula to estimate the transition probabilities via maximum likelihood

Having learned about first-order discrete Markov processes, in the next section, we will build upon and extend the idea to having the transition probabilities depend upon a longer amount of history than just the preceding state. This introduces what are known as higher-order discrete Markov processes. Since we will be building upon the concepts and terminology of first-order Markov processes, a lot of the next section will be very similar, and so it will be a much shorter section.

Higher-order discrete Markov processes

First-order discrete Markov processes are extremely flexible, but there may be situations where we feel that the assumption that the transition probabilities depend only on the current state is an unrealistic one. There are plenty of situations where what happens next depends on more than just the preceding state in history. Take a sequence of words, for example. It would be a crude model that said that the probability of the next word in this sentence only depended on the immediately preceding word – see point *2* in the *Notes and further reading* section at the end of the chapter.

Can we improve upon this simple assumption? Can we make our transition probabilities depend upon longer stretches of history? Yes, we can. We can take the simplest case of the transition probabilities depending not only on the current state but also on the preceding state. This means the probability of a state depends upon the previous two states. For obvious reasons, this is called a second-order discrete Markov process.

Second-order discrete Markov processes

Since the transition probabilities in a second-order process depend upon the current state and the state before that, this means that we have a transition matrix of the following form:

$$p_{ij,k} = \text{Prob}(i,j \to k) = \text{Prob}\left(X_{t+1} = k \mid X_t = j, X_{t-1} = i\right)$$

Eq. 22

If we have N states, then we interpret *Eq. 22* as saying that the transition matrix is a $N^2 \times N$ matrix. The N^2 rows are indexed by the combination ij, with $i,j \in 1, 2, \ldots, N$. Each row gives a probability distribution. The row corresponding to ij gives us the probabilities of the next state given we are currently in state j and were in state i before that.

Many of the properties of our new transition matrix elements are the same or similar to the transition probabilities for a first-order discrete Markov process. For example, the probabilities in each row must sum to 1, so we have the following:

$$\sum_{k=1}^{N} p_{ij,k} = 1 \quad \text{for all } i \text{ and } j$$

Eq. 23

To illustrate this, let's consider a scenario where we have $N = 3$ states. The transition probabilities $p_{ij,k}$ can be represented in matrix form, as shown in *Figure 11.2*:

Columns denote the next state

	$k = 1$	$k = 2$	$k = 3$
$i = 1, j = 1$	$p_{11,1}$	$p_{11,2}$	$p_{11,3}$
$i = 1, j = 2$	$p_{12,1}$	$p_{12,2}$	$p_{12,3}$
$i = 1, j = 3$	$p_{13,1}$	$p_{13,2}$	$p_{13,3}$
$i = 2, j = 1$	$p_{21,1}$	$p_{21,2}$	$p_{21,3}$
$i = 2, j = 2$	$p_{22,1}$	$p_{22,2}$	$p_{22,3}$
$i = 2, j = 3$	$p_{23,1}$	$p_{23,2}$	$p_{23,3}$
$i = 3, j = 1$	$p_{31,1}$	$p_{31,2}$	$p_{31,3}$
$i = 3, j = 2$	$p_{32,1}$	$p_{32,2}$	$p_{32,3}$
$i = 3, j = 3$	$p_{33,1}$	$p_{33,2}$	$p_{33,3}$

Rows denote the preceding two states

Figure 11.2: Transition probabilities for a second-order discrete Markov process in matrix form

The transition probabilities $p_{ij,k}$ are static, just as for first-order processes. Similarly, once we know the current state j and the preceding state i we can use the probabilities in row ij to sample the next state. Repeating this, we can generate an example trajectory from our second-order discrete Markov process. How we use second-order processes isn't really that different from how we use first-order processes.

Extending to third-order processes, fourth-order processes, and so on follows a similar pattern. For an nth process, the probability of the next state is dependent on the preceding n states. Our transition matrix is of the following form:

$$p_{i_1 i_2 \cdots i_n, i_{n+1}} = \text{Prob}(i_1 i_2 \cdots i_n \to i_{n+1}) = \text{Prob}\left(X_{t+1} = i_{n+1} \mid X_t = i_n, X_{t-1} = i_{n-1}, \ldots, X_{t-n+1} = i_1\right)$$

Eq. 24

We interpret *Eq. 24* as saying that the transition matrix is a $N^n \times N$ matrix. The N^n rows are indexed by the combination $i_1 i_2 \ldots i_n$. Each row gives a probability distribution. The row corresponding to $i_1 i_2 \ldots i_n$ gives us the probabilities of the next state given we are currently in state i_n, were in state i_{n-1} before that, were in state i_{n-2} two timepoints previously, and so on, up to being in state i_1 n timepoints ago.

As always, the transition probabilities $p_{i_1 i_2 \cdots i_n, i_{n+1}}$ are static. Again, normalization of the transition probabilities means we have the following:

$$\sum_{i_{n+1}=1}^{N} p_{i_1 i_2 \cdots i_n, i_{n+1}} = 1$$

Eq. 25

One thing to note about higher-order models is that we have a lot of transition probabilities: N^{n+1}, in fact. Ultimately, we will have to estimate these probabilities from data. With N^{n+1} transition probabilities to estimate that is a lot of training data we require. Consequently, it may be preferable to construct a simplified model of the transition probabilities themselves. We won't go into how to do that here, as it is a chapter in itself. Instead, we aim to make you aware of the issue and of the generic approach to solving the issue – see point 3 in the *Notes and further reading* section at the end of the chapter for details of an example of this approach.

Evolution of the state probability distribution in higher-order models

One thing you may have noticed is that when talking about second-order and n^{th} discrete processes, we haven't discussed how the state probability distribution evolves. We haven't introduced the equivalent of *Eq. 9* that was used in the evolution of first-order discrete Markov processes. Why is that? Well, let's restrict ourselves to second-order discrete Markov processes and try to formulate the equivalent of *Eq. 9*.

To determine the probability of being in state k at time t, we need to do the equivalent of *Eq. 8* and sum over all the rows in the transition matrix, picking out the probability in column k. That means we have the following,

$$\text{Prob}(X_t = k) = \sum_{i=1}^{N} \sum_{j=1}^{N} \text{Prob}(X_{t-1} = j, X_{t-2} = i) p_{ij,k}$$

Eq. 26

Let's introduce the following shorthand notation:

$$\pi_t^{(2)}(i,j) = \text{Prob}(X_t = j, X_{t-1} = i)$$

Eq. 27

We can then write *Eq. 26* in matrix algebra terms as follows:

$$\underline{\pi}_t = \underline{\pi}_{t-1}^{(2)} \underline{P}$$

Eq. 28

In *Eq. 28*, the matrix $\underline{P}$ is the matrix of the second-order transition probabilities $p_{ij,k}$ as shown in *Figure 11.2*. Also, in *Eq. 27* and *Eq. 28*, the superscript (2) in $\pi_t^{(2)}$ is used to indicate that the probabilities $\pi_t^{(2)}(i,j)$ depend upon two state values, i and j.

You'll notice that the left-hand side of *Eq. 28* is a vector of probabilities that depends only on a single state, while the right-hand side has a vector of probabilities that depends on two states. This means that REF _Ref151061948 \ h *Eq. 28* does not have the pleasing symmetry of *Eq. 9* where both the left-hand and right-hand sides are single-state probability distributions. Is there a way around this? Can we make our second-order discrete Markov process look more like a first-order Markov process? More generally, can we make an nth-order discrete Markov process look like a first-order process? With a bit of re-defining of what we mean by a state, the answer is yes. In the next sub-section, we'll show how.

A higher-order discrete Markov process is a first-order discrete Markov process in disguise

In a second-order discrete Markov process, we use a transition matrix to model transitions of the form $ij \rightarrow k$. The transition $ij \rightarrow k$ means, for example, the neighboring pair of states $X_{t-2} = i$, $X_{t-1} = j$ lead to state $X_t = k$. After that transition has occurred, we can take the neighboring pair of states $X_{t-1} = j, X_t = k$ and use them along with the transition probabilities to generate the state for the next timepoint X_{t+1}.

What we are doing here is taking states two at a time. This suggests re-defining our states as pairs of states. So instead of having states $i \in 1, 2, ..., N$, we have state pairs $ij \in 11, 12, ..., 1N, 21, 22, ...2N, ..., NN$. We have N^2 possible state pairs. At this stage, the notation is getting a bit cumbersome, so we'll just re-index those N^2 states using the variable $l \in 1, 2, ..., N^2$. A bit of simple math shows that for state pair ij, the index value $l = (i - 1)N + j$. So, now we have transitions from one state pair to another. We can think of l being equivalent to a state pair, and so we have transitions from l to l' where $l, l' \in 1, 2, ..., N^2$. Because we have the same number, N^2, of "from" states as "to" states in our new transition structure, this is beginning to look more like a first-order process. We now have transition probabilities $p_{ll'}$ that represent the probability of transition from state pair l to state pair l'. But what does the transition matrix look like? It is clearly going to be a $N^2 \times N^2$ matrix, but what are the matrix elements? Here is where we need to think carefully. If we transition from state pair $l \equiv ij$, then our "to" state must be a pair that starts with j. For example, we can get a transition $12 \rightarrow 23$ but not a transition $12 \rightarrow 13$. Consequently, many of our transition probabilities are going to be zero in this new expanded way of writing things. To make it more concrete, we'll take the second-order process transition matrix in *Figure 11.2* and write it in this new form using "from" and "to" state pairs. The resulting transition matrix is shown in *Figure 11.3*:

Columns are the 'to' state-pairs, indexed by l'

	$l' = 1$ ($jk = 11$)	$l' = 2$ ($jk = 12$)	$l' = 3$ ($jk = 13$)	$l' = 4$ ($jk = 21$)	$l' = 5$ ($jk = 22$)	$l' = 6$ ($jk = 23$)	$l' = 7$ ($jk = 31$)	$l' = 8$ ($jk = 32$)	$l' = 9$ ($jk = 33$)
$l = 1$ ($ij = 11$)	$p_{11,1}$	$p_{11,2}$	$p_{11,3}$	0	0	0	0	0	0
$l = 2$ ($ij = 12$)	0	0	0	$p_{12,1}$	$p_{12,2}$	$p_{12,3}$	0	0	0
$l = 3$ ($ij = 13$)	0	0	0	0	0	0	$p_{13,1}$	$p_{13,2}$	$p_{13,3}$
$l = 4$ ($ij = 21$)	$p_{21,1}$	$p_{21,2}$	$p_{21,3}$	0	0	0	0	0	0
$l = 5$ ($ij = 22$)	0	0	0	$p_{22,1}$	$p_{22,2}$	$p_{22,3}$	0	0	0
$l = 6$ ($ij = 23$)	0	0	0	0	0	0	$p_{23,1}$	$p_{23,2}$	$p_{23,3}$
$l = 7$ ($ij = 31$)	$p_{31,1}$	$p_{31,2}$	$p_{31,3}$	0	0	0	0	0	0
$l = 8$ ($ij = 32$)	0	0	0	$p_{32,1}$	$p_{32,2}$	$p_{32,3}$	0	0	0
$l = 9$ ($ij = 33$)	0	0	0	0	0	0	$p_{33,1}$	$p_{33,2}$	$p_{33,3}$

Rows are the 'from' state-pairs, indexed by l

Figure 11.3: Transition probabilities for a second-order discrete Markov process written as a first-order process transition matrix

In *Figure 11.3*, we can clearly see all the zero transition probabilities that correspond to transitions that can't occur when we construct overlapping "from" and "to" state pairs from a sequence of states.

We have also written the matrix elements $p_{ll'}$ in *Figure 11.3* in terms of the matrix elements $p_{ij,k}$ in *Figure 11.2*. Because of this, we can see that on any row of the matrix in *Figure 11.3*, we only have the probabilities from a single row of the matrix in *Figure 11.2*. This means that since each row of the matrix in *Figure 11.2* is a properly normalized probability distribution, each row of the matrix in *Figure 11.3* is too. In math terms, we have the following:

$$\sum_{l'=1}^{N^2} p_{ll'} = \sum_{k=1}^{N} p_{ij,k} = 1 \text{ for all } l \equiv ij$$

Eq. 29

As we hoped, our second-order discrete Markov process can be thought of as just a first-order discrete Markov process that has been written in a more compact form. If we unpack the second-order process, we reveal the first-order process inside, albeit with a larger and sparse transition matrix.

What this means is that all the results we proved about first-order processes, such as how to find stable and limiting distributions, apply to second-order processes as well.

You'll not be surprised to find that the idea of expanding the states into overlapping sequences extends to n^{th}-order discrete Markov processes, and we can write an nth-order process as a first-order process. But be aware – for an nth-order process, we use overlapping tuples of n consecutive states. Our new state tuples are indexed by $l \in 1, 2, ..., N^n$, and our expanded transition matrix is $N^n \times N^n$. Even for $n = 3$ or $n = 4$, the transition matrices can be very large, with most of the matrix elements being zero. Being able to write an nth-order process as an expanded first-order process is useful for proving mathematical results, but sometimes it is more computationally efficient to use the original n^{th}-order form of the process for certain calculations; for example, generating example state trajectories.

Higher-order discrete Markov processes are still memoryless

Since a higher-order discrete Markov process is just a first-order discrete Markov process in disguise, it must also be memoryless. By making transition probabilities depend upon more of the state history, we are making more of that history influence the next state that is produced, but we still have conditional independence. This means that for an nth-order process, once we have specified the values of X_{t-n}, $X_{t-n+1}, ..., X_{t-1}$, we have everything we need to produce the next state X_t. Anything of the older state history before X_{t-n} has no influence on the outcome of X_t. Just like for first-order Markov models, we have forgotten this older history.

Having made the important connection between higher-order and first-order discrete Markov models, we'll wrap up this section and summarize what we have learned.

What we learned

In this section, we have learned the following:

- In higher-order discrete Markov processes, the transition probabilities depend upon more of the preceding state history than just the current state.

- In higher-order discrete Markov processes, the transition probability matrix is non-square. This is because there are many more combinations of states that we transition from than the number of single states we transition to. The transition matrix is of size $N^n \times N$ for an nth-order process.

- By extending the definition of the "from" and "to" states in our transition matrix, we can write our n^{th}-order discrete Markov process as a first-order Markov process, but at the expense of making the transition matrix of size $N^n \times N^n$ and very sparse.

Having learned about higher-order discrete Markov processes, we'll now move on to another variant of Markov models, namely Hidden Markov Models (HMMs). These models are useful when we have differences between the states of the model and the data we see.

Hidden Markov Models

In this section, we will give a brief outline of what an HMM is. The mathematics behind HMMs is extensive and beyond what we can cover here. Instead, we will focus on describing what they are and how they work conceptually.

An HMM is a first-order discrete Markov process. This means it has a set of states between which it transitions, governed by a transition matrix. The difference from the first-order discrete Markov processes of the previous sections is that we don't directly observe those states. The states are hidden from us. They are **latent**, hence the term *hidden* in an HMM.

What we observe is a sequence of symbols that are emitted at each point along the state sequence. Why is this useful? One of the main benefits is that we can model situations where we think there are distinctly different phases or modes of behavior that we want to understand, but we can't directly observe those different modes. Instead, we can only observe the symbols. A concrete example will make this clearer.

The example we will use is that of wanting to understand a shopper's intent when they are browsing an e-commerce website. We can observe their actions such as which pages they are clicking on; for example, are they looking at the details of a product or are they reading a review, are they jumping to another similar product, or are they about to put the product in their cart and check out? What we'd like to understand is whether the shopper's intent is serious or whether they just browsing casually.

If we were able to observe a shopper's sequence of "intent" states, we might have a transition matrix network diagram like that shown in *Figure 11.4*:

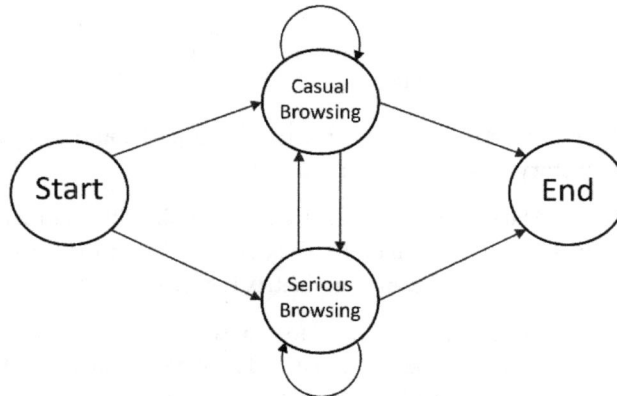

Figure 11.4: Transition structure for e-commerce shopper intent

In *Figure 11.4*, there are two main states, *Casual Browsing* and *Serious Browsing*. A shopper can switch between the two states but also remain within each state as time progresses. Here, our time variable will be the page clicks that the shopper makes. In the transition structure shown in *Figure 11.4*, we haven't given any transition probabilities. That is deliberate. At this stage, we just want to focus on the transition structure.

You'll also notice that in *Figure 11.4* we also have two additional states, *Start* and *End*. These states are included to allow us to start the state sequence off at a definite point – in this case, the *Start* state. So, every shopper browsing session is considered to start in the *Start* state. You'll notice that the *Start* state only has edges leaving it. A shopper starts at the *Start* state and moves either to *Casual Browsing* or *Serious Browsing*, but they can never move back to the *Start* state. A state with only out-edges is called a **source state** (or **source node** in network terminology) because it serves as the starting source of all paths. The *End* state plays a similar role. It only has in-edges, and so once we are in the *End* state, we can't go anywhere else – the browsing session has finished. A state with only in-edges is called a **sink state** or **absorbing state** (**sink node** or **absorbing node** in network terminology) because it acts as a state where paths vanish or sink or are effectively absorbed.

Emission probabilities

The transition structure in *Figure 11.4* is deliberately simplistic to make the explanation easier, but it retains enough realism of the real-world problem to illustrate the data science issue we are trying to solve. The problem is that we don't observe the shopper's state sequence. At any point in time (the click sequence), we don't know whether a shopper is just casually browsing or is more serious about buying. We'd like to, because we could potentially target the shopper with a relevant promotion to increase that buy probability. What we do observe is what page the shopper is on at each click. We know if the shopper has jumped to another product or is reading a customer review. We can record which action the shopper performed and so we have a sequence of recorded "symbols" telling us which action the shopper performed at each click. An individual browsing session might look something like that shown in *Figure 11.5*:

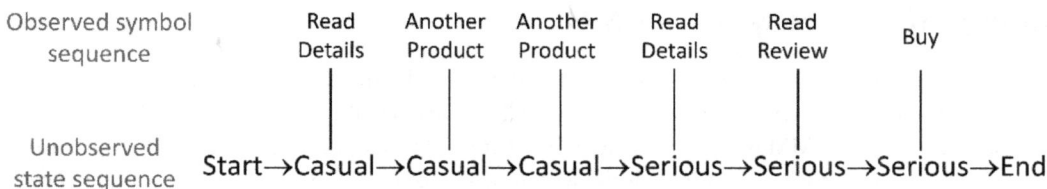

Figure 11.5: An example browsing session showing the state sequence and the symbol sequence

In *Figure 11.5*, we have shown the unobserved sequence of "intent" states that the shopper went through, along with the symbol or action the shopper took in each state. Note that no symbol or action was generated in the *Start* or *End* states, as their purpose is only to start and end the browsing session. For all other steps in the state sequence, the state has produced a symbol, which in this case represents the observed action the shopper took. In HMM terminology, we say the state has **emitted** a symbol. With each of the four states in *Figure 11.4*, we have a set of **emission probabilities** that determine how likely we are to emit each symbol given the state we are in. *Figure 11.6* shows the same transition structure as in *Figure 11.4* but with the emission probabilities (in the square boxes) for each state:

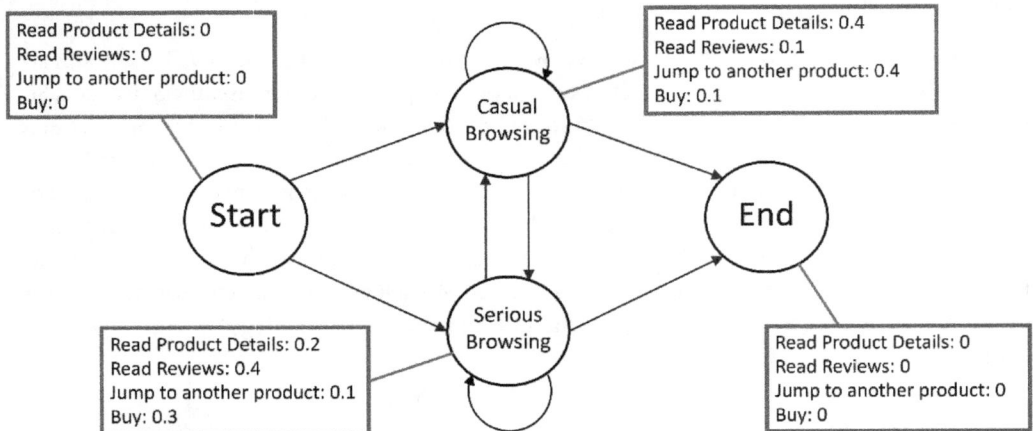

Figure 11.6: Shopper transition structure with the associated emission probabilities

From *Figure 11.6*, we can see that the *Casual Browsing* state has a higher probability of emitting the *Jump to another product* action compared to the *Serious Browsing* state, as the shopper jumps around looking at multiple products, not being able to work out which one they want to buy, if at all. Similarly, the *Serious Browsing* state has a higher emission probability for the *Read Reviews* action, as the shopper gets serious about evaluating a specific product. Ultimately, the *Serious Browsing* state also has a higher emission probability for the *Buy* action, as a shopper who is doing more than just an initial casual browse is more likely to buy a product. Note also that the *Start* and *End* states have zero emission probabilities as no actual shopper action is performed in these states.

Making inferences with an HMM

HMMs are extremely useful. They allow us to take an observed sequence of symbols or actions and model it in terms of a sequence of latent states that ordinarily we wouldn't have access to. Ideally, we'd like to use our shopper HMM in *Figure 11.6* to infer what was the shopper's intent at each point in the session. We'd like to infer what was the underlying state sequence from the observed symbol sequence.

Inferring the state sequence using the Viterbi algorithm

The state at each point in a sequence is a random variable. There are many different state values that can emit the symbol observed at a given timepoint. Consequently, there are many state sequences that are compatible with the observed symbol sequence. Which state sequence should we choose? If we don't know which one to choose, how can we infer the underlying state sequence?

As usual, when we make inferences using probabilistic models, we use the likelihood. If we have the model parameters (transition probabilities and emission probabilities), we could in principle take a sequence of symbols and use the likelihood to infer what is the most likely state sequence given the symbol sequence. The algorithm that allows us to make that inference is the **Viterbi algorithm**.

Let's think about that for a moment. Using the Viterbi algorithm, we can take a sequence of symbols or actions, such as what a shopper clicks on, and infer what their most likely intent was at each timepoint. That is a very powerful capability to have.

Estimating the transition probabilities using the Baum-Welch algorithm

However, there is a problem. To use the Viterbi algorithm, we need to know the state transition probabilities. But we can't use the maximum likelihood estimates of the transition probabilities in *Eq. 21* because to do so requires us to know the state sequence, which is hidden from us. What do we do? The answer is to infer the model parameters at the same time as we infer the expected state transitions in the state sequence. The algorithm that does that for us is the **Baum-Welch algorithm**.

Note that the Baum-Welch algorithm estimates the probability of each transition type for each timepoint *t*. An alternative, known as **Viterbi training**, uses an iterative process whereby the most probable state sequence, obtained via the Viterbi algorithm, is used to update the estimates of the transition and emission probabilities. So, in Viterbi training, we get a single definite state path estimated along with estimates of the transition and emission probabilities, while in the Baum-Welch algorithm, we get the probability of each possible state path.

The Viterbi algorithm and the Baum-Welch algorithm are both sophisticated algorithms, and we do not have space here to go through their derivation. Our main purpose here is to highlight that algorithms exist to estimate both the HMM parameters and the likely latent state sequence, making HMMs more than just a theoretical curiosity but a genuinely useful modeling tool. For references giving more details behind the Viterbi and Baum-Welch algorithms, see point *4* in the *Notes and further reading* section at the end of the chapter.

That explanation of how we do inference with HMMs brings us to the end of this section on HMMs, so let's recap what we have learned about HMMs and then summarize the chapter overall.

What we learned

In this section, we have learned the following:

- That an HMM is a discrete Markov process but one where the state sequence is hidden from us.

- What we observe in an HMM is a set of symbols emitted by the hidden states.

- The emitted symbols tell us information at each step of the hidden state sequence.

- The probabilities of seeing different symbols are given by the emission probabilities of the HMM. The emission probabilities are parameters of the HMM.

- Given the parameters of the HMM, we can infer the most likely (hidden) state sequence from the observed symbol sequence using the Viterbi algorithm.

- If we don't know the parameters of the HMM, we can use the Baum-Welch algorithm to estimate them and infer the expected state transition at each point in the observed symbol sequence.

Summary

This chapter has been about dynamical systems, but from a data science perspective. That means we have focused on those dynamical systems that are heavily used in data science modeling. Consequently, we have spent most of the chapter focusing on Markov chain models – first-order and higher-order discrete Markov processes. Despite the outward simplicity of discrete Markov models, they are a very powerful tool for modeling real-world scenarios. To help understand discrete Markov models, in this chapter, we have covered the main concepts underlying their behavior. Those concepts are the following:

- A dynamical system has a state, and that state changes over time.

- For some dynamical systems the time variable is continuous, while for other dynamical systems the time variable is discrete.

- An evolution equation determines how a dynamical system evolves.

- First-order discrete Markov processes are probabilistic discrete-time models that specify how a system evolves from one state to another over time.

- For a first-order discrete Markov process, the probability distribution from which the next state is drawn depends only on the current state and is given by the corresponding row of the transition probability matrix.

- The rate matrix of a first-order discrete Markov process determines the evolution of the state probability distribution.

- Stable distributions and limiting distributions of first-order discrete Markov processes.

- First-order discrete Markov processes are memoryless.

- In higher-order discrete Markov processes, the transition probabilities depend upon more of the preceding state history than just the current state.

- A higher-order discrete Markov process can be written as a first-order Markov process.

- An HMM is a discrete Markov process but one where the state sequence is hidden from us. Instead, we observe a sequence of emitted symbols.

- Given the parameters of an HMM, we can infer the most likely (hidden) state sequence from the observed symbol sequence using the Viterbi algorithm.

- If we don't know the parameters of the HMM, we can still use the observed symbol sequence along with either Viterbi training or the Baum-Welch algorithm to estimate model parameters and appropriate state sequences.

Our next chapter covers another advanced topic, about a specific but highly used class of data science algorithms called kernel methods.

Exercises

1. We have a three-state second-order discrete Markov process whose states are labeled 1, 2, and 3. The transition probability matrix is given (in non-square form) in *Table 11.1*. The rows represent the two preceding "from" states of each transition type, while the columns represent the "to" state of each transition type. The transition probability matrix is in the same form illustrated in *Figure 11.2*.

    We want to generate a sequence of state values, X_i, $i = 1, 2, ..., N$, of length $N = 200$. Starting the sequence with $X_1 = 1$, $X_2 = 1$, use the transition probability matrix to sample the remaining 198 state values in the sequence:

	k=1	k=2	k=3
i=1, j=1	0.27	0.33	0.40
i=1, j=2	0.12	0.67	0.21
i=1, j=3	0.30	0.30	0.40
i=2, j=1	0.08	0.56	0.36
i=2, j=2	0.72	0.19	0.09
i=2, j=3	0.43	0.27	0.30
i=3, j=1	0.16	0.16	0.68
i=3, j=2	0.45	0.45	0.10
i=3, j=3	0.25	0.38	0.37

Table 11.1: Transition probability matrix for second-order Markov process in Q1

2. By the end of the 200-length sampled sequence that we generated in *Q1*, the state distribution $\pi_{t=200}$ will have almost completely forgotten the influence of the starting states X_1 and X_2. This means that effectively, the last state in the sampled sequence, X_{200}, will be a good approximation of a sample of a state from the limiting distribution π_∞ of the Markov process. By repeating the process of sampling an $N = 200$-length sequence 10000 times, and each time recording the final state value, X_{200}, calculate a numerical estimate of the limiting distribution $[\pi_\infty(1), \pi_\infty(2), \pi_\infty(3)]$. For each of the 10000 $N = 200$ sequences you generate, start them off by sampling X_1 and X_2 independently from the discrete uniform distribution $[\frac{1}{3}, \frac{1}{3}, \frac{1}{3}]$.

3. By writing the transition probability matrix in *Table 11.1* in square form, as we showed in *Figure 11.3*, identify the stationary distribution of the second-order Markov process (there is only one) and compare it to your numerical estimate for the limiting distribution from *Q2*.

Notes and further reading

1. The extra complication is that we can have more than one limiting distribution; that is, there is more than one solution π_∞ to *Eq. 12*. Which limiting distribution we end up in after iterating for an infinite amount of time will then depend upon the details of the starting distribution π_0.

2. Very high-order discrete Markov models have been used to model language generation. This approach to modeling language also has some similarities with how **large language models (LLMs)** work in that LLMs also use a large number of the preceding states to predict the next word. However, the inclusion of the self-attention mechanism in LLMs makes them significantly different and superior to high-order Markov models for language generation.

3. One of the earliest approaches to doing this was by *A.E. Raftery, A Model for High-Order Markov Chains, Journal of the Royal Statistical Society, Series B, 47(3):528-539 (1985)*.

4. I learned about HMMs in the context of DNA sequence analysis. The book I learned from was *Biological sequence analysis* by *R. Durbin, S. Eddy, A. Krogh*, and *G. Mitchison, Cambridge University Press, Cambridge, 1998*. I still think it gives one of the clearest and most succinct explanations of HMMs, the Viterbi algorithm, and the Baum-Welch algorithm.

5. The tutorial paper by *L.R. Rabiner, A tutorial on hidden Markov models and selected applications in speech recognition, Proceedings of the IEEE. 77 (2): 257–286 (1989)*, is an excellent explanation of HMMs, the Viterbi and Baum-Welch algorithms, as well as explaining different variants of HMMs (auto-regressive HMMs and HMMs that allow for explicit inclusion of a user-specified distribution for the contiguous time spent in a single state). The paper can be found at https://ieeexplore.ieee.org/document/18626. You may also be able to find non-paywalled copies of the paper on the internet.

6. I also like the tutorial introduction to HMMs by *G. Slade, The Viterbi algorithm demystified*. This is in the form of an unpublished technical report but still widely cited. You can find a copy of the report at https://www.researchgate.net/publication/235958269_The_Viterbi_algorithm_demystified. Be aware that the transition probability matrix is defined as the transpose of the way I have defined it in this chapter; that is, Slade has "to" states in the rows of the transition probability matrix and "from" states in the columns.

12

Kernel Methods

Our remaining chapters will now focus on more advanced topics. Due to their advanced nature, we will not attempt to cover them in the same level of detail as we have done for the topics of earlier chapters. Instead, we will focus on getting the essential concepts and ideas behind these topics across. The aim is not to make you an expert in these topics but to introduce you to them so that you can recognize them when you see them again, or if you want to learn more at a later date. This focus on the essentials means that each of these chapters on advanced topics will be shorter than previous chapters.

The first of our advanced topics is **kernel methods**. Kernel methods, or **kernelized learning algorithms**, are very widely used. The math they are based on is both advanced and elegant. That math relates to machine learning algorithms that make use of similarities between feature vectors. To understand kernel methods, we will need to understand why similarity-based learning algorithms are common. We will also need to understand the main principles behind the elegant math of kernel functions. To do that we will cover the following topics:

- *The role of inner-products in common learning algorithms*: In this section, we will see why inner products are at the heart of many machine learning algorithms

- *The kernel trick*: In this section, we will learn about kernel functions and how kernel functions allow us to compute inner products in new feature spaces implicitly and very simply

- *An example kernelized learning algorithm*: In this section, we will see, using a code example, the simplicity and power of a kernelized learning algorithm

Technical requirements

All code examples given in this chapter can be found at the GitHub repository at `https://github.com/PacktPublishing/15-Math-Concepts-Every-Data-Scientist-Should-Know/tree/main/Chapter12`. To run the Jupyter notebooks you will need a full Python installation, including the following packages:

- `pandas` (>=2.0.3)
- `numpy` (>=1.24.3)
- `scikit-learn` (>=1.3.0)
- `matplotlib` (>=3.7.2)

The role of inner products in common learning algorithms

In *Chapter 3*, we introduced the **Principal Component Analysis (PCA)** unsupervised learning algorithm and showed how all the calculations in PCA could be expressed in terms of inner products between the feature vectors of the different points in the training data.

In *Chapter 3*, we also explained that the inner product $\underline{x}^\top \underline{y}$ between vectors $\underline{x}$ and $\underline{y}$ gives a measure of how similar vectors $\underline{x}$ and $\underline{y}$ are to each other. Since many learning algorithms are based on the idea that similar datapoints behave similarly, it is not surprising that PCA and indeed many other learning algorithms make use of inner products.

Like PCA, many classical statistical and machine learning algorithms can be expressed solely in terms of inner products, such as **Linear Discriminant Analysis (LDA)**, **Fisher Discriminant Analysis (FDA)**, **Canonical Correlation Analysis (CCA)**, and **Support Vector Machines (SVMs)**. We will refer to these types of algorithms as **inner-product based learning algorithms**.

Given the prevalence of inner-product based learning algorithms, it is natural to ask whether inner products between the feature vectors in a training dataset are all we need. The answer is more subtle.

Sometimes we need new features in our inner products

We can only calculate inner products between datapoints using the features we already have. OK, but we already know this. What is the big deal? Sometimes this is OK and we don't need to construct any new features. Sometimes it is not.

Imagine that we are trying to construct a linear discriminant. A linear discriminant is a simple algorithm that constructs a decision rule in the form of a linear combination of the features. The linear discriminant is essentially a line drawn in the feature space. If a point falls on one side of the line, we classify it as being in one class, and if it falls on the other side of the line, we classify it as being in the other class. *Figure 12.1* shows a simple example. We have just two features, x_1 and x_2. Our two classes are the red class and the blue class. You can easily see from *Figure 12.1* that the class that a given point is in is completely determined by which side of the straight line $x_2 = -x_1$ the point is on. In other words, knowing whether the value of the $x_1 + x_2$ linear combination is greater than or less than 0 is enough to determine the class. In this case, we would say that the classes are **linearly separable**. The existing features, x_1 and x_2, that we have on our dataset are enough to construct an accurate discriminant function from any training data. We don't need to construct any additional features.

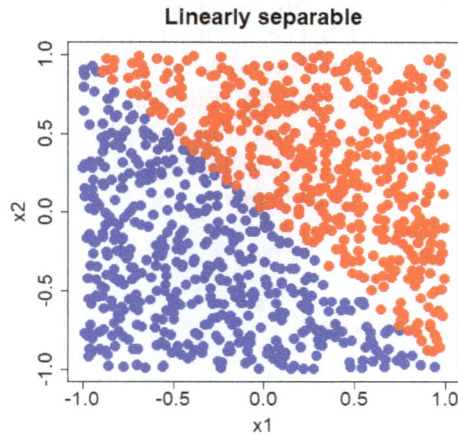

Figure 12.1: Existing features can linearly separate the red and blue classes

This is not always the case. Take the example in *Figure 12.2*. Clearly, the two classes, red and blue, are still separable by a simple boundary. However, that boundary is not a line. We would say that the two classes in this second example are not linearly separable.

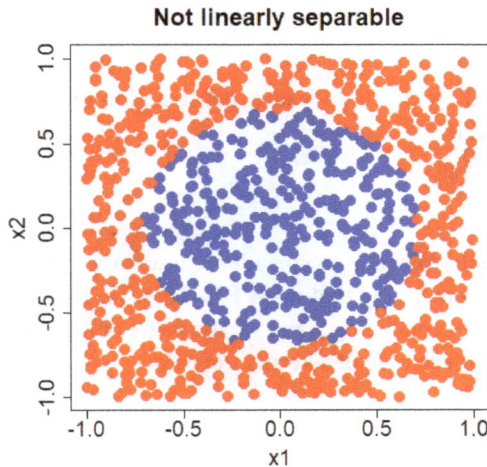

Figure 12.2: Two classes that cannot be linearly separated by existing features

You may be able to notice that the boundary separating the red and blue dots in *Figure 12.2* is a circle and is given by the $x_1^2 + x_2^2 = \frac{1}{2}$ relation. So, knowing whether the value of $x_1^2 + x_2^2$ is greater than or less than 0.5 is enough to tell us which class each point is in. We can do this calculation if we have the values of x_1^2 and x_2^2 in our training dataset. OK, calculating x_1^2 and x_2^2 from the values of x_1 and x_2 is not difficult. However, it does illustrate that to correctly separate the two classes, we would have to construct some new feature values. Moreover, the example we have shown in Figure 12.2 is a very simple one. In real-world datasets, we will not be able to simply visualize the data and spot the new features required to separate the classes. In real-world datasets, we must go through a lengthy process of constructing new features and trying them out. This can be inefficient. Fortunately, using the **kernel trick** can help us perform this feature construction process implicitly and allows us to explore whole families of new features in a very simple way. We will learn about the kernel trick in a moment, but for now, let's summarize what we have learned in this section.

What we learned

In this section we have learned the following:

- Many machine learning algorithms are based on computing the inner product between datapoints in the training dataset.

- The existing features in a dataset may not be sufficient to construct an accurate learning algorithm and so new features need to be constructed.

Having highlighted the issue that we may need to construct new features to make our inner-product based learning algorithms work, in the next section, we will learn about the kernel trick and how it can allow us to do feature construction implicitly.

The kernel trick

To learn how the kernel trick allows us to do feature construction implicitly and efficiently, we will first have to learn what a kernel is.

What is a kernel?

The simplest way to think about a kernel is to consider it as a mapping that takes two vectors as input and returns a scalar. It is a mapping that maps $\mathbb{R}^d \times \mathbb{R}^d \to \mathbb{R}$. This means that a kernel is a function $f(x, y)$, with the input vectors being x and y. The value of f is a real number. This means that the inner product $x^\top y$, is an example of a kernel function.

That is a high-level mathematical definition of what a kernel is, but what is the intuition behind this? An $f(x, y)$ kernel function applied to the vectors x and y is typically used to measure the similarity between those vectors. Consequently, we usually want our kernel function to have its largest values when x and y are most similar and its lowest values when x and y are least similar. We want the function to decrease smoothly and monotonically in between those two scenarios.

Commonly used kernels

You may recall from *Chapter 3* that the simple inner product $x^\top y$ between two vectors x and y is also called the **dot-product** because we can also write it as $x \cdot y$. Consequently, the $f(x, y) = x^\top y$ function is an example of a **dot-product kernel**. More generally, a dot-product kernel is any function of $x \cdot y$. So, the function that follows, where g is any univariate function, is a dot-product kernel:

$$f(x, y) = g(x \cdot y)$$

Eq.1

Commonly used forms of dot-product kernels are where the function $g(s) = (1 + cs)^p$. These are called polynomial dot-product kernels because the resulting kernel function f is a polynomial of the dot-product $x \cdot y$.

Another commonly used class of kernel functions is the **translationally invariant** kernels. These are kernels of the following form:

$$f(x, y) = h(x - y)$$

Eq.2

They are called translationally invariant because translating the x and y vectors by adding a constant vector a to both does not change the value output by the kernel function. One of the most used kernels of this type is the **squared-exponential kernel**, as seen here:

$$f(x, y) = \exp\left(-\frac{1}{2b^2}\|x - y\|^2\right)$$

Eq.3

This kernel is also known as the **Radial Basis Function** (RBF) kernel. It is also sometimes called the **Gaussian kernel**. The b parameter in *Eq.3* provides a scale on which kernel function f measures the differences between the x vector and the y vector. The smaller the value of b, the smaller the output value of f for the given vectors $x \neq y$.

You'll also realize that if in addition, we restrict the vectors x, y to have a fixed length (say 1) so that they live on the surface of the unit-sphere, then the kernel function in *Eq.3* will look as follows:

$$f(x, y) = \exp\left(-\frac{1}{2b^2}[2 - 2x \cdot y]\right) = \exp\left(-\frac{1}{b^2}\right) \times \exp\left(\frac{1}{b^2} x \cdot y\right)$$

Eq.4

The right-hand side of *Eq.4* is an example of a dot-product kernel. By restricting the space from which the x and y vectors come, we can change the properties of the kernel function. In the example in *Eq.4*, we have changed a translationally invariant kernel to a dot-product kernel. In fact, using the usual Taylor series representation of the exponential function, we can write the right-hand side of *Eq.4* as follows:

$$\exp\left(-\frac{1}{b^2}\right) \times \exp\left(\frac{1}{b^2} x \cdot y\right) = \exp\left(-\frac{1}{b^2}\right) \times \sum_{k=0}^{\infty} \frac{1}{k!} \left(\frac{x \cdot y}{b^2}\right)^k$$

Eq.5

So, when the x and y vectors lie on the surface of the unit-sphere, we can think of the squared-exponential kernel as being the same as a polynomial dot-product kernel of infinite order.

Perhaps more interesting is the fact that the kernel in *Equations 3 and 4* has a parameter, b, that we as a user specify and can vary. In fact, our simple polynomial dot-product kernel $f(x, y) = (1 + c x \cdot y)^p$ had user-specified parameters. We can think of both c and p as being parameters of the polynomial dot-product kernel that we can vary. It is the fact that our kernel functions have parameters that we can vary that allows us to explore whole classes of new features efficiently and implicitly. We will learn about this later in this section.

Kernel functions for other mathematical objects

Can we generalize kernels further? Yes, we can. We can usually reduce any mathematical object to a set of numbers, whether real or complex. For example, a matrix is just a set of matrix elements, that is, a set of numbers, but they are usually arranged in a rectangular structure. It is the rectangular arrangement of the matrix elements that gives the matrix its overall interesting properties, but it is still just a set of numbers. That means we can store or represent the matrix as one long vector of numbers. Consequently, we can in principle construct a kernel function that takes two matrices as inputs and returns a scalar value. The precise details of how that matrix kernel function is constructed are up to us. For example, if we had two $N \times N$ symmetric real matrices, A and B, we could define a function that measures the distance or dissimilarity between them. An example dissimilarity function might take the following form:

$$\text{dissimilarity}(A, B) = \text{tr}\left[(A - B)^{\top}(A - B)\right]$$

Eq.6

The expression in *Eq.6* is the square of the **Frobenius norm** of the $A - B$ matrix. The Frobenius norm is a standard matrix norm or way of measuring the size of a matrix. So, in *Eq.6*, we are measuring the size of the difference between A and B. From the dissimilarity measure in *Eq.6*, we can define similarity as 1 minus the dissimilarity. Since we tend to use kernel functions for measuring similarity that would give us a kernel function for matrices of the following form:

$$f(A, B) = 1 - \text{tr}\left[(A - B)^{\top}(A - B)\right]$$

Eq.7

What this illustrates is that we can construct kernel functions on spaces of any mathematical objects. We can consider a kernel function to be, more generally, a function that takes two objects, m_1 and m_2, as input and returns and scalar value. The mathematical objects m_1 and m_2 could be two matrices, two graphs, two strings, or two documents, as long as we can ultimately represent them mathematically. This means that when we use the kernel trick to implicitly construct new features, we can use it to implicitly construct new features for matrices, graphs, strings, documents, and so on. This gives us a powerful tool for extending learning algorithms that we typically think of as only applying to vectors and getting those learning algorithms to work for more exotic classes of objects.

Combining kernels

The kernel functions in *Eq.1 – 3* are very simple. This does not mean that all kernels have a mathematically simple form. In fact, sometimes we want to construct more complex kernel functions from simpler ones by applying mathematical operations to combine several simple kernel functions.

The simplest mathematical operation we can consider is to just combine them linearly. That is, given two kernel functions, $f_1(x,y)$ and $f_2(x,y)$, that operate on mathematical objects x and y, we can construct a new function:

$$f_3(x,y) = af_1(x,y) + bf_2(x,y)$$

Eq.8

If the f_1 and f_2 functions are valid kernel functions, then so is the f_3 function. Linear combination of kernel functions is also usually the most complex mathematical combination operation we are likely to perform.

Positive semi-definite kernels

Having introduced what a kernel function is, we'll now highlight a specific subset of kernels: the positive semi-definite kernels. These are the kernel functions that we will use in the kernel trick.

What is a positive semi-definite kernel?

We have already said that we can think of a kernel as taking mathematical objects x and y that live in some mathematical space and mapping them to a scalar value. We can also use a kernel function to perform a mapping within the space from which x and y come. Consider the calculation that follows:

$$\int f(x,y)v(x)\rho(x)dx$$

Eq.9

The integral in *Eq.9* is over the space from which the mathematical objects x and y are drawn. The result of the integration in *Eq.9* is a function of y, so we can think of the integration in *Eq.9* as mapping the $v(x)$ function to some function of y. The $\rho(x)$ function is a weighting or measure function. Often, we take $\rho(x)$ to be some probability density function that gives the probability of getting the object x in the mathematical space we are dealing with.

Now, what if the output function of y that we got from the integration in *Eq.9* was just proportional to the $v(x)$ input function? We would have the specific result that follows:

$$\lambda v(y) = \int f(x,y)v(x)\rho(x)dx$$

Eq.10

Hang on a minute! We recognize *Eq.10*. It is our old friend, the **eigenfunction equation**. This means that if we can find particular functions $v(x)$ that satisfy *Eq.10*, then $v(x)$ is said to be an **eigenfunction** of the $f(x,y)$ kernel and the constant of proportionality, λ, in *Eq.10* is the corresponding **eigenvalue**.

How does this help us? The eigenfunctions and eigenvalues help us characterize the kernel. There is one characteristic we are interested in for the kernel trick, and that is positive definiteness. A kernel function is said to be **positive definite** if all its eigenvalues are greater than zero, and **positive semi-definite** if all its eigenvalues are non-negative. From the fact that all the eigenvalues are non-negative, we can show that the property of positive semi-definiteness is equivalent to the relationship that follows:

$$\int f(x,y)v(x)v(y)\rho(x)\rho(y)dxdy \geq 0 \quad \text{for any function } v$$

Eq.11

The relationship in *Eq.11* is the more fundamental definition of positive semi-definiteness, but the standard way to prove that a kernel function $f(x,y)$ is positive semi-definite is to determine its eigenfunctions and eigenvalues and see whether any of the eigenvalues are negative.

Great. We now have all the mathematical pieces in place to introduce the kernel trick.

Mercer's theorem and the kernel trick

The kernel trick is based upon **Mercer's theorem**. Mercer's theorem says the following: if we have a positive semi-definite kernel function $f(x,y)$ then the value of f represents the value of the inner product between $\Phi(x)$ and $\Phi(y)$, where the mathematical object $\Phi(x)$ is some mapping of the mathematical object x to a new mathematical space.

Okay, that definition of Mercer's theorem is a bit abstract, so we'll make it more concrete by going back to using d-dimensional vectors $\underline{x}$ and $\underline{y}$. If we have a positive semi-definite kernel function, $f(\underline{x},\underline{y})$, then the value of f represents the inner product between the $\underline{\Phi(x)}$ and $\underline{\Phi(y)}$ vectors, where $\underline{\Phi(x)}$ is the mapping of the vector $\underline{x}$ to a new vector $\underline{\Phi(x)}$, and likewise $\underline{\Phi(y)}$ is the mapping of the vector y

to a new vector $\Phi(y)$. This means that from x, we have constructed new features $\Phi(x)$ and calculated the inner product $\Phi(x) \cdot \Phi(y)$ using the $f(x, y)$ kernel function. In math, we are saying the following:

$$f(x, y) = \Phi(x) \cdot \Phi(y)$$

Eq.12

However, here is the subtlety: at no point did we say what the new features $\Phi(x)$ were. Mercer's theorem told us that we didn't have to. Mercer's theorem told us that using the $f(x, y)$ kernel function is equivalent to implicitly computing inner products in some new feature space. This is the kernel trick.

Why is this useful? Remember that many of the kernels that we have already introduced have some parameters in them. By varying the kernel function parameters, we are effectively varying the new features that we are implicitly creating. Varying the parameters is the same as varying or exploring across a whole set of new feature spaces.

Mercer's theorem – an example

Let's look at a simple but explicit example of Mercer's theorem in action. We'll return to our features x_1, x_2 in *Figures 12.1* and *12.2*. If we have two points, $x = (x_1, x_2)$ and $y = (y_1, y_2)$, in our original feature space, then the inner product between them is as follows:

$$x \cdot y = x_1 y_1 + x_2 y_2$$

Eq.13

Now we'll explicitly construct a new feature space, Φ. From the existing $x = (x_1, x_2)$ feature vector, we define our feature vector $\Phi(x)$ to be as follows:

$$\Phi(x) = \left(x_1^2, x_2^2, \sqrt{2}\, x_1 x_2 \right)$$

Eq.14

Similarly, the $y = (y_1, y_2)$ point in the original feature space maps to the following:

$$\Phi(y) = \left(y_1^2, y_2^2, \sqrt{2}\, y_1 y_2 \right)$$

Eq.15

The inner product, $\Phi(x) \cdot \Phi(y)$, in the new feature space is then given by the following:

$$\Phi(x) \cdot \Phi(y) = x_1^2 y_1^2 + x_2^2 y_2^2 + 2 x_1 x_2 y_1 y_2 = (x_1 y_1 + x_2 y_2)^2$$

Eq.16

Comparing the right-hand side of *Eq.16* to *Eq.13*, we can see that it is the same as $(x \cdot y)^2$. This means that, for this example, we have the following:

$$\Phi(x) \cdot \Phi(y) = (x \cdot y)^2$$

Eq.17

In this example, we have explicitly specified what the mapping from the original feature space x to the new feature space Φ was. However, we didn't have to. *Eq.17* tells us that if all we want or need to do is compute inner products in the new feature space, then we don't have to know what the mapping from the original to the new feature space is. We can just use the $(x \cdot y)^2$ polynomial dot-product kernel to compute inner products in the new feature space. This is the kernel trick in action.

For our explicit example, the $\Phi(x)$ mapping only produced a finite number of features. We went from a two-dimensional feature vector to a three-dimensional feature vector. Is this always the case? You can probably guess by looking at *Eq.17* that the fact that using the $(x \cdot y)^2$ polynomial dot-product kernel is equivalent to constructing a finite-dimensional feature vector stems from the fact that the order of the polynomial is finite, that is, we just have a quadratic dot-product kernel. You'd be right. Using a higher-order polynomial dot-product, such as $(1 + 3x \cdot y)^4$, will lead to more implicit features, but still a finite number of them.

Wait, I hear you say. *Eq.5* tells us that when x and y are on the surface of the unit-sphere, then the squared-exponential kernel is equivalent to a polynomial dot-product kernel of infinite order. So, does using a squared-exponential kernel mean that we would implicitly be creating an infinite number of new features? Yes, it does. By using different kernel functions, we can explore a wide range of different new feature spaces. This is the power of the kernel trick – it allows us to efficiently explore new feature spaces.

So, how do we use Mercer's theorem and the kernel trick in a learning algorithm? This is what we will cover next.

Kernelized algorithms

Mercer's theorem tells us that we can take a positive semi-definite kernel function f and use it to calculate the inner-product values in our inner-product based learning algorithm. This is often called **kernelizing** the algorithm.

Take PCA as an example of an inner-product based learning algorithm. We know from Mercer's theorem that computing the inner products using a kernel function will be equivalent to doing the PCA in some new feature space, even though we do not (necessarily) know what that new feature space is. The new kernelized version of the learning algorithm is called **kernel PCA**.

The great thing about using Mercer's theorem in this way is that if our kernel function has some parameters, we can treat those parameters as hyper-parameters and vary them until we minimize some loss function, or until we achieve maximal prediction accuracy on a validation set. Each kernel function parameter value is equivalent to performing a PCA in some new feature space. Therefore, varying the parameters is equivalent to performing a whole family of PCAs across a whole family of new feature spaces until we find the feature space that gives us the best prediction accuracy.

In general, the approach to kernelizing any inner-product based learning algorithm is simple. It is as follows:

1. Take your inner-product based learning algorithm.
2. Replace the inner product calculations with kernel function evaluations.

That's it. You're good to go.

In practice, thinking about the original learning algorithm solely in terms of inner product calculations and learning how to express all calculations solely in terms of inner products can take a bit of getting used to, but once you do, the preceding steps are all there is to it.

If kernelizing an inner-product based learning algorithm is simple, you might ask: is there much of a difference between the original learning algorithm and its kernelized version? The answer is no. We can think of the original learning algorithm as just a special case of its kernelized version that uses a linear kernel function. For example, our original PCA algorithm, explained in *Chapter 3*, is just an example of kernel PCA that uses a linear dot-product kernel, and so uses a kernel function of the following form:

$$f(x, y) = x \cdot y$$

Eq.18

That short section on how we can, in principle, use the kernel trick concludes this section on kernels, so let's recap what we have learned about kernels and the kernel trick.

What we learned

In this section, we have learned the following:

- A kernel function maps two mathematical objects to a scalar value
- About commonly used kernels, such as dot-product kernels and translationally invariant kernels, that are applied to vectors
- How kernel functions can be combined in linear combinations
- About positive definite and positive semi-definite kernel functions
- How Mercer's theorem tells us that a positive semi-definite kernel function corresponds to an inner product calculation in some unknown (implicit) new feature space

- The kernel trick uses Mercer's theorem to replace the inner products in an inner-product based learning algorithm with a kernel function to create a kernelized version of the learning algorithm

Having learned about the ideas behind how to create kernelized learning algorithms, in the next section, we will see an example of a kernelized algorithm in action.

An example of a kernelized learning algorithm

To illustrate the simplicity of kernelized algorithms we'll demonstrate with a code example for a specific inner-product based learning algorithm. The algorithm we'll use is **Fisher Discriminant Analysis (FDA)**, which is an algorithm for assigning points to class labels. The standard version of FDA is a form of **Linear Discriminant Analysis (LDA)**. When we run the kernelized version of the FDA, we will be doing **Kernel Fisher Discriminant Analysis (kFDA)**.

kFDA code example

We will start with the example data in *Figure 12.1*. The classes are linearly separable, so we'll use a linear Fisher discriminant to construct a classifier. A linear discriminant for a two-class problem uses the orthogonal distance of a point x from a line w to determine which class the point is in. If the point x is one side of the line, we say it is in class 1, while if it is on the other side of the line, we say it is in class 2. The line w is our predictive model that we need to train.

Measuring how far a point is from the line w is equivalent to measuring how far the point x is along the line β, which is orthogonal to the line w. This means that we can express the classifier as the mathematical condition $\beta^\top x > c$, where c is some constant. If the point x satisfies this condition the point is in one class, if it doesn't satisfy this mathematical condition it is in the other class.

Training the linear discriminant is the process of determining the optimal line w which minimizes the classification error on the training dataset. Determining the optimal line w is equivalent to finding the optimal line β. Since we can see from *Figure 12.1* that the two classes (the red and the blue points) can be separated by a straight line, we know that a linear discriminant using just the features we have, x_1 and x_2, will be sufficient to achieve a high-level of accuracy.

I have written a Python class, `KFDA_Poly`, which allows us to do kFDA. I have kept things simple; it only allows us to do kFDA using pure polynomial dot-product kernels of the form that follows:

$$k(x, y) = (x \cdot y)^p$$

Eq.19

The `KFDA_Poly` class is defined in the `kernel_fda.py` module, which can be found in the Chapter12 directory of the GitHub repository. When calling the constructor for the `KFDA_Poly` class, we specify the degree p of the kernel that we want to use. To do linear FDA, we specify the degree $p = 1$.

The data for *Figure 12.1* is in the `lda_ex1.csv` file in the `Data` directory of the GitHub repository. We have labeled the classes 1 and 2 rather than `red` and `blue`. Class 1 corresponds to the blue points, while class 2 corresponds to the red points. The code example that follows can be found in the `Code_Examples_Chap12.ipynb` Jupyter notebook in the GitHub repository. First, we'll read in the data:

```
import pandas as pd
import kernel_fda
# Read in the data
df_LDA_ex1 = pd.read_csv('../Data/lda_ex1.csv')
```

Let's look at the data:

```
# Take a quick look at the dataframe
df_LDA_ex1
```

This gives the table that follows, from which we can see that we have 1,000 datapoints, each consisting of the two features x_1, x_2 and the class label:

```
        x1          x2        class
0     -0.297128    0.477975   2
1     -0.575548   -0.274354   1
2     -0.793637   -0.681858   1
3      0.842911   -0.766655   2
4     -0.566261    0.621195   2
...    ...         ...        ...
995   -0.797437    0.922392   2
996   -0.247480    0.526261   2
997   -0.562740   -0.328817   1
998   -0.376398    0.211520   1
999    0.799909    0.486870   2
```

Now we'll build a linear Fisher discriminant. We'll instantiate a `KFDA_Poly` object with a linear kernel (`degree=1`). Specifying a linear kernel is saying that we are going to do kFDA but with a linear kernel, so this is equivalent to linear FDA:

```
# Create the linear classifier
linear_classifier_ex1 = kernel_fda.KFDA_Poly(degree=1)
```

Now we'll fit the linear classifier using the training data we have just read in:

```
# Fit the linear classifier
linear_classifier_ex1.fit(X=df_LDA_ex1[['x1','x2']],
                          y=df_LDA_ex1['class'])
```

We can then score the trained linear classifier on the training set using the built-in score function:

```
linear_classifier_ex1.score(X=df_LDA_ex1[['x1','x2']],
                            y_true=df_LDA_ex1['class'])
```

This gives an output of 0.998. We can see that the model scores very well on the training set. The proportion of the training points that the classifier correctly classifies is 0.998, that is, nearly 100% accuracy on the training set. This is to be expected, as we know just by looking at *Figure 12.1* that the two classes are separable by a straight line, so a properly trained linear classifier should be capable of fitting the training data nearly perfectly. We also know that this trained classifier would predict any hold-out datapoints accurately provided they are also drawn from the same distribution as the training data. Therefore, for the purposes of this example, there is no need to test our classifier on a holdout sample.

Now we'll repeat the process using the data from *Figure 12.2*. We know from looking at *Figure 12.2* that a straight line can't separate the two classes perfectly. Consequently, a trained linear Fisher discriminant should score poorly on the training data in *Figure 12.2*. Let's check.

The data for *Figure 12.2* is in the `lda_ex2.csv` file in the `Data` directory of the GitHub repository. It is in the same format as the previous example. First, we'll read in the data:

```
# Read in the data
df_LDA_ex2 = pd.read_csv('../Data/lda_ex2.csv')
```

Now we'll repeat the process we went through with the first example and train a linear Fisher discriminant on this data. We will start by instantiating the linear classifier:

```
# Create the linear classifier
linear_classifier_ex2 = kernel_fda.KFDA_Poly(degree=1)
```

Next, we'll fit it to the training data from *Figure 12.2*:

```
# Fit the linear classifier
linear_classifier_ex2.fit(X=df_LDA_ex2[['x1','x2']],
                          y=df_LDA_ex2['class'])
```

Now, we'll score the trained linear classifier on the training set data:

```
linear_classifier_ex2.score(X=df_LDA_ex2[['x1','x2']],
                            y_true=df_LDA_ex2['class'])
```

This gives an output of 0.502. We can see the score on the training set is close to 0.5, that is, only about 50% accuracy. This is a lot lower than in our first example. This is to be expected. No straight line can separate the two classes in *Figure 12.2*.

Can you think why the accuracy on the training set was close to 0.5, even though we have trained (or rather, optimized) this linear classifier on the training data?

We know that the red and blue points in *Figure 12.2* are separated by the $x_1^2 + x_2^2 = \frac{1}{2}$ boundary. So, if a point $\underline{x}$ has $x_1^2 + x_2^2 > \frac{1}{2}$, it is in the red class, while if $\underline{x}$ has $x_1^2 + x_2^2 < \frac{1}{2}$, it is in the blue class. This tells us that if we had a perfect classifier for this dataset, we could write our classifier condition as $x_1^2 + x_2^2 > \frac{1}{2}$. This classifier condition can also be written as $\beta^T\underline{\Phi} > \frac{1}{2}$, where $\underline{\Phi} = \left(x_1^2, x_2^2, \sqrt{2}\,x_1 x_2\right)$ and $\underline{\beta} = \left(1,1,0\right)$. This is in the form of a linear discriminant classifier, but one where we are using a new feature vector $\underline{\Phi}$. However, the $\underline{\Phi}$ vector is precisely the new feature vector that was implicitly created when we used a quadratic dot-product kernel in our Mercer's theorem example in the previous section. This suggests that if we train a kernel Fisher discriminant classifier using a quadratic dot-product kernel $f(\underline{x},\underline{y}) = (\underline{x} \cdot \underline{y})^2$, the trained classifier should be capable of perfectly separating the red and the blue points in the training data shown in *Figure 12.2*. Let's see.

First, we will instantiate a kernel classifier object by specifying a polynomial dot-product kernel of degree 2:

```
# Create a quadratic dot-product kernel Fisher Discriminant
kernel_classifier = kernel_fda.KFDA_Poly(degree=2)
```

Next, we will fit the kernel Fisher discriminant to the training data from *Figure 12.2*:

```
# Fit the kernel classifier to the training data
kernel_classifier.fit(X=df_LDA_ex2[['x1','x2']],
                      y=df_LDA_ex2['class'])
```

Finally, we'll score the trained kernel Fisher discriminant classifier on the training data. We should get something a lot higher than 0.5 and much closer to 1:

```
# Score the trained classifier on the training data
kernel_classifier.score(X=df_LDA_ex2[['x1','x2']],
                        y_true=df_LDA_ex2['class'])
```

This gives an output of 0.911, so we do get a trained classifier that fits the training data much better than a standard linear Fisher discriminant. The reason the trained classifier doesn't fit the training data perfectly, that is, that the accuracy proportion is not 1, is simply due to sampling variation. If we increased the size of the training data, we would get closer and closer to a score of 1.

The code example shows how easy it is to use kernelized algorithms in practice. It is usually as simple as specifying the kernel that we want to use and then running the algorithm as we would normally run the un-kernelized version. The use of non-linear kernels in the algorithm means we can learn the non-linear structure present in the data by implicitly creating new features. Learning this non-linear structure is not something that the un-kernelized version of the algorithm is capable of.

In the code example, we were able to deduce which kernel to use a priori. In real-world situations, the choice of kernel is something we would typically experiment with or optimize as a hyper-parameter of the algorithm. Due to the simplicity of using the kernelized algorithm, varying the parameters of the kernel is not difficult.

That concludes the demonstration of a real kernelized algorithm. We'll recap what we have learned from in this section and then wrap up the chapter overall.

What we learned

In this section, we have learned the following:

- How running a kernelized algorithm is as simple as selecting a kernel
- How a kernelized algorithm can correctly learn the non-linear structure present in a dataset, while the standard linear version of the algorithm cannot

Summary

This chapter has been focused on kernel methods, which are also called kernelized algorithms. The chapter has been short so that we can focus on the most important concepts underpinning kernel methods. Those concepts are as follows:

- Inner-product based learning algorithms are very common because an inner product captures the similarity between feature vectors, and learning by similarity is a natural basis for many machine learning algorithms.
- Inner products calculated from the existing features on a dataset may not be sufficient to learn the non-linear structure present in the dataset.
- Construction of new features can be necessary to make our learning algorithms accurate.
- Mercer's theorem tells us that positive semi-definite kernel functions implicitly construct new features and calculate inner products in those new feature spaces.
- There are different types of kernel functions.
- We can use the kernel trick to kernelize any inner-product based learning algorithm.
- Using kernelized algorithms in practice can be as simple as specifying a choice of kernel and its parameters.
- By varying the parameters of a kernel, we can effectively explore many different new feature spaces. This can be an efficient way to learn the non-linear structure in a dataset.

Our next chapter is focused on another advanced topic. It is one you have probably heard of but may not be that familiar with. The topic of the next chapter is **information theory**.

Exercises

We have already given a lengthy code example of a simple kernelized algorithm in the main part of this chapter. Therefore, for the exercises, we will demonstrate some of the more complex aspects of kernel methods. Due to this increase in complexity, we will only ask a single question. It is intentionally challenging, so don't be surprised if you don't manage to complete it fully. Have a go at answering the exercise and then compare your answer to the one in the `Answers_to_Exercises_Chap12.ipynb` Jupyter notebook in the GitHub repository.

- The data in the `kernel_PCA_matrix_data.csv` file in the `Data` directory of the GitHub repository contains the matrix elements of `N=200`, 4x4 matrices. Each matrix corresponds to a single row of the `.csv` file. The column headings are of the form `i_j`, where i and j are integers, representing the i,j matrix element. For example, the column with the `1_3` heading holds the `1,3` matrix elements of each matrix. Use the data to perform a kernel PCA of the matrix data, where each matrix is a single datapoint. Use the matrix kernel function in *Eq.7* to calculate inner products between any two matrices. You should plot the datapoints in a PCA score plot and comment on what you see.

  - You will find it useful to reshape the data in the `.csv` file into a Python list of matrices, that is, square arrays.

  - You can use the `numpy.linalg.norm` NumPy function to calculate the Frobenius norm used in *Eqs. 6* and *7*.

  - Review the material in *Chapter 3* on how to do PCA using just the Gram matrix of the centered data matrix. The i,j matrix element of the Gram matrix is the inner product between the ith centered datapoint and the jth centered datapoint.

  - You will need to center the data, but this centering needs to be done in the new feature space that is implicitly created by our kernel function. Unfortunately, you don't know what the new features are. Fortunately, you don't have to. If the $N \times N$ Gram matrix of the uncentered data has matrix elements G_{ij}, then the matrix elements, F_{ij}, of the Gram matrix of the centered data can be calculated via the following equation:

$$F_{ij} = G_{ij} - \frac{1}{N} \sum_j G_{ij'} - \frac{1}{N} \sum_i G_{i'j} + \frac{1}{N^2} \sum_{i,j} G_{i'j'}$$

Eq.20

  - You will need to do an eigen-decomposition of the centered Gram matrix. The centered Gram matrix is real and symmetric, so you should use the `numpy.linalg.eigh` NumPy function to do this eigen-decomposition.

13

Information Theory

Information theory and information-theoretic concepts are very useful, but it is unlikely that you will have to formally make use of them as a data scientist. By this, we mean that you are unlikely to have to use detailed information-theoretic mathematical proofs or techniques in your work. But – and it's an important but – the ideas and ways of thinking that information theory introduces are worth understanding. And that is what this chapter aims to achieve. To do that, we will cover the following topics:

- *What is information and why is it useful?*: Here, we'll define precisely what we mean by information and how we quantify it mathematically

- *Entropy as expected information*: Here, we'll introduce the concept of the average information associated with a random variable and its probability distribution

- *Mutual information*: Here, we'll extend our information theory concepts to multiple random variables

- *The Kullback-Leibler divergence*: Here, we'll extend our information theory concepts to multiple distributions and show how we can use them to build optimal approximations of distributions

Technical requirements

All the code examples provided in this chapter can be found in this book's GitHub repository at `https://github.com/PacktPublishing/15-Math-Concepts-Every-Data-Scientist-Should-Know/tree/main/Chapter13`. To run the Jupyter notebooks, you will need a full Python installation that includes the following packages:

- `pandas` (>=2.0.3)

- `numpy` (>=1.24.3)

- `scipy` (>=1.11.1)

- `scikit-learn` (>=1.3.0)

- `matplotlib` (>=3.7.2)

What is information and why is it useful?

As we said in the introduction, information theory and information-theoretic ideas are very useful. To understand those ideas, one of the first things we must address is what we mean by *information*. I am talking conceptually here. Once we have nailed down our conception of what information is about, writing down a mathematical definition will be easier.

The concept of information

Part of the difficulty in introducing information theory as a mathematical subject is that different people use the word "information" and apply it to different concepts. For example, the word "information" could apply to the following:

- The semantic content or meaning of an action or event, such as a particular word being spoken or written. This is the most common conception that people have of what information should be about. It is the idea that the information associated with a thing should somehow be related to what that thing means.

- The intended effect or purpose of an action or event. When speaking a particular word, or triggering a specific event, we usually have an intended outcome that we would like to achieve or engineer. The association between the event and outcome may not be perfect, but we might think that the information associated with the event measures something about the desired outcome.

- How well an action or event encodes or identifies the thing it was intended to communicate. This is like the previous possible interpretation of "information," but rather than concerning ourselves with how signal and outcome are related, in this interpretation, we focus solely on the efficiency and precision of the communication process. This is a narrower interpretation of what "information" is about.

As you can imagine, the last of those interpretations of what "information" should be about is the narrowest of the three, and so also the easiest to define mathematically. Therefore, you will not be surprised that this is the conceptual definition of what "information" is about that mathematicians and scientists have settled on. Indeed, the founder of modern information theory, Claude Shannon, titled his 1948 landmark paper on information theory *A mathematical theory of communication*. For mathematicians, statisticians, scientists, engineers, and data scientists, information theory concerns itself with quantifying the efficiency of communication. For a longer discussion of the various possible interpretations of "information", see [1] in the *Notes and urther reading* section at the end of this chapter.

So, we've nailed down conceptually what we mean when we use the word "information," but how do we define it mathematically so that we can measure it?

The mathematical definition of information

In the UK, when I was growing up, there was a popular game called "Guess Who." The game consisted of two players, who would each have an identical set of character playing cards, such as those shown in *Figure 13.1*. Each player would choose a character in secret. Each player had to guess the character chosen by their opponent by asking questions about the attributes of their opponent's chosen character – for example, "Are they wearing a hat?" or "Do they have black hair?" – with just "yes" or "no" answers to the questions. The winner of the game was the person who could identify their opponent's chosen character first.

Figure 13.1: The game of identifying a person from their attributes

Now, imagine that you and I were playing the game and you asked me if my chosen character was working at a computer, and I said yes. With that answer, you could immediately identify who my chosen character was from the characters in Figure 13.1. The information content of the answer was very high because it communicated very efficiently (precisely in fact) which character we were talking about. The answer, "Yes, the person is working at a computer" encodes very precisely the signal about which character I selected in secret.

Alternatively, if you asked the question, "Is your selected character a person," I would also reply yes. It would have been a very silly question to ask, but for illustration purposes, we'll stick with it. All the characters in Figure 13.1 are people, so my answer, "Yes, they are a person" does not help you narrow down the identity of my selected character. My answer does not efficiently encode the signal about who I have chosen.

From this, we can see that the information content of my answers, how efficiently it communicated to you who I had chosen, was dependent on how rare the attribute you were asking about was – having a computer was rare (only one character did) while being a person was common (all the characters were people). Since the rarity of an attribute is defined by the probability distribution of the attribute, such as Prob(Has Computer) or Prob(Has Black Hair), this also tells us that the mathematical definition of information is related to probability and that information-theoretic concepts are probabilistic. This is another reason why we emphasized the importance of understanding probability in *Chapter 2*.

We now know that the information associated with a thing or event is related to the probability of that thing or event occurring. We also know that the lower the probability of the thing or event occurring, the higher the amount of information associated with it.

The formal mathematical definition of the amount of information associated with an event, *x*, occurring is as follows:

$$\text{Information} = -\log P_X(X = x)$$

Eq. 1

We have been very formal in Eq. 1 and used the $P_X(X = x)$ notation to emphasize that we are dealing with a random variable, X, and we are talking about the information associated with X taking a value of x. Information itself is specifically concerned with the probability of a specific outcome, x, not the full distribution, $P_X(X = x)$. We will encounter information-theoretic concepts such as entropy, which are associated with probability distributions, later. Because information increases as x becomes rarer, and hence more surprising if we do see it, it is also said to represent the **degree of surprise** of x.

One of the first things you may notice about Eq. 1 is that we haven't said which base we're using when taking the logarithm. Most times, when we have a logarithm in a formula, the choice of base does not affect the **decision** we make based on the formula. Here, however, we are using the logarithm not to make a decision, but to directly quantify the amount of information associated with the outcome, $X = x$. The choice of base will have an effect here.

So, which base should we use? It is up to you, but there are some commonly used bases. The most common is base 2, in which case we define the information associated with the outcome, $X = x$, to be as follows:

$$\text{Information} = -\log_2 P_X(X = x)$$

Eq. 2

When we use base 2 for measuring information, the amount of information is said to be measured in **bits**. This means that Eq. 2 tells us the number of bits of information associated with the outcome, $X = x$. Less frequently, you may see the term **Shannon** used instead of **bit** for the unit of information when using base 2. Of course, we could choose a different base to use for our logarithm in Eq. 1. The other most common choices are as follows:

- Base e, so that we use the natural logarithm, and the information associated with $X = x$ is calculated as $-\ln P_X(X = x)$. Unsurprisingly, when we use the natural logarithm, we say the information is measured in *nats*.

- Base 10, so that the information associated with $X = x$ is calculated as $-\log_{10} P_X(X = x)$. In this case, we say the information is measured in *Hartleys*, or *Harts*, named after the electronics pioneer Ralph Hartley, who contributed to the foundations of information theory. You may also see the words *ban* and *dit* used when measuring information with base 10 logarithms.

The second thing you should notice about Eq. 1 is that it is monotonic in terms of probability. Figure 13.2 shows a plot of the information (measured in bits) associated with an event and the probability of that event:

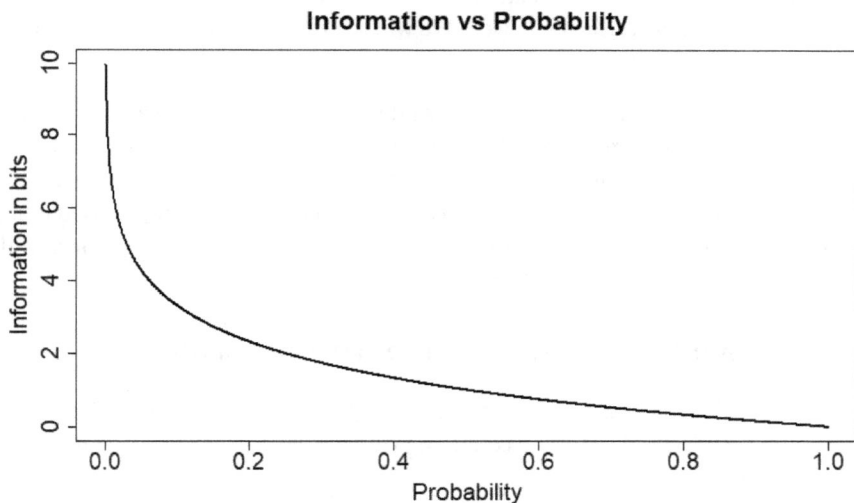

Figure 13.2: A plot of information content, measured in bits, of an event, and the probability of that event

The monotonic nature of the curve in Figure 13.2 confirms what we outlined earlier, namely that the rarer an event is, the more insight or information it gives us, with very rare events having high information content. For example, an event that occurs only 1 in 100 times, and so has a probability of occurrence of 0.01, has an information content of $-\log_2(0.01) \approx 6.64$ bits. At this point, it is also worth highlighting that an event that has a 50% probability of occurring has an information content in bits of $-\log_2\left(\frac{1}{2}\right) = 1$. This is a good reason for measuring information in bits because we can easily remember that an event that is equally likely to occur as not has 1 bit of information associated with it.

Information theory applies to continuous distributions as well

We introduced the mathematical definition of information by talking about the probability of an event or outcome. Implicit in that is that we are talking about an outcome from a discrete random variable. This was also explicit in our "Guess Who" game example, where we spoke about the probability of a character having a particular attribute, such as black hair. Can we generalize the definition of information to continuous probability distributions? The answer is yes. For a continuous random variable, X, with probability density, $p_X(x)$, we can always define a particular event that X takes a value between x and $x + dx$. So, from the density, $p_X(x)$, we can always create discrete events and an associated probability distribution. That means with a bit more mathematical work, we can extend information-theoretic concepts to continuous random variables as well. We will see this in action when we introduce the concept of entropy in the next section.

It may be tempting to think that from what we just said we can assign an information value to the value, x, of a continuous random variable. Surely we can just define information as $-\log p_X(x)$? Strictly speaking, no. $p_X(x)$ is a probability density, not a probability, and information can only be measured on probabilities. However, you may see expressions such as $-\log p_X(x)$ loosely or sloppily used in informal proofs of information-theoretic formulae. In these circumstances, it is almost always the case that the proof can easily be made more rigorous and we arrive at the same formulae.

Why we measure information on a logarithmic scale

You may have looked at Eq. 1 and asked why we measure information on a log scale. Why did we use the logarithm function in Eq. 1 when we could, in principle, have used any other monotonically increasing function of $P_X(X = x)$?

The simplest answer is the one Shannon gave in his original paper on *A mathematical theory of communciation*. Since we think of the information associated with an event with how much it helps us narrow down possibilities, a probability of $\frac{1}{4}$ helps us narrow down possibilities twice as much as a probability of $\frac{1}{2}$. In our game of "Guess Who", an attribute that only 25% of characters have, such as possessing a stethoscope, could narrow down the field of possible characters twice as much as an attribute that 50% of the characters have, such as possessing a hat. Likewise, a probability of $\frac{1}{20}$ helps us narrow down possibilities twice as much as a probability of $\frac{1}{10}$. Therefore, it is natural for us to want to measure information in such a way that the difference in information between probabilities of $\frac{1}{4}$

and $\frac{1}{2}$ is the same as the difference in information between probabilities of $\frac{1}{20}$ and $\frac{1}{10}$. This means using a monotonic function, f, so that we have the following:

$$f\left(\tfrac{1}{4}\right) - f\left(\tfrac{1}{2}\right) = f\left(\tfrac{1}{20}\right) - f\left(\tfrac{1}{10}\right)$$

Eq. 3

Or more generally, the function, f, must satisfy the following equation:

$$f(\alpha p_1) - f(p_1) = f(\alpha p_2) - f(p_2) \quad \text{for any } p_1, p_2 \in [0,1] \text{ and } \alpha \in [0,1]$$

Eq. 4

Eq. 4 is only satisfied by the logarithm function. So, if we want halving the probability of occurrence to always lead to a linear increase in information, we must measure information on a logarithmic scale.

Why is quantifying information useful?

In short, information theory is useful because it enables us to quantify how efficiently a signal is encoded when being transmitted. That means we can use information theory to work out better encodings and ultimately optimize the encoding, and in doing so minimize any signal loss at the receiving end. The practical applications of information theory to signal analysis and communications design are huge.

In information theory, the person transmitting the signal is called the **transmitter**, and the person or process wanting to receive the signal is the **receiver**. The transmitter transmits the signal via some communication channel. When transmitting the signal via the channel, the signal is encoded. This scenario is represented schematically in Figure 13.3:

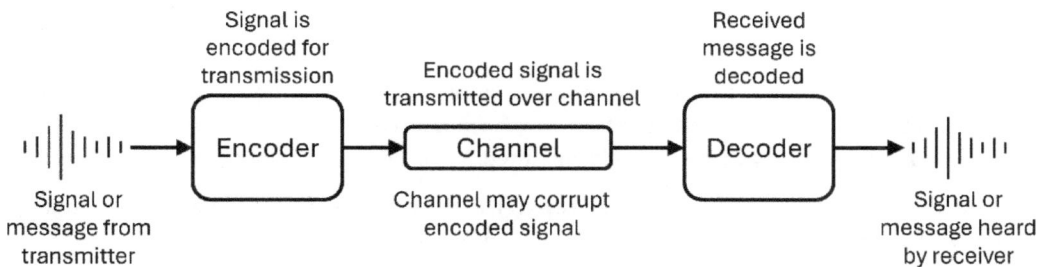

Figure 13.3: Schematic of the signal transmission and receiving process

For a specific example, think about when you have a conversation with a friend using your cell phone. Your voice signal – changes in air pressure caused by you speaking – is encoded digitally and transmitted via radio waves, then decoded, and sent as electrical signals to the speaker in your friend's cell phone.

In our "Guess Who" game, I was (reluctantly) transmitting the signal of which character I had selected. I encoded that signal (the chosen character's identity) by sending details of attributes the character had – for example, they had a stethoscope. The information associated with that single attribute, as measured by $-\log P$(Character has stethoscope), quantitatively tells me how well the attribute has encoded the signal (which character I chose). In this case, the encoding is not perfect as it only narrows down the possibilities in Figure 13.1 to two people. It is lossy. I could make it better. For example, I could increase the message length and include more attributes, such as that the character has a stethoscope and black hair, which picks out just one person in Figure 13.1, or I could use an attribute that encodes the signal more efficiently, such as that they have a computer, which picks out one person.

Signal loss in the communication process may also occur because the communication channel is itself noisy – that is, it corrupts the encoded signal as it is transmitted. Information theory also focuses on how to design robust encoding processes that can handle noisy imperfect communication channels, so information theory is a particular focus of companies building and running cell phone networks.

Having explained what we mean by information, and how and why we quantify it, this is a good place to wrap up this section with a recap of what we have learned.

What we've learned

In this section, we've learned the following:

- Information theory concerns itself with the communication of signals and the efficiency of encoding those signals.

- Information-theoretic concepts are probabilistic concepts.

- The smaller the probability of an event or outcome occurring, the higher the information associated with that event or outcome.

- We measure information on a logarithmic scale.

- When we use logarithms to base 2, the resulting information is measured in bits. When we use logarithms to base e, the resulting information is measured in nats. When we use logarithms to base 10, the resulting information is measured in Hartleys, Harts, bans, or dits.

- When transmitting to another user (the receiver), the information associated with the encoding used tells us how efficient that encoding is in communicating the signal to the receiver.

Having learned that the information associated with a single outcome is defined from the probability of that single outcome occurring, in the next section, we will learn about information-theoretic concepts that relate to the whole probability distribution of possible outcomes.

Entropy as expected information

For our "Guess Who" game, there are several attributes that a character can have. Here, I have listed the complete (for this purpose) set of possible attributes:

- Has a hat

- Has a stethoscope

- Has black hair

- Has a computer

- Has a chemical flask

From what we've learned about information theory so far, we've seen that if I tell you my chosen character has a computer, then you can narrow down the possibilities to a single character. This might suggest that asking if my chosen character has a computer is the most efficient question you can ask me. This isn't quite true. There is no guarantee that my chosen character does have a computer. To identify the best question to ask, we should look at the **expected information** you'll get by asking a question.

How do we calculate the expected amount of information you get from asking a single question? We'll use a character having a stethoscope to illustrate this. Two out of the eight characters in Figure 13.1 have a stethoscope, so $P(\text{Has Stethoscope}) = \frac{1}{4}$. So, $P(\text{Doesn't have stethoscope}) = \frac{3}{4}$. If you ask me the question, "Does your chosen character have a stethoscope?", there are two possible outcomes: "yes" or "no." If the answer is "yes," that narrows my chosen character down to two possibilities and gives you $-\log_2(\frac{1}{4})$ bits of information. If the answer is "no," that narrows my chosen character down to six possibilities and gives you $-\log_2(\frac{3}{4})$ bits of information.

Now, how likely is my answer to be "yes," and how likely is my answer to be "no?" We already know these probabilities. They are $\frac{1}{4}$ and $\frac{3}{4}$, respectively. That means we can easily calculate the average or expected amount of information you are going to get from me in response to your question. It is as follows:

$$-\frac{1}{4}\log_2(\tfrac{1}{4}) - \frac{3}{4}\log_2(\tfrac{3}{4}) \approx 0.811 \text{ bits}$$

Eq. 5

Let's write Eq. 5 more symbolically:

$$P(\text{Has stethoscope}) \times -\log_2 P(\text{Has stethoscope}) + P(\text{Doesn't have stethoscope}) \times -\log_2 (\text{Doesn't have stethoscope})$$

Eq. 6

Eq. 6 makes the average information nature of the calculation more explicit. We can generalize Eq. 6 to any attribute and since our answers have only two possibilities, "yes" or "no," we can always write $P(\text{Doesn't have attribute}) = 1 - P(\text{Has attribute})$. For any attribute, we can calculate the expected information you'll get back from me when you ask about that attribute. For convenience, we'll shorten "Has attribute" to just "A." This expected information is then given by the following equation:

$$\text{Expected information} = -P(A)\log_2 P(A) - (1 - P(A))\log_2(1 - P(A))$$

Eq. 7

Table 13.1 shows the probabilities and expected information for each of the five attributes we listed earlier:

Attribute	P(Has attribute)	Expected Information
Has hat	0.5	1 bit
Has stethoscope	0.25	0.811 bits
Has black hair	0.375	0.954 bits
Has computer	0.125	0.544 bits
Has chemical flask	0.125	0.544 bits

Table 13.1: Character attributes and their expected information

You'll recall from the *What is information and why is it useful?* section that the higher the information associated with an answer, the more it allowed us to narrow down the possibilities of which character I had secretly selected. Table 13.1 shows us that the question that allows us to narrow down the characters most effectively is asking if they have a hat since it gives us the most information on average. Asking whether they have a computer is considerably less effective. This doesn't match what we were initially thinking. How come? It should be clear from the formula in Eq. 7 what has happened. Although knowing a character has a computer narrows down the possibilities efficiently to just one person, it is unlikely that a person does have a computer, and this has reduced the information we get back on average when we ask if they have a computer.

We can re-write the equation in Eq. 7 like so:

$$\text{Expected information} = -p\log_2 p - (1 - p)\log_2(1 - p)$$

Eq. 8

Here, we just plug in the value of $P(\text{Attribute})$ for the value of p. As a function of p, we can ask, "What is the value of p that gives the highest expected information?" We do this by finding the maximum value of the right-hand side of Eq. 8 with respect to p. We'll use differential calculus to do this. We'll differentiate the right-hand side of Eq. 8 with respect to p, set the derivative to zero, and then solve for the resulting value of p. Doing this give us a condition:

$$-\log_2 p + \log_2(1 - p) = 0$$

Eq. 9

Solving Eq. 9 gives us $p = \frac{1}{2}$. This means asking about an attribute whose probability is $\frac{1}{2}$ is the most efficient question we can ask. From Table 13.1, we can see that having a hat occurs with a probability of $\frac{1}{2}$, so asking whether a character has a hat is as efficient as it is possible to get in this game. Each time we ask a question where $P(\text{Has Attribute}) = \frac{1}{2}$, we remove half the remaining characters. Each time we divide the remaining characters into two groups and ask which group the chosen character is in, we are playing the game as efficiently as possible.

Pro tip

Performing a task by dividing it into two parts, removing one part, and then repeating, is a very general technique for efficiently performing that task. For example, it can be used for efficiently sorting objects (merge sort) and finding the roots of an equation (the bisection method).

Entropy

In Eq. 7 and Eq. 8, we calculated expected information. Expected information is the average information across the possible outcomes of a random variable. You may have seen a formula like that in Eq. 7 or Eq. 8 before and seen it referred to as **entropy**. That is because entropy and expected information are two names for the same thing, although entropy is the more commonly used name. By now, you'll also realize that entropy is something we calculate about random variables.

For each attribute in our example, we had just two possible outcomes – either "yes, the character has the attribute," or "no, the character doesn't have the attribute." The outcome is a random variable with two values, "yes" and "no," which occur with probabilities $P(A)$ and $1 - P(A)$. To calculate the entropy in Eq. 7, all we needed was those two probabilities. This is also true when we have a random variable with more than two outcomes. We only need its probability distribution to calculate its entropy. For a discrete random variable, X, with probability distribution, P_X, the entropy is calculated as follows:

$$\text{Entropy} = \text{Expected information} = -\sum_x P_X(X = x)\log P_X(X = x)$$

Eq. 10

Here, x represents an outcome of the random variable, X, so the summation in Eq. 10 is over all the possible outcomes we can get for the random variable.

Since entropy is just expected information, it is measured in the same units as we measure information. As before, those units are determined by the base we use for the logarithm in Eq. 10. The choice of base is up to you.

We typically denote entropy by the symbol $H(X)$. This looks like we are applying a function, H, to X. This is deliberate, to emphasize the fact that entropy is a quantity associated with random variables, so we can think of entropy as a function of X.

The entropy of continuous random variables

Entropy is one of those quantities we mentioned in the previous section that we can generalize from discrete random variables to continuous random variables. We can do so by considering discrete outcomes, $x \leq X \leq x + \Delta x$, using the formula for the entropy of a discrete random variable, and finally using the usual calculus trick of reducing the size of our intervals, Δx, to zero. Doing so, for a continuous random variable, X, with probability density, $p_X(x)$, we find the entropy is given by the following equation:

$$\text{Entropy} = H(X) = -\int p_X(x)\log p_X(x)dx + \text{Constant}$$

Eq. 11

Unfortunately, the constant in Eq. 11 becomes infinite when we take $\Delta x \to 0^+$. However, the constant is just that – a constant. It does not depend on $p_X(x)$ in any way, so it is the integral term in Eq. 11 that contains the interesting and relevant behavior of X. Because the integral term in Eq. 11 is the difference between the entropy and the constant, it is called the **differential entropy**. That is, for a continuous random variable, X, with density, $p_X(x)$, we have the following:

$$\text{Differential Entropy} = -\int p_X(x)\log p_X(x)dx$$

Eq. 12

Because the expression in Eq. 12 is what we use when discussing the entropy of continuous random variables, you will frequently see the prefix **differential** dropped, and the integral on the right-hand side of Eq. 12 referred to as the entropy, $H(X)$, of the continuous random variable, X. I will do so as well. When I refer to the entropy of the continuous random variable, X, and I use the symbol $H(X)$, I mean its differential entropy, as defined in Eq. 12.

The integration in Eq. 12 is over the support of the random variable, X. If X is a two-dimensional real random variable, the integration will be over a two-dimensional real space, while if X is one-dimensional and real, the integration will be over the real line.

Calculating the entropy for some different continuous distributions can give us more insight into how entropy behaves and what it represents. We'll look at two common continuous distributions – the uniform distribution and the Gaussian distribution. The formulae for the density functions and entropies of these two distributions are given here.

Uniform distribution $U(a, b)$:

$$\text{Density} = \begin{cases} \frac{1}{b-a} & \text{for } x \in [a, b] \\ 0 & \text{otherwise} \end{cases} \quad , \quad \text{Entropy} = \log(b-a)$$

Eq. 13

Gaussian distribution $N(\mu, \sigma^2)$:

$$\text{Density} = \frac{1}{\sqrt{2\pi\sigma^2}} \exp\left(-\frac{1}{2\sigma^2}(x-\mu)^2\right) \quad , \quad \text{Entropy} = \frac{1}{2}\log 2\pi e \, \sigma^2$$

Eq. 14

Again, in Eq. 13 and Eq. 14, we haven't specified which base we are taking logarithms to – it is your choice. The entropy formulae in Eq. 13 and Eq. 14 are correct for any choice of base.

Because we have been able to calculate the entropy for these two distributions in terms of the parameters of their density functions, from Eq. 13 and Eq. 14, we can see how the entropy behaves as we change those parameters and hence change the shape of the distributions.

If we increase the width, $b - a$, of the uniform distribution, the entropy increases. Similarly, if we make our Gaussian distribution wider by increasing its standard deviation, σ, we increase its entropy. In both cases, the higher the variance of the distribution, the higher the entropy. Also, note that in both cases, the entropy is independent of the mean of the distribution. It doesn't matter where we locate our distribution; the entropy is only dependent on how dispersed the distribution is.

What does entropy tell us?

Because entropy increases with the variance of a probability distribution, many people tend to think of entropy as a measure of **uncertainty** or **disorder**. This may be how you have encountered entropy before – in physics. In physics, entropy is a concept associated with thermodynamics. In thermodynamics, heating a system increases the disorder (think molecules moving about more rapidly with increasing temperature). The physics formulae for entropy are the same as those from information theory, namely the formulae in Eq. 10 and Eq. 12.

Yet entropy is just average information, and we have tended to think of an increase in information as increasing the certainty with which we can identify the underlying cause of the information. How are these two viewpoints compatible with each other?

A large entropy tells us that our random variable, X, has a large variance and so tells us that the range of likely values we'd get from a single observation of that random variable is large. It tells us about the uncertainty in that single observation before we make it. It tells us about $P_X(x)$. In contrast, the information associated with an observation, $X = x$, tells us how well that observation narrows down the underlying state of the system that gave rise to the value x. It tells us about $P(\text{state} \mid x)$. Since entropy is just average information, it tells us how much we can narrow down possibilities on average. It tells us how much, on average, we can reduce uncertainty about our knowledge of a system from a single observation, x. To reduce the uncertainty a lot with a single observation of X, there must be a lot of uncertainty, **a priori**, in X. So, entropy tells us about the **a priori** uncertainty in a random variable, X, and the certainty with which we can identify the associated underlying state from observations of X. The two viewpoints of entropy are compatible with each other. They are two sides of the same coin.

The Maximum Entropy technique

The physics view of entropy leads to an interesting inference technique called **Maximum Entropy** or **MaxEnt** for short.

The physics view says that entropy tells us about the number of underlying states compatible with a density function, $p_X(x)$. Let's suppose we have some random variable, X, but I don't know what its distribution is. However, I do know its mean value, μ, and its variance, σ^2 (or I have good estimates of them, say from a sample of data). I want to model the distribution of X and do some calculations with that distribution. What distribution should I use? One that is compatible with the values of μ and σ^2.

A reasonably logical choice is to use the most probable distribution, but what do we mean by that? We can say that the most probable distribution of X is the one that has the highest number of possible underlying states compatible with its density, $p_X(x)$. But that number is the entropy. So, it turns out that a reasonable choice for $p_X(x)$ is one that maximizes the entropy subject to the constraints that the mean of X is μ and the variance of X is σ^2.

How do we do that maximization calculation? By using calculus – that is, using Lagrange multipliers as usual to impose the constraints. The objective function we must maximize is as follows:

$$-\int p_X(x)\log p_X(x)dx + \lambda_1 \int p_X(x)dx + \lambda_2 \int p_X(x)xdx + \lambda_3 \int p_X(x)\,x^2\,dx$$

Eq. 15

The first term in our objective function in Eq. 15 is the entropy. The next three terms impose the constraints through the Lagrange multipliers $\lambda_1, \lambda_2, \lambda_3$. The constraints are that we must match the specified mean and variance, and we must have a properly normalized distribution, $p_X(x)$.

Technically, to maximize the expression in Eq. 15, we must use a technique called **calculus of variations**, albeit a relatively simple version of it in this instance. However, because of this extra complexity, we will just quote the answer. The density, $p_X(x)$, that maximizes the entropy for a specified mean, μ, and variance, σ^2, is as follows:

$$p_X(x) = \frac{1}{\sqrt{2\pi\sigma^2}} \exp\left(-\frac{1}{2}(x-\mu)^2\right)$$

Eq. 16

This is just the density function of the Gaussian distribution. What does this mean? It tells us that of all the distributions, $p_X(x)$, that have mean, μ, and variance, σ^2, the Gaussian distribution is the one that has the highest entropy. It tells us that for mean, μ, and variance, σ^2, the Gaussian distribution is the most probable, in the sense that it maximizes the entropy – it has the highest number of underlying states compatible with a mean, μ, and variance, σ^2. This is another reason for studying the Gaussian distribution. It is the most probable distribution if we only know the mean and variance.

What happens if we don't know the mean and variance? What happens if we only know the minimum and maximum values of X? What is the most probable distribution then? Repeating the MaxEnt calculation, we find the properly normalized distribution that has the highest entropy is the uniform distribution. You can now see why we suggested looking at the entropy of both the uniform and Gaussian distributions – they are exceptional distributions.

In this section, we learned about the different aspects of entropy. So, let's recap.

What we've learned

In this section, we learned the following:

- The expected information tells us the average amount of information we get from an observation of a random variable, averaged across all the possible outcomes of the random variable
- The expected information is more commonly known as entropy
- Entropy is measured in the same units as information
- Entropy can be defined and calculated for both discrete probability distributions and continuous distributions
- Entropy increases with an increase in the variance of a distribution
- The information theory definition of entropy is the same as the physics definition of entropy and they encapsulate the same concept
- The MaxEnt technique can be used to determine the most probable distribution compatible with specified constraints
- The Gaussian distribution has the highest entropy for a given specified mean and variance
- The uniform distribution has the highest entropy for a given finite support

Having learned about the expected information (entropy) of a single random variable, in the next section, we'll learn about the information associated with multiple random variables.

Mutual information

In this section, we're going to look at information-theoretic concepts relating to multiple random variables. We'll focus on the case of just two random variables, X and Y, but you'll soon realize that the new calculations and concepts we'll introduce generalize easily to more than two random variables. As usual, we'll start with the discrete case first before introducing the continuous case later.

Because we're looking at two discrete random variables, X and Y, we'll need their joint probability distribution, $P_{X,Y}(x, y)$, which we'll use as shorthand for $P_{X,Y}(X = x, Y = y)$. The joint distribution is just a probability distribution so we can easily measure its entropy, which we'll denote by $H(X, Y)$. Applying the usual rules that entropy is expected information, $H(X, Y)$ is given by the following formula:

$$H(X, Y) = -\sum_x \sum_y P_{X,Y}(x, y) \log P_{X,Y}(x, y)$$

Eq. 17

To understand Eq. 17, let's look at what would happen if X and Y were independent of each other. The joint distribution would be given by $P_{X,Y}(x, y) = P_X(x) P_Y(y)$. Plugging this into Eq. 17 and remembering the rules of logarithms of products from *Chapter 1*, we'll find that we have the following here:

$$H(X, Y) = -\sum_x P_X(x) \log P_X(x) + -\sum_y P_Y(y) \log P_Y(y) = H(X) + H(Y)$$

Eq. 18

Eq. 18 tells us that when X and Y are independent, the entropy, $H(X, Y)$, is just the sum of the entropies of the separate variables, X and Y. We also know that if X and Y are independent, then knowing the value of X tells us nothing about the value of Y. We gain no information about Y if we know X. There is no information in common between X and Y.

But what would happen if X and Y weren't independent and X and Y did have some information in common? Knowing the value of X would tell us information about Y and vice versa. What would be the size of this common or **mutual information**? Well, we can take a simple approach and define this common or mutual information to be the difference between the average information we get from the random variables separately and the average information we get from the random variables together. This means that mutual information, $I(X, Y)$, is defined as a difference between entropies and is given by the following formula:

$$I(X, Y) = H(X) + H(Y) - H(X, Y)$$

Eq. 19

Although $I(X, Y)$ is called mutual information, it is defined from entropies of the random variables, X and Y, not the specific outcomes, x and y, of those random variables. This is why the symbol we use for mutual information, $I(X, Y)$, looks like a function that's applied to X and Y. And being defined as a difference between entropies, $I(X, Y)$ is measured in whatever units you use for the base of your logarithms – bits if you're taking logarithms to base 2, nats if you're taking logarithms to base e, and so on.

Conditional entropy

To get an intuition about what the mutual information, $I(X, Y)$, tells us, we'll use Bayes' theorem from *Chapter 5* to re-write our formula:

$$P_{X,Y}(X = x, Y = y) = P_{Y|X}(Y = y|X = x)P_X(X = x)$$

Eq. 20

The notation, $P_{Y|X}$, represents the conditional distribution of Y – that is, the distribution of Y once we know the value of X. From Eq. 20, we can calculate the entropy, $H(X, Y)$, and hence the mutual information, $I(X, Y)$. By doing so, we get the following:

$$I(X, Y) = H(Y) - H(Y|X)$$

Eq. 21

Here, $H(Y|X)$ is the entropy defined from the conditional distribution, $P_{Y|X}(Y = y|X = x)$, averaged over all possible values of X. So, $H(Y|X)$ is given by the following formula:

$$H(Y|X) = -\sum_x \sum_y P_{X,Y}(X = x, Y = y)\log P_{Y|X}(Y = y|X = x)$$

Eq. 22

Since the entropy, $H(Y|X)$, is calculated from the conditional distribution, $P_{Y|X}$, it is known as the **conditional entropy**. You'll recall from the previous section that entropy measures the uncertainty of a random variable. So, $H(Y|X)$ is the average uncertainty in Y after we know X. And since $H(Y)$ measures the uncertainty in Y when we don't know X, the expression in Eq. 21 now gives us a way to understand the mutual information. Eq. 21 tells us that $I(X, Y)$ is the reduction in uncertainty about Y that we get on average from knowing the value of X.

We can use Bayes' theorem to write the following:

$$P_{X,Y}(X = x, Y = y) = P_{X|Y}(X = x|Y = y)P_Y(Y = y)$$

Eq. 23

This means we can also write the mutual information, $I(X, Y)$, as follows:

$$I(X, Y) = H(X) - H(X|Y)$$

Eq. 24

Here, the entropy, $H(X|Y)$, is the conditional entropy calculated from the conditional distribution, $P_{X|Y}$. Eq. 24 means we can also interpret the mutual information, $I(X, Y)$, as the reduction in uncertainty about X that we get on average from knowing the value of Y. Because we have two different ways of interpreting $I(X, Y)$, they must be equivalent. This means we can use the following formula:

Average reduction in uncertainty about Y from knowing X = Average reduction in uncertainty about X from knowing Y

Eq. 25

This reduction in uncertainty is always non-negative. If X and Y are independent, then we gain no information about Y from knowing the value of X, while if X and Y are correlated, then knowing the value of X does tell us something about the value of Y:

$$I(X, Y) \geq 0$$

Eq. 26

By comparing Eq. 21 and Eq. 24, we can also see that we have $I(X, Y) = I(Y, X)$, as we would expect for a measure of the mutual information between X and Y.

Mutual information for continuous variables

As we did with entropy, we can define mutual information for continuous random variables, X and Y. We can do this by constructing discrete random variables using small intervals, Δx and Δy, of x and y, and then taking the limit, $\Delta x \to 0^+, \Delta y \to 0^+$. As before, we must deal with constants that become infinite as $\Delta x \to 0^+, \Delta y \to 0^+$. However, once we do the formula for the (differential) entropy, $H(X, Y)$, is given in terms of the joint density, $p_{X,Y}(x, y)$, and is as follows:

$$H(X, Y) = -\iint p_{X,Y}(x, y) \log p_{X,Y}(x, y) dx dy$$

Eq. 27

Here, mutual information is still defined via the following formula:

$$I(X, Y) = H(X) + H(Y) - H(X, Y) = H(Y) - H(Y|X) = H(X) - H(X|Y)$$

Eq. 28

The conditional entropies, $H(Y|X)$ and $H(X|Y)$, are defined from the conditional densities, $p_{Y|X}(y|x)$ and $p_{X|Y}(x|y)$, in the way you would expect:

$$H(Y|X) = -\iint p_{X,Y}(x,y)\log p_{Y|X}(y|x)dxdy$$

Eq. 29

$$H(X|Y) = -\iint p_{X,Y}(x,y)\log p_{X|Y}(x|y)dxdy$$

Eq. 30

Mutual information as a measure of correlation

Since the mutual information, $I(X, Y)$, measures the information that X and Y have in common, it is a measure of the strength of the relationship between X and Y, just like the Pearson correlation coefficient. The advantage of the mutual information is that it is invariant to the monotonic transformation of X and Y. This means that we could create new random variables, $X' = f(X)$ and $Y' = g(Y)$, and so long as the functions, f and g, are monotonic, we have the following:

$$I(X', Y') = I(X, Y)$$

Eq. 31

This tells us that if we have a non-linear relationship between our random variables, X and Y, the mutual information will be just as confident at identifying and quantifying that relationship as when the relationship is linear. The mutual information is a sort of "non-linear correlation coefficient." In contrast, the Pearson correlation coefficient value would be different for a non-linear relationship between X and Y compared to a linear relationship. The Pearson correlation coefficient may not spot that there is a strong relationship between X and Y, if that relationship is non-linear.

This makes mutual information a popular tool for feature selection in machine learning, by filtering out features that have low mutual information scores with the target variable. Because many machine learning algorithms can build predictive models of non-linear relationships between features and targets, mutual information is excellent at identifying features a machine learning algorithm can make good use of.

Let's illustrate Eq. 31 with a code example. First, we'll introduce the data we'll use in the code example. The plots in Figure 13.4 show scatter plots of two variables, y and z, against a third variable, x:

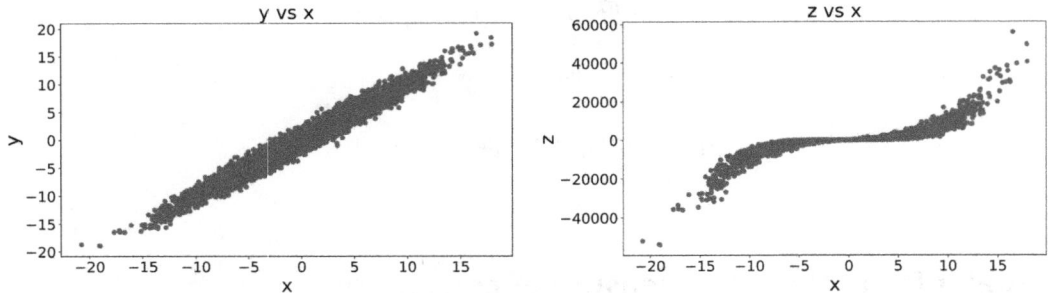

Figure 13.4: Plots of y and z against x

There is a clear strong linear relationship between y and x, while the relationship between z and x is non-linear. It won't surprise you to learn that the z values have been constructed from the y values via $z = 8y^3$. If we wanted to build a model of y or z, then x would be a good feature to use. Let's look at a code example showing how the mutual information is just as good at identifying that x is a good feature for predicting z as it is at identifying x is a good feature for predicting y.

Mutual information code example

You'll find this code example in the `Code_Examples_Chap13.ipynb` Jupyter notebook in this book's GitHub repository.

The first problem we must address is that the definition of mutual information in Eq. 19 is for random variables. In Figure 13.4, we have data – that is, a sample from random variables. This means we can only estimate the mutual information and we must construct an estimation algorithm to do that. Fortunately, someone has already done that for us. We can use the `scikit-learn` library's `mutual_info_regression` function from `sklearn.feature_selection` to estimate the mutual information from samples of two variables. We'll use the data that is plotted in Figure 13.4. The data is stored in the `mutual_information_data.csv` file in the `Data` directory of this book's GitHub repository. The data contains three columns called x, y and z. First, we must read in the data:

```
import numpy as np
import pandas as pd
import matplotlib.pyplot as plt
from sklearn.feature_selection import mutual_info_regression
from scipy.stats import pearsonr
# Read in the data
df_mutual = pd.read_csv('../Data/mutual_information_data.csv')
```

We'll calculate the Pearson correlation coefficient between x and y and between x and z. We'll use the `scipy.stats.pearsonr` function to calculate the Pearson correlations for us:

```
print("Pearson correlation between y and x = ",
        pearsonr(df_mutual['x'], df_mutual['y'])[0])
print("Pearson correlation between z and x = ",
        pearsonr(df_mutual['x'], df_mutual['z'])[0])
```

This gives us the following output:

```
Pearson correlation between y and x =   0.9801371625555979
Pearson correlation between z and x =   0.7569629714461419
```

There is a 26% difference between the two Pearson correlation coefficient values. Using the Pearson correlation coefficient to identify relationships between variables, from samples of those variables, works well when the relationship is linear. However, when the relationship between two variables is non-linear, the Pearson correlation coefficient is not so good at identifying that a relationship is present.

Now, for comparison, we'll estimate the mutual information between x and y and between x and z:

```
print("Mutual information between y and x = ",
        mutual_info_regression(df_mutual['x'].values.reshape(-1,1),
                                df_mutual['y'])[0])
print("Mutual information between z and x = ",
        mutual_info_regression(df_mutual['x'].values.reshape(-1,1),
                                df_mutual['z'])[0])
```

This gives us the following output:

```
Mutual information between y and x =   1.6219665917644344
Mutual information between z and x =   1.6154808774857417
```

There is a 0.4% difference between those two estimates of mutual information. This says that mutual information is just as good at identifying a relationship between two variables when that relationship is non-linear as when the relationship is linear.

This code example concludes this section on mutual information, so let's recap what we've learned.

What we've learned

In this section, we've learned the following:

- The idea of entropy can be generalized to multiple random variables
- The mutual information, $I(X, Y)$, tells us how much reduction in uncertainty about Y we get on average from knowing the value of X, and vice versa

- Random variables that are independent of each other have zero mutual information

- The conditional entropy, $H(Y|X)$, tells us the average entropy of Y when we know the value of X, and so tells us about the average uncertainty in Y after we know X

- Mutual information and conditional entropy can be calculated for continuous random variables

- The mutual information between X and Y gives us a measure of the correlation between X and Y, even when the relationship between X and Y is non-linear

Having learned how to quantify the information that is common between two different random variables, in the next section, we'll learn how to quantify the information that is in common between two different distributions of the same single random variable.

The Kullback-Leibler divergence

In the previous section, we learned about the similarities between random variables from an information theory perspective. But let's return to just a single random variable, X, and ask what would happen if we had two different distributions for that variable, X. As ever, we'll start with the discrete case.

Wait a moment! How can we have two different distributions for the same random variable? The variable is either random, and so has a given distribution, or it is not random. Yes, that's correct, but imagine that we have its true distribution, $P_X(x)$, and an approximation to its true distribution. We'll call the approximation $Q_X(x)$. It may be that the true distribution is too complex to practically work with in a data science algorithm, so we want to replace it with a more tractable distribution, $Q_X(x)$. Ideally, we'd like some way of measuring the difference between $P_X(x)$ and $Q_X(x)$ so that we can tell how good an approximation $Q_X(x)$ is. We need a measure of the "distance" between $P_X(x)$ and $Q_X(x)$. Even better, if $Q_X(x)$ had some parameters, such as hyperparameters, we could use them to minimize that distance and make $Q_X(x)$ look as much like $P_X(x)$ as we possibly can. That way, we'd get the benefit of a tractable and easy-to-use distribution, $Q_X(x)$, that is as close as possible to the true distribution.

So, how do we measure the difference between $P_X(x)$ and $Q_X(x)$? Using information, of course!

Relative entropy

For any outcome value, x, we can calculate the difference in information that we'd get according to $P_X(x)$ and according to $Q_X(x)$. This information difference is as follows:

$$\text{Information difference} = -\log Q_X(x) + \log P_X(x) = \log\left(\frac{P_X(x)}{Q_X(x)}\right)$$

Eq. 32

But this is the information difference at a single value, x, so to construct a difference between P_X and Q_X, we must calculate the average of Eq. 32 over all possible values of x, weighted with the true probabilities, $P_X(x)$. Doing so gives us the following formula:

$$\text{Average Information Difference} = \sum_x P_X(x) \log\left(\frac{P_X(x)}{Q_X(x)}\right)$$

Eq. 33

Because Eq. 33 measures the difference in two average information values – that is, a difference in entropies – it is called the **relative entropy**. However, it has another more commonly used name, as we shall see.

The expression on the right-hand side of Eq. 33 is not a measure of distance between P_X and Q_X because it is not symmetric – that is, if we swap P_X and Q_X around in Eq. 33, we get a slightly different mathematical expression. Instead, we refer to Eq. 33 as a **divergence**. In this case, it is called the Kullback-Leibler divergence, or KL-divergence for short, and given the symbol $D_{KL}(P_X || Q_X)$. So, formally we have the following:

$$D_{KL}(P_X || Q_X) = \sum_x P_X(x) \log\left(\frac{P_X(x)}{Q_X(x)}\right)$$

Eq. 34

We also have the following formula:

$$D_{KL}(Q_X || P_X) = \sum_x Q_X(x) \log\left(\frac{Q_X(x)}{P_X(x)}\right)$$

Eq. 35

Although we can see from Eq. 34 and Eq. 35 that $D_{KL}(Q_X || P_X) \neq D_{KL}(P_X || Q_X)$ and so we cannot take $D_{KL}(P_X || Q_X)$ as a distance measure, we can still use it as a measure of the difference between P_X and Q_X. This means we can still use $D_{KL}(P_X || Q_X)$ for our original goal of finding an optimized approximation to the distribution, P_X. In particular, $D_{KL}(P_X || Q_X)$ has the following useful properties:

$$D_{KL}(P_X || Q_X) = 0 \text{ if and only if } Q_X = P_X$$

Eq. 36

$$D_{KL}(P_X || Q_X) > 0 \text{ if } Q_X \neq P_X$$

Eq. 37

These properties tell us that the KL-divergence is only zero if our approximation is perfect. By adjusting any parameters in Q_X to make the KL-divergence smaller, we will be making our approximation Q_X closer to P_X.

KL-divergence for continuous variables

As you have probably already guessed, all the results about KL-divergences for discrete random variables can be generalized to continuous random variables. The formulae are those you would guess, so we will just state them here. For a continuous random variable, X, the KL-divergence between probability densities $p_X(x)$ and $q_X(x)$ is given by the following formula:

$$D_{KL}(p_X \| q_X) = \int p_X\left(x\right)\log\left(\frac{p_X(x)}{q_X(x)}\right)dx$$

Eq. 38

Using the KL-divergence for approximation

To finish this chapter, we'll give an example of how we can use the KL-divergence for approximating a distribution. This example is deliberately simple to help illustrate the ideas. Don't worry – there is a more complex example that has been set as an exercise at the end of this chapter.

We will use a continuous random variable, X, whose true density is a Laplace distribution with density given by the following formula:

$$p_X(x) = \frac{1}{2\lambda}\exp\left(-\frac{|x|}{\lambda}\right)$$

Eq. 39

Here, X has a mean of zero and a variance of $2\lambda^2$. We are going to approximate $p_X(x)$ by a Gaussian distribution with mean zero and variance, σ^2, so we'll write the following:

$$q_X(x) = \frac{1}{\sqrt{2\pi\sigma^2}}\exp\left(-\frac{x^2}{2\sigma^2}\right)$$

Eq. 40

The variance, σ^2, is the parameter we will adjust to make q_X as close as possible to p_X by minimizing the KL-divergence, $D_{KL}(p_X \| q_X)$, given in Eq. 38. We can simplify the calculation by first re-writing $D_{KL}(p_X \| q_X)$, as follows:

$$D_{KL}(p_X \| q_X) = -\int p_X(x)\log q_X(x)dx + \int p_X(x)\log p_X(x)\,dx$$

Eq. 41

The second integral on the right-hand side of Eq. 41 only depends on p_X, so it does not depend on our parameter, σ^2. To determine the optimal value of σ^2, we only need minimize the first integral on the right-hand side of Eq. 41. The first integral is as follows:

$$-\frac{1}{2\lambda}\int_{-\infty}^{\infty}e^{-\frac{|x|}{\lambda}}\left(-\frac{1}{2}\log 2\pi\sigma^2 - \frac{x^2}{2\sigma^2}\right)dx = \frac{1}{2}\log 2\pi\sigma^2 + \frac{\lambda^2}{\sigma^2}$$

Eq. 42

So, to determine the optimal value of σ^2, we must minimize the right-hand side of Eq. 42 with respect to σ^2. We can do this by differentiating with respect to σ^2 and setting the derivative equal to zero. By doing this, we get the following formula:

$$\frac{\partial}{\partial \sigma^2}\left[\frac{1}{2}\log 2\pi\sigma^2 + \frac{\lambda^2}{\sigma^2}\right] = \frac{1}{2\sigma^2} - \frac{\lambda^2}{\sigma^4} = 0$$

Eq. 43

Solving Eq. 43 for σ^2, we find that $\sigma^2 = 2\lambda^2$.

This result is somewhat trivial. It says that the optimal value of σ^2, the variance of our Gaussian approximation, is equal to the variance of our true distribution. We could have guessed that. There are also other simpler methods, such as moment matching, for setting the variance parameter of an approximation. However, our main goal here was to illustrate, in as simple a way as possible, the main steps in using the KL-divergence to derive an optimal approximation to another distribution. There is another reason why we chose to use this very simple example, and why our exercises are still relatively simple at the end of this chapter. It is because, for these examples, we can calculate $D_{KL}(p_X||q_X)$ exactly and minimize it exactly. This is not always the case.

Variational inference

When we use KL-divergences in real data science algorithms, we are typically constructing some approximation, q_θ, to the Bayesian posterior probability, $p_\theta = \text{Prob}(\underline{\theta}|\text{Data})$, where $\underline{\theta}$ represents the parameters of a probabilistic model we have used to model our training data. The practice of constructing an approximation to a model posterior by varying the parameters of the approximation until a KL-divergence is optimal, is part of a wider field of machine learning methods known as **variational inference techniques**, or **variational inference**.

In general, calculating and minimizing $D_{KL}(p_\theta||q_\theta)$ can be hard and the resulting optimized approximation, q_θ is not always an accurate approximation, or as accurate as we would like it to be. Instead, in variational inference, it is more common to work with $D_{KL}(q_\theta||p_\theta)$. Here, we minimize $D_{KL}(q_\theta||p_\theta)$ with respect to the parameters of the approximate density, q_θ.

Minimizing $D_{KL}(q_\theta||p_\theta)$ has some nice properties. Deriving those properties is beyond the scope of this short section on the KL-divergence, so we will just state the main property at a high level. Minimizing the KL-divergence, $D_{KL}(q_\theta||p_\theta)$, is equivalent to maximizing a lower bound on the Bayesian evidence of the model. You may recall from *Chapter 8* that the Bayesian evidence is a quantity that measures how good our probabilistic model form is, given the data. From this, we can rigorously show that minimizing $D_{KL}(q_\theta||p_\theta)$ does indeed obtain the best possible approximation, q_θ, and this motivates us to work with $D_{KL}(q_\theta||p_\theta)$.

Minimizing $D_{KL}(q_\theta||p_\theta)$ when working with real data science algorithms can also require using sophisticated optimization techniques, so we won't go into the details here. However, it does explain why our introductory examples needed to be simple and why we focused on minimizing $D_{KL}(p_\theta||q_\theta)$ instead.

This section provided a concise introduction to the KL-divergence. So, let's summarize what we've learned about it before wrapping up this chapter.

What we've learned

In this section, we've learned the following:

- Relative entropy measures the average difference in information between two different distributions for the same random variable

- Relative entropy is also known as the KL-divergence and can be used as an asymmetric measure of the difference between two distributions

- We can use the KL-divergence to optimize how well one distribution approximates another

Summary

This chapter was about information theory. Although you are less likely to directly use the calculations demonstrated in this chapter compared to the material from other chapters, the concepts and ideas behind information theory can be invaluable. Information-theoretic concepts give us a different way to think about probability, distributions, and what is conveyed when we observe a piece of data. Those concepts are as follows:

- Information theory concerns itself with the communication of signals and the efficiency of encoding those signals.

- The smaller the probability of an event or outcome occurring, the higher the information associated with that event or outcome.

- We measure information on a logarithmic scale.

- The expected information tells us the average amount of information we get from an observation of a random variable. The expected information is more commonly known as entropy.

- Entropy increases with the variance of a distribution, so it quantifies the uncertainty about an outcome before we have measured it.

- The MaxEnt technique can be used to determine the most probable distribution compatible with specified constraints.

- The idea of entropy can be generalized to multiple random variables.

- The mutual information, $I(X, Y)$, tells us how much reduction in uncertainty about Y we get on average from knowing the value of X, and vice versa.

- The conditional entropy, $H(Y|X)$, tells us the average entropy of Y when we know the value of X, so it tells us about the average uncertainty in Y after we know X.

- The mutual information between X and Y gives us a measure of the correlation between X and Y, even when the relationship between X and Y is non-linear.

- Relative entropy measures the average difference in information between two different distributions for the same random variable.

- Relative entropy is also known as the KL-divergence and can be used as an asymmetric measure of the difference between two distributions.

- We can use the KL-divergence to optimize how well one distribution approximates another.

This was a very brief introduction to a very large topic. For more extensive introductions to information theory, see the books suggested in [2] of the *Notes and further reading* section.

Like this chapter on information theory, our next chapter will also make use of probabilistic concepts. The next chapter will also cover an advanced topic: Bayesian non-parametric modeling.

Exercises

This section contains a series of exercises. The answers to all these can be found in the `Answers_to_Exercises_Chap13.ipynb` Jupyter notebook in this book's GitHub repository.

We have a composite random variable, X, that consists of three binary random variables, A_1, A_2, A_3. We denote this as $X = (A_1, A_2, A_3)$. We'll use a_1 for the outcome for A_1, a_2 for the outcome of A_2, and a_3 for the outcome of A_3. This means $a_1, a_2, a_3 \in \{0,1\}$.

We can write the outcome, x, for the overall random variable, X, as a three-digit bit-string. For example, $x = 010$ –to represent the outcome, $a_1 = 0, a_2 = 1, a_3 = 0$. There are $2^3 = 8$ possible values for x; these are 000,001,010,011,100,101,110,111. We can also denote the true probability distribution, $P_X(x)$, as $P_{A_1,A_2,A_3}(a_1, a_2, a_3)$; it corresponds to eight numbers (between 0 and 1) that all add up to 1.

Now, let's introduce our approximation, Q_X. We will use a product approximation, so we'll write the following:

$$Q_{A_1,A_2,A_3}(a_1, a_2, a_3) = P_{A_1}^{(\text{approx})}(a_1)\, P_{A_2}^{(\text{approx})}(a_2)\, P_{A_3}^{(\text{approx})}(a_3)$$

Eq. 44

We've put the superscript "approx" on the distributions on the right-hand side of Eq. 44 to emphasize that we're constructing an approximation and that $P_{A_1}^{(\text{approx})}(a_1)$ is not the true marginal distribution, $P_{A_1}(a_1)$, but an approximation to it.

Here are the exercises concerning this question:

1. What is the only form the approximation, $P_{A_1}^{(\text{approx})}(a_1)$, can take? What are the parameters of this approximation?

2. Using this approximation for $P_{A_1}^{(\text{approx})}(a_1)$, and similar approximations for $P_{A_2}^{(\text{approx})}(a_2)$ and $P_{A_3}^{(\text{approx})}(a_3)$, substitute these into Eq. 44 to write down the full mathematical form for the approximation, $Q_{A_1,A_2,A_3}(a_1, a_2, a_3)$.

3. Using the mathematical expression for $Q_{A_1,A_2,A_3}(a_1, a_2, a_3)$ that you wrote in question 2, derive an expression for the KL-divergence, $D_{KL}\left(P_{A_1,A_2,A_3} \middle\| Q_{A_1,A_2,A_3}\right)$, in terms of the parameters of $Q_{A_1,A_2,A_3}(a_1, a_2, a_3)$ and the true marginal distributions, $P_{A_1}(a_1)$, $P_{A_2}(a_2)$, and $P_{A_3}(a_3)$.

4. Minimize the expression for the KL-divergence you derived in question 3 concerning the parameters of $Q_{A_1,A_2,A_3}(a_1, a_2, a_3)$. Comment on the solution.

Notes and further reading

To learn more about the topics that were covered in this chapter, take a look at the following resources:

1. For a short but very readable discussion on the broader aspects of what information is, I like the book by L. Floridi, *Information: A Very Short Introduction*, 1st Edition (2010), Oxford University Press, Oxford, UK. ISBN: 978-0199551378.

2. For additional texts on the mathematical aspects of information theory, see the following:

    A. For a modern readable introduction to the mathematical theory of information, I like the book by J.V. Stone, *Information Theory: A Tutorial Introduction*, 1st Edition (2015), Sebtel Press. ISBN: 978-0956372857.

    B. Another accessible and well-established account of mathematical information theory is the book by J.R. Pierce, *An Introduction to Information Theory: Symbols, Signals and Noise*, Revised 2nd Edition (2003), Dover Publications, New York, USA. ISBN: 978-0486240619.

    C. The most authoritative textbook on information theory is probably *Elements of Information Theory*, 2nd Edition (2006), by T.M. Cover and J.A. Thomas. It is published by Wiley-Interscience, Hoboken, New Jersey, USA. It is a lengthy textbook (nearly 800 pages).

    D. A modern and well-known book linking aspects of information theory and machine learning is that by D.J.C. MacKay, *Information Theory, Inference and Learning Algorithms*, 1st Edition (2003), Cambridge University Press, Cambridge, UK. ISBN: 978-0521642989. A PDF copy of the book can be found online at `https://www.inference.org.uk/itila/book.html`.

14

Non-Parametric Bayesian Methods

Building a predictive model requires us to make assumptions. For example, we often need to assume some fixed mathematical form for the relationship between our predictive features and the response variable. It is the parameters within that mathematical form that we usually vary and optimize through a training process, not the mathematical form. If those parametric assumptions are incorrect, we get a poorly performing model. Often it would be better to not make those parametric assumptions and to use a non-parametric modeling approach. That is what we do in this chapter. We do so by putting Bayesian priors on the functions and relationships that we model. This makes the methods we use non-parametric Bayesian methods. To learn about them we must introduce some new modeling ideas and concepts. We do that by covering the following topics:

- *What are non-parametric Bayesian methods?*: This is where we learn about the key concept of not making parametric assumptions about the relationship between features and response variables

- *Gaussian processes*: This is where we learn what **Gaussian processes** (**GPs**) are and how they can be used as a prior on the relationship between our features and our response variable

- *Dirichlet processes*: This is where we learn what **Dirichlet processes** (**DPs**) are and how they can be used as a prior on a probability distribution

Technical requirements

All code examples given in this chapter can be found in the GitHub repository at `https://github.com/PacktPublishing/15-Math-Concepts-Every-Data-Scientist-Should-Know/tree/main/Chapter14`. To run the Jupyter Notebooks, you will need a full Python installation including the following packages:

- `pandas` (>=2.0.3)
- `numpy` (>=1.24.3)
- `scikit-learn` (>=1.3.0)
- `matplotlib` (>=3.7.2)

What are non-parametric Bayesian methods?

As the name suggests, non-parametric Bayesian methods are Bayesian, so it may be a good time to review the section in *Chapter 5* on Bayesian probabilistic modeling. Because we are using Bayesian methods, we will be using priors. As usual with Bayesian methods, there is a subjective element to setting the prior. The prior is something we choose. Choose a slightly different prior and we will get slightly different inferences. However, it is what we use the prior for that is the interesting aspect of non-parametric Bayesian methods.

In *Chapter 5*, when we were using Bayesian methods to build probabilistic models, we had the data points, (x_i, y_i), $i = 1, 2, \ldots, N$. We modeled the target variable values, y_i, as random variables whose probability density function was given by some function, $f(x_i|\theta)$, with θ being the model parameters. We'd write this in statistical modeling notation as follows:

$$y_i \sim f(x_i|\theta) \quad i = 1, 2, \ldots, N$$

Eq. 1

What *Eq. 1* says is that the observations of the target (response) variable, y, are random variables whose distribution changes with the predictive features, x. For example, the mean of y changes with x, and how those changes occur is controlled by the model parameters, θ. *Eq. 1* allows us to calculate the likelihood of the observations, $y_1, y_2, \ldots, y_N$. We would then put a prior, $P(\theta)$, on the parameters, θ. Combining the prior and the likelihood gives us the posterior, the probability of the model parameters, θ, given the data. This is **Bayesian parametric modeling**.

In *Eq. 1*, we have a fixed mathematical form for f and it is the model parameters, θ, that control the variation in the target values, $y = (y_1, y_2, \ldots, y_N)$. Typically, the more model parameters we have, the more flexibility we have in the possible variation of y. However, because the number of model parameters is finite, we always have a finite-dimensional model. Because of this, we are restricting the possible variation in y that our model can capture or learn.

An alternative approach is to allow the function to be infinite dimensional. In doing so, we are effectively taking the parameters, θ, out of the problem. Instead, we just have the function, f, which describes a surface over x. We still want to put some restrictions on the possible shapes f can take. For example, we may want it to be smoothly varying and not have spiky up and down sections to it. In other words, we want to draw or sample f from some space of (reasonably behaved) functions that we specify. We do this by putting a prior on f. Once we have a prior on f, we can again use it to calculate a posterior probability for the function, f, given the data. This is **Bayesian non-parametric modeling**.

We still have parameters

The name *Bayesian non-parametric modeling* can be confusing when you first encounter it. We still have some parameters in the overall calculation that are part of our prediction algorithm. It would be difficult to run a machine learning algorithm that had no parameters whatsoever. We still have parameters in our algorithm, and we will even optimize those parameters on a training set. However, those parameters relate to a different part of the overall calculation – typically, they are part of the mathematical specification of the prior on the function, f. The word "non-parametric" in Bayesian non-parametric modeling refers to the fact that we have no explicit parameters in the relationship that links the response, y_i, to the features, x_i.

The different types of non-parametric Bayesian methods

The idea we have introduced is a very general one. It is the idea that instead of specifying a model function, f, by specifying some fixed parametric form (with parameters, θ) and putting a prior on the function parameters, instead we introduce flexibility into the choice of f by drawing it from some space of functions by putting a prior directly on f. Because of the generality of the idea, it can be applied in many ways. Two of the most common types of priors we put on f come from the following:

- Gaussian Processes
- Dirichlet Processes

These are the priors we shall focus on in this chapter, with most of our focus being on **Gaussian process regression** (GPR) since that is a very intuitive way to understand the power of non-parametric Bayesian methods.

The pros and cons of non-parametric Bayesian methods

Non-parametric Bayesian methods specify how the function, f, in *Eq. 1* can vary. Since f controls the distribution of y, this means that non-parametric Bayesian methods focus on how the data, y, varies from one data point to another given changes in the feature vector, x, from one data point to another. This makes Bayesian non-parametric models about smoothing or interpolating between the training data observations. This also makes Bayesian non-parametric models extremely flexible and relatively easy to use. Instead, using the parametric approach in *Eq. 1*, we'd probably have to bake domain-specific knowledge into the model, $f(x_i|\theta)$. In contrast, in non-parametric Bayesian models, we'd typically just get the data and let the model interpolate between the observations for us. Non-parametric Bayesian models are very data driven.

The downside to that is, i) Non-parametric Bayesian models do not scale well with increases in N, the size of the training dataset, since they are directly using all the training data to make the predictions, ii) non-parametric Bayesian models are less able to extrapolate compared to say a parametric model whose mathematical form has been constructed using explicit domain knowledge. Finally, as with all Bayesian methods, there is some art and subjectivity in specifying the choice of any priors.

That's the introduction complete, so, let's summarize what we have learned.

What we learned

In this section, we have learned the following:

- Non-parametric Bayesian methods focus on specifying a modeling function as coming from a prior distribution over functions, instead of being of a fixed mathematical form

- Non-parametric Bayesian models still have parameters

Having learned in general what non-parametric Bayesian methods are, in the next section, we'll look at a specific method, that of GPR.

Gaussian processes

We're going to look in detail at a specific non-parametric Bayesian method that makes use of GPs. As the name suggests, GPs involve a Gaussian distribution. In fact, the Gaussian distribution is the prior that we put on our function, f. This makes GPs widely used in non-parametric Bayesian methods. We can use them for constructing both regression models and classification models. To keep this chapter short, we'll only illustrate GPR, but many of the concepts and ideas are the same for both GPR and GP classification. Personally, I also find GPR the easiest non-parametric Bayesian method to understand, so it is a good place to start.

To start we need to set up our model. We'll model our observations, y_i, as follows:

$$y_i = f(x_i) + \varepsilon_i \quad , \quad \varepsilon_i \sim Normal(0, \sigma^2) \quad i = 1, 2, \ldots N$$

Eq. 2

For a fixed choice of f, Eq. 2 says that our observation, y_i, is a Gaussian noise corrupted version of $f(x_i)$. We will also assume that the noise values, ε_i, from different observations are independent of each other.

Eq. 2 looks like *Eq. 1* but we don't have any parameters, $\underline{\theta}$, because we're not specifying a fixed parametric form for f. All we're saying in *Eq. 2* is that f is some function. If we were to calculate the joint distribution of the observations, $y_1, y_2, \ldots, y_N$, it would be boring, as it is a multivariate Gaussian with a diagonal covariance matrix. It would be the following:

$$\underline{y} \sim Normal(\underline{f}, \sigma^2 \underline{I}_N)$$

Eq. 3

In *Eq. 3*, the vector, $\underline{f}$, is given by $\underline{f} = \left(f(\underline{x}_1), f(\underline{x}_2), \ldots, f(\underline{x}_N) \right)$. So far, in *Eq. 3*, we have only considered the variation in the random noise variables, ε_i, when deriving the distribution of $\underline{y}$. Now, we introduce our prior on the function, f. In GPR, we put a GP prior on f. This is denoted by saying the following:

$$f \sim \mathcal{GP}(0, K)$$

Eq. 4

What *Eq. 4* says is that the function, f, has some distribution. It is like an infinite dimensional random variable. Its random variation is controlled by the function, K – more on that in a moment.

In practical terms, what this means is that as we vary the possible choice of f, the values of $f(\underline{x}_1)$ and $f(\underline{x}_2)$ will be correlated. In fact, the value of $f(\underline{x}_i)$ and $f(\underline{x}_j)$ will be correlated for any values of i and j. A GP prior on f says that **a priori**:

$$\underline{f} \sim Normal(\underline{0}, \underline{K})$$

Eq. 5

The $N \times N$ matrix, $\underline{K}$, has matrix elements, K_{ij}, $i, j = 1, 2, \ldots, N$, and these are calculated from the function, K, via the following:

$$K_{ij} = K(\underline{x}_i, \underline{x}_j)$$

Eq. 6

Now, if we combine the variation of $\underline{f}$ in *Eq. 5* with the variation of $\underline{y}$ in *Eq. 3*, we get that the overall variation of $\underline{y}$ is given by the following:

$$\underline{y} \sim Normal(\underline{0}, \underline{K} + \sigma^2 \underline{I}_N)$$

Eq. 7

Eq. 7 gives us the distribution of y once we have incorporated both the variation over the random noise variables, ε_i, and the variation over the possible functions, f.

Our model for y in *Eq. 7* is very simple. It says y is a multivariate Gaussian with a mean vector of $\underline{0}$ and covariance matrix, $\underline{K} + \sigma^2 \underline{I}_N$. The zero mean vector relates to the 0 in the notation, $f \sim \mathcal{GP}$ $(0, K)$. It means that our prior expectation for $f(\underline{x})$ was zero for any value of $\underline{x}$, and so once we've incorporated the variation over f, we're assuming that the expected value of y_i is zero. Without loss of generality, we can always mean center our data sample, $y_1, y_2, \ldots, y_N$, before we start modeling, so that our GP prior in *Eq. 4* is always appropriate. It is also easy to generalize the GP prior to include a mean function, $\mu(\underline{x})$, by writing the following:

$$f \sim \mathcal{GP}(\mu, K)$$

Eq. 8

In this case, *Eq. 7* would become the following:

$$\underline{y} \sim Normal\left(\underline{\mu}, \underline{K} + \sigma^2 \underline{I}_N\right)$$

Eq. 9

where the vector, $\underline{\mu}$, is given by $\underline{\mu} = (\mu(\underline{x}_1), \mu(\underline{x}_2), \ldots, \mu(\underline{x}_N))$. For simplicity of illustration, from now on we're going to assume $\mu(\underline{x}) = 0$ for any $\underline{x}$ (i.e., our target values, y_i, have expectations of zero).

The kernel function

To model our data using a (zero mean) GP, we just need to know the value of σ^2 and the choice of function, K. What is this function, K? What does it do? From *Eq. 7*, we can see that K determines the covariance between y_i and y_j. A high covariance value means we expect that y_i and y_j vary in a similar way (i.e., they are similar). That similarity is quantified by $K(\underline{x}_i, \underline{x}_j)$.

In *Chapter 12*, we encountered functions that measure the similarity between feature vectors, $\underline{x}_i$ and $\underline{x}_j$. They were called **kernel functions**. That is what the function, K, is. K is a kernel function. Unsurprisingly, the types of kernel functions that we use in GPR are the same as we used in kernel methods in **Chapter 12**. Common choices for K are the RBF (squared-exponential) kernel, dot-product kernels, and the **Matérn kernel**.

When we use an RBF kernel, we are using a function, K, of the following form:

$$K(\underline{x}_i, \underline{x}_j) = A \times \exp\left(-\frac{\left|\underline{x}_i - \underline{x}_j\right|^2}{2b^2}\right)$$

Eq. 10

This is the RBF kernel we encountered in *Chapter 12* but with the addition of the extra parameter, A. From this, it should be clear that GPR is not parameter free. If we were to use the RBF kernel in *Eq. 10* in combination with *Eq. 7*, then our GPR model has three parameters, A, b, and σ^2. But how do we determine appropriate values for those parameters? By fitting the GPR model to training data. We do that next.

Fitting GPR models

The parameters of our GPR model are σ^2 and whatever parameters we have in our kernel function, K. *Eq. 7* tells us the probability of the data, y, given the parameters. We can view this as a likelihood. It is the marginal likelihood because we have integrated over all the possible functions, f, coming from our GP prior. We can use the marginal likelihood to estimate the GPR model parameters by maximum likelihood. Alternatively, if we put priors on those GPR parameters, we can estimate them via **maximum a posteriori** (**MAP**) estimation, or even sample directly from the Bayesian posterior of the model parameters – see *Chapter 5* for a reminder of these techniques. For simplicity, we'll stick to maximum likelihood estimation. We'll write out the log-likelihood explicitly as follows:

$$\text{log-likelihood} = -\frac{N}{2}\log 2\pi - \frac{1}{2}\text{logdet}\left(\underline{K} + \sigma^2 \underline{I}_N\right) - \frac{1}{2}\underline{y}^\top \left(\underline{K} + \sigma^2 \underline{I}_N\right)^{-1}\underline{y}$$

Eq. 11

Fitting our GPR model by maximum likelihood requires us to calculate derivatives of the log-likelihood in *Eq. 11* with respect to its parameters.

In practice, there are several existing Python packages for doing GPR, so we don't have to do the differentiation and equation solving ourselves, however, instead, we just specify what sort of kernel function we want to use in our model. In a moment, we'll demonstrate GPR using the `scikit-learn` package, but before we do, we'll explain how we use a fitted GPR model to make predictions.

Prediction using GPR models

Okay, so imagine we've fitted our GPR model using maximum likelihood to obtain optimal values for σ^2 and the kernel function parameters. How do we use the fitted model to make predictions? We just use *Eq. 7* again. Imagine we are trying to make a prediction of the value of y at the point, $\underline{x}_*$. We'll denote that value by y_*. By extending the derivation behind *Eq. 7*, we can derive the joint distribution of y_* and y. In doing so, we get the following:

$$(\underline{y}, y_*) \sim Normal\left(\underline{0}, \underline{K}_* + \sigma^2 \underline{I}_{N+1}\right)$$

Eq. 12

In *Eq. 12*, we are modeling $N + 1$ data points, whilst in *Eq. 7*, we were modeling N. In *Eq. 12*, the $(N + 1) \times (N + 1)$ matrix, $\underline{K}_*$ is of the following block form:

$$\underline{K}_* = \begin{pmatrix} \underline{K} & \underline{k}_* \\ \underline{k}_*^\mathsf{T} & K(\underline{x}_*, \underline{x}_*) \end{pmatrix}$$

Eq. 13

The matrix, $\underline{K}$, is the same as in *Eq. 7*. The N-dimensional column vector, $\underline{k}_*$, is given by the following:

$$\underline{k}_*^\mathsf{T} = \left(K(\underline{x}_1, \underline{x}_*), K(\underline{x}_2, \underline{x}_*), \dots, K(\underline{x}_N, \underline{x}_*) \right)$$

Eq. 14

To calculate the joint distribution in *Eq. 12*, all we need is our kernel function and the value of σ^2. From the Gaussian distribution in *Eq. 12*, we can calculate the conditional density, $p(y_*|\underline{y})$, by applying Bayes' theorem. For a Gaussian distribution, it is a straightforward calculation, so we'll just give the following result:

$$y_*|\underline{y} \sim Normal\left(\underline{k}_*^\mathsf{T} \left(\underline{K} + \sigma^2 \underline{I}_N \right)^{-1} \underline{y} \;,\; K(\underline{x}_*, \underline{x}_*) + \sigma^2 - \underline{k}_*^\mathsf{T} \left(\underline{K} + \sigma^2 \underline{I}_N \right)^{-1} \underline{k}_* \right)$$

Eq. 15

What you'll notice from *Eq. 15* is that we get a probabilistic prediction. We get a distribution. We can calculate the expected value of y_*. For the univariate Gaussian distribution in *Eq. 15*, we can easily read off the expectation value of y_*. It is as follows:

$$\mathbb{E}(y_*|\underline{y}) = \underline{k}_*^\mathsf{T} \left(\underline{K} + \sigma^2 \underline{I}_N \right)^{-1} \underline{y}$$

Eq. 16

Let's unpack *Eq. 16* in more detail. We can re-write it as follows:

$$\mathbb{E}(y_*|\underline{y}) = \sum_{j=1}^{N} w_j y_j \;,\; w_j = \sum_{i=1}^{N} K(\underline{x}_i, \underline{x}_*) \left(\underline{K} + \sigma^2 \underline{I}_N \right)^{-1}_{ij}$$

Eq. 17

Eq. 17 says the prediction is given by a weighted sum of the observations, $y_1, y_2 \dots, y_N$. This is what we meant when we said non-parametric Bayesian methods are very data driven and smooth the training data. Our model of y_* is calculated directly from all the training data, $y_1, y_2 \dots, y_N$, not from a fixed mathematical form whose parameters are determined by the data. This is what we mean by a non-parametric method. The downside to this is that as the amount of data increases, the computational cost of calculating $\mathbb{E}(y_*|\underline{y})$ increases. The matrix inversions of the $N \times N$ matrices in *Eq. 17* scale as $O(N^3)$ and so become very computationally costly as N becomes large.

If we return to the prediction distribution, $y_*|y$, in *Eq. 15*, we see that not only do we get the expectation of y_*, but we also get the standard deviation. That means we automatically get an estimate of the uncertainty around the prediction. We can easily calculate 95% confidence intervals around the expectation value in *Eq. 16*. So, GPR not only makes predictions of y_*, it also tells us when it is confident about those predictions.

To see the power of GPR in action, we'll now look at a code example.

GPR code example

The following code example can be found in the `Code_Examples_Chap14.ipynb` Jupyter Notebook in the GitHub repository.

Since GPR is a commonly used technique, it is available in the `scikit-learn` package. We'll use the `sklearn.gaussian_process.GaussianProcessRegressor` class to do our GPR. For our code example, we're going to use a one-dimensional feature, x. The training data can be found in the `gp_data.csv` file in the `Data` directory of the GitHub repository. The training data has two columns, the x and y values. There are 25 data points in the training data. First, we'll read in the data:

```
import pandas as pd
import matplotlib.pyplot as plt
from sklearn.gaussian_process import GaussianProcessRegressor
from sklearn.gaussian_process.kernels import RBF, WhiteKernel

# Read in the data
df_gp = pd.read_csv('../Data/gp_data.csv')
```

We then take a quick look at the data:

```
# Look at the data
df_gp.head(5)
```

This then gives the following output:

```
       x            Y
  -3.213717    15.410904
  -2.583426    22.534045
  -2.147617     4.547179
  -1.634624     9.537378
  -1.50614      8.213919
```

It is also instructive to plot the training data, which we do next:

```
# Plot the training data
plt.scatter(df_gp['x'], df_gp['y'])
```

```
plt.title('Gaussian Process Regression Training Data')
plt.xlabel('x')
plt.ylabel('y')
plt.show()
```

This gives us the plot in *Figure 14.1*.

Figure 14.1: Plot of training data for GPR

There is clearly some relationship between *x* and *y*, but it is not obvious what it is. It could be a linear relationship, but it could be more complex. This is a good example of where to apply GPR. We will use an RBF kernel for our GPR; that is, we choose our function K to be the following:

$$K(x_i, x_j) = A\exp\left(-\frac{1}{2b^2}(x_i - x_j)^2\right)$$

Eq. 18

We can do this using the RBF constructor from the `sklearn.gaussian_process.kernels` module. We'll set $b = 1$ but `scikit-learn` will optimize this value for us. The value, $b = 1$, is just an initial guess. We specify the RBF kernel using the following syntax:

```
1*RBF(length_scale=1.0)
```

The prefactor of 1 in front of the RBF constructor call introduces an extra parameter into our RBF kernel, in this case, it is the parameter, A, and we have initialized its value to 1. Again, `scikit-learn` will optimize this parameter for us by fitting to the data. We must also add a noise component to our kernel. This is the $\sigma^2 I_N$ part of the overall covariance matrix in *Eq. 7*. We do this by adding a `scikit-learn` `WhiteKernel` object to our RBF kernel. Again, `scikit-learn` will optimize the noise level parameter, σ^2, when it fits to the data:

```
# Specify our kernel
kernel = 1*RBF(length_scale=1.0) + WhiteKernel(
    noise_level_bounds=(1.0, 20.0))
```

So, now that we have our kernel, we can create a GPR object by calling the GaussianProcessRegressor constructor, passing in the kernel object we already created:

```
# Create GaussianProcessRegressor object which we will use to do our
# Gaussian Process Regression
gaussian_process = GaussianProcessRegressor(kernel=kernel)
```

Now; we can optimize the parameters; A, b, σ, by calling the fit method of the Gaussian ProcessRegressor object, into which we pass our training data:

```
# Optimize the model parameters
gaussian_process.fit(df_gp['x'].values.reshape(-1,1), df_gp['y'])
```

Let's look at the optimized parameters. We can do this by just looking at the kernel object:

```
# Look at the kernel object after we have optimized it
gaussian_process.kernel_
```

This gives us the following output:

```
9.24**2 * RBF(length_scale=0.412) + WhiteKernel(noise_level=8.33)
```

We can see the optimized kernel is of the following form:

$$K(x_i, x_j) = 9.24^2 \exp\left(-\frac{1}{2 \times 0.412^2}(x_i - x_j)^2\right) + 8.33\,\delta_{ij}$$

Eq. 19

This means `scikit-learn` has chosen the optimal values as $A = 9.24^2$, $b = 0.412$, and $\sigma^2 = 8.33$. Having got our optimal `GaussianProcessRegressor` object, let's use it to make some predictions. We will make predictions for regularly spaced values of x between the maximum and minimum values of x seen in the training dataset. The predictions are the expectation values of the response (target) variable at each of the prediction points. We'll also get the standard deviations of the response variable at the prediction points so we can plot confidence intervals around our predictions:

```
# Create a range of x values for prediction
x_predict = np.arange(
    df_gp['x'].min(), df_gp['x'].max(), 0.01).reshape(-1,1)

# Get the prediction (expectation) and standard deviation.
mean_prediction, std_prediction = gaussian_process.predict(x_predict,
    return_std=True)
```

We'll also calculate the true expectation values of the response variable at each of the prediction points. I can do this because I know the formula used to create the true expectation values of the response variable in the training data:

```
# Calculate true expectation value
y_true = [(-3.4 - 2.0*x +(x-1.0)**2 + 5.5*np.sin(5.0*x-4.0))
         [0] for x in x_predict]
```

Finally, we plot the predictions (estimates of the expectation values), the 95% confidence intervals, the true expectation values, and the training data points:

```
# Plot the predictions
plt.plot(x_predict, y_true, label="True value", linestyle="dotted")
plt.scatter(df_gp['x'], df_gp['y'], label="Observations")
plt.plot(x_predict, mean_prediction, label="Mean prediction")
plt.fill_between(
    x_predict.ravel(),
    mean_prediction - 1.96 * std_prediction,
    mean_prediction + 1.96 * std_prediction,
    alpha=0.5,
    label=r"95% confidence interval",
)
plt.title('Gaussian Process Regression Predictions')
plt.xlabel('x')
plt.ylabel('y')
plt.legend()
plt.show()
```

This gives us the plot in *Figure 14.2*.

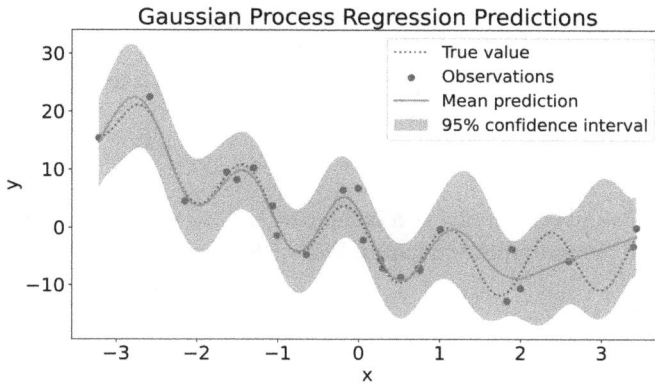

Figure 14.2: Plot of GPR predictions

We can see from *Figure 14.2* that the estimated expectation value (the mean prediction) follows the true expectation value reasonably closely, including the oscillatory pattern present. Despite having only 25 data points in the training set, we have been able to uncover a lot of the true structure present. In contrast, had we used a parametric model, we probably would have fitted a linear or quadratic relationship to the data, causing us to miss this oscillatory structure.

Where the estimated expectation value deviates from the true expectation value, it is where there are fewer training data points, but in these regions, the 95% confidence interval is also relatively wider, telling us that the estimated expectation value may be less accurate in these regions. This again highlights the data-driven nature of non-parametric Bayesian methods; where we have lots of observations, our predictions will be accurate, and where we have fewer observations, our predictions will be less accurate.

That has been a lengthy code example and a lengthy section overall, so, let's wrap up the section by summarizing what we have learned.

What we learned

In this section, we have learned the following:

- A GP is specified by a kernel (covariance) function and a mean function.
- The kernel function of a GP has parameters.
- In GPR, our model is the mean function of the response (target) variable and we put a GP prior on that mean function.
- In GPR, we can calculate the marginal likelihood of the data conditional on the kernel function parameters.

- In GPR, we can fit the kernel function parameters to the training data by maximizing the marginal likelihood.

- We make predictions in GPR by calculating the expectation value and variance of the response variable conditional on the training data. The conditional expectation value can be expressed as a weighted sum of the training data.

- We can use the `scikit-learn` package to do GPR.

Having learned about GPs and GPR, in the next section, we will learn about another type of stochastic process used in Bayesian non-parametric analysis. We will learn about DPs.

Dirichlet processes

GPs are not the only type of process used in non-parametric Bayesian methods, although they are possibly the most used. In this section, we will introduce another type of stochastic process, the DP. As with GPs, we will use DPs as priors for functions that we want to make inferences about.

Since the last section was a lengthy one with a lengthy code example, we will keep this section short and only give a high-level view of DPs.

How do DPs differ from GPs?

As you might have guessed, we use a DP as a prior on a function. However, GPs already did that for us, so how do DPs differ from GPs? When we used GPs in GPR, the GP provided a prior for a generic function, f. There were no restrictions on the type of function, f, could be. Sometimes, we will want to model a particular type of function. For example, we might need to build a model of a probability distribution. In this case, we use a DP to construct our prior.

Let's make that more explicit. I want to model a probability distribution, F, of a discrete random variable, X. A DP provides me with a prior for F. Sampling from the DP gives me an instance of F. We can think of a DP as a distribution of distributions.

The DP notation

A DP is specified by two quantities, α and H, so we denote the DP using the notation, $DP(\alpha, H)$. If our distribution, F, has a DP prior, then we write the following:

$$F \sim DP(\alpha, H)$$

Eq. 20

But what do α and H mean? H is usually referred to as the **base distribution**. You can think of it as the distribution about which the DP is centered. In fact, H is the expectation value of F in *Eq. 20*, meaning the following:

$$\mathbb{E}(F) \;=\; H$$

Eq. 21

So, H is the distribution function we expect to get on average if we were to repeatedly generate functions, F, according to *Eq. 20*.

The base distribution, H, can be discrete or continuous. It is up to us to specify and will depend on what sort of problem we are modeling. You may also see the notation, H_0, used for the base distribution.

The value of α controls how much variation (dispersion) we get around the base distribution, H. A larger value of α means we will get less variation around the base distribution and so the functions, F, generated will be very close to H. In the limit, $\alpha \to \infty$, we would effectively only get the base distribution, H, if we were to sample F according to *Eq. 20*. In contrast, a low value of α means we get a lot of potential variation away from the base distribution, and instances of F sampled according to *Eq. 20* could look very different from H. You can see that the higher the value of α, the more concentrated F will be around H. Hence, α is known as the **concentration parameter**.

We have defined a DP, but how do we begin to use one? How do we use it as a prior for our distribution F? To answer these questions, we must first learn how to sample from a DP.

Sampling a function from a DP

We've specified that distribution, F, is distributed according to $DP(\alpha, H)$, but how do we generate an example of F? How do we sample from the DP? One way to do that is to make use of what is known as the **stick-breaking algorithm**, which is given here:

1. Draw values, $x_1, x_2, x_3, \ldots$, from H.
2. Draw values, $v_1, v_2, v_3, \ldots$, from the Beta$(1, \alpha)$ distribution.
3. Set weight, $w_1 = v_1$, and calculate subsequent weights, $w_i = v_i \prod_{j=1}^{i-1}(1 - v_j)$.
4. Construct the probability distribution that has probability mass, w_i, at x_i. This is our sampled distribution, F, for our discrete random variable, X.

From the last step, we can see that our sampled distribution, F, is given the following:

$$F \;=\; \sum_{i=1}^{\infty} w_i \delta_{x, x_i}$$

Eq. 22

The Kronecker delta-function notation, δ_{x,x_i}, represents a unit probability mass located at x_i, and so is 1 if x and x_i are the same value and zero otherwise. The expression on the right-hand side of *Eq. 22* gives the mathematical equation for the probability that our discrete random variable, X, has a value of x.

The distribution in *Eq. 22* is like the empirical distribution function that we met in *Chapter 2*. You'll also notice that it is discrete. The distribution, F, in *Eq. 22* is a sum of point masses. The only possible values we can get for X from the distribution in *Eq. 22* are the distinct values, $x_1, x_2, x_3\ldots$ Wait? We get a discrete distribution, F, even if our base distribution, H, was continuous? Yes. However, we still have the following:

$$\mathbb{E}(F) \;=\; H$$

Eq. 23

This means that if we generated lots and lots of discrete distributions according to the recipe in *Eq. 22*, the average of all those discrete distributions would be close to H even if no single one of the discrete distributions looked that much like H.

You will also have noticed by now that the sampled distribution, F, in *Eq. 22* consists of an infinite number of point masses. This makes the stick-breaking algorithm more of a useful theoretical construction than a practical tool. However, the weights, w_i, in *Eq. 22* get smaller and smaller as i increases, and so the contributions to F from larger values of i become insignificant. So-called **truncated stick-breaking (TSB)** algorithms take advantage of this and replace the representation of F in *Eq. 22* with a finite sum of point masses.

In practice, we will often be more interested in generating a finite sample of observations from F, not sampling F itself. Fortunately, there are algorithms for generating a finite sample of N values from F when $F \sim DP(\alpha, H)$. We will meet one in a moment. For now, we'll keep with the view that *Eq. 22* is a useful way of representing what a sample from a DP looks like.

Generating a sample of data from a DP

In the recipe that led to *Eq. 22*, we sampled an instance of F from its DP prior, $DP(\alpha, H)$. The distribution, F, is the distribution from which we consider our data variable, X, to be drawn. However, when we used GP priors in GPR, we focused on the data by integrating over all possible functions coming from the prior to get the marginal joint distribution of the random variables, $X_1, X_2, \ldots, X_N$, that represent the data. Can we do the same thing here? Can we sample an instance of $X_1, X_2, \ldots, X_N$ from the marginal joint distribution once we have marginalized over all values of F from the DP prior?

The answer is yes. In principle, one approach would be to just sample lots of instances of F using the recipe in *Eq. 22*, and from each instance of F, sample some values of X. However, there is a more efficient and practical algorithm to sample from the marginal distribution of X. The algorithm to do so goes by the unusual name of the **Chinese restaurant process**. It is also called the **infinite Pólya urn sampling process**. To sample N values from the marginal joint distribution, the Chinese restaurant process is defined as follows:

1. Draw a value of x_1 from H.
2. For $i \geq 2$, with probability, $\alpha/(i + \alpha - 1)$, draw a value of x_i from H, and with probability, $(i - 1)/(i + \alpha - 1)$, draw a value of x_i from the empirical distribution function formed from the values, $x_1, x_2, \ldots, x_{i-1}$.
3. Set $i \rightarrow i + 1$ and repeat *Step 2* whilst $i \leq N$.

Bayesian non-parametric inference using a DP

So far, we have learned a lot about DPs and how to sample from them, but we haven't discussed how we combine them with observations (i.e., real data). We haven't discussed how we would use a DP prior in a data science algorithm. Since we want to use a DP prior to help us model how our data is distributed, there are two kinds of tasks we are interested in:

- Modeling the probability distribution of a discrete random variable
- Modeling the probability density of a continuous random variable

We'll now look at each of those tasks in turn.

Using a DP to model a probability distribution

To show how to combine a DP prior on a discrete probability distribution with our data, we first need to set up our data science problem.

We have a set of observations, $x_1, x_2, \ldots, x_N$, that we model as being samples of i.i.d. random variables, $X_1, X_2, \ldots, X_N$, which all have a probability distribution, F. This means we write the following:

$$X_1, X_2, \ldots, X_N \sim F$$

Eq. 24

Since we are taking a non-parametric Bayesian approach to our modeling, we say the distribution, F, is distributed **a priori** according to a DP prior:

$$F \sim DP(\alpha, H)$$

Eq. 25

What we now want is the posterior distribution of F conditional on the values, $x_1, x_2, \ldots, x_N$. For a DP prior of the form in *Eq. 25*, it turns out that the posterior of F is very simple. It is also a DP. The posterior of F is given by the following:

$$F\left(X = x \right)\Bigg| x_1, x_2, \ldots, x_N \sim DP\left(\alpha + N \ , \frac{\alpha}{\alpha + N} H + \frac{N}{\alpha + N} \frac{1}{N} \sum_{i=1}^{N} \delta_{x, x_i} \right)$$

Eq. 26

We'll take a moment to delve into *Eq. 26*. From the form of the DP in *Eq. 26*, we can see that the base distribution is of the following form:

$$\text{Posterior Base Distribution} = \frac{\alpha}{\alpha + N} H + \frac{N}{\alpha + N} \text{Empirical Distribution}$$

Eq. 27

The posterior base distribution is a weighted sum of our original prior base distribution, H, and the empirical distribution formed from the data. As the amount of data increases and N becomes very much bigger than α, the posterior base distribution tends toward the empirical distribution function. But as N grows, so does the posterior concentration parameter, $\alpha + N$. Overall, this means that as N becomes big, the posterior distribution of F becomes concentrated around the empirical distribution (i.e., our posterior says F is effectively just the empirical distribution). In contrast, at small N or when N is still comparable to α, the posterior distribution of F will show some variation about a posterior base distribution that is still a mix of the original base distribution, H, and the empirical distribution function. This is the usual Bayesian analysis phenomenon. When we have lots of data, our inferences are driven by the data, but when we have limited amounts of data, our inferences are still affected by our prior. From this, we also get an idea of how to set a suitable value for α. It should be the amount of data above which we want the data to dominate our inferences.

Eq. 26 tells us that the posterior distribution of F, conditional on the data, is a DP, but how does that help us in practical terms? How can we use it to make inferences from the data? Well, we already know how to generate samples from a DP using the Chinese restaurant process algorithm. If we use a TSB algorithm, we can generate an example distribution from the posterior DP. This means we can generate instances of F that are compatible with the training data, and we can generate new samples of data compatible with the training data. *Q2* in the exercises at the end of the chapter is designed to illustrate this aspect of using DPs in practice.

Using a DP to model a probability density

Because the sampled distribution, F, generated in *Eq. 22* is a discrete distribution, it means that a DP is not directly a good choice for a prior of a probability density. If we want to use non-parametric Bayesian methods to model probability density functions, we need a different approach. Fortunately, we can do so with only a small modification. We do so by modeling the probability density as a mixture model. We use a mixture of parametric density functions of a specific form, but we put a DP prior on the discrete distribution from which we get the parameters of the mixture component densities. This is called a **Dirichlet process mixture (DPM)**.

If we want to model a set of observations, $x_1, x_2, \ldots, x_N$, using a DPM, we again consider the observations to be drawn from random variables, $X_1, X_2, \ldots, X_N$. We then model those random variables by writing the following:

$$X_i \sim f(\underline{\theta}_i) \quad i = 1, 2, \ldots, N$$

$$\underline{\theta}_i \sim F \quad i = 1, 2, \ldots, N$$

$$F \sim DP(\alpha, H)$$

Eq. 28

Eq. 28 says that our random variable, X_i, is distributed according to some parametric distribution, f. For example, f might be a Gaussian. The distribution, f, is characterized by parameters, $\underline{\theta}_i$, for example, its mean and variance. The vector of parameters, $\underline{\theta}_i$, is distributed according to the distribution, F, and F is drawn from a DP. Since F is drawn from $DP(\alpha, H)$, the stick-breaking representation in *Eq. 22* tells us that $\underline{\theta}_i$ could be any of an infinite number of different possible values corresponding to the point masses from which F is made. This means we are modeling our data using an infinite mixture distribution. DPMs are infinite mixtures.

We can also see from *Eq. 28* that since the parameter vectors, $\underline{\theta}_i$, for the different values of i are all drawn from the same discrete distribution, F, we will inevitably get some parameter vectors, $\underline{\theta}_i$, having the same value as each other. The number, K, of distinct parameter vectors, will be less than N. In effect, the model in *Eq. 28* naturally clusters the data points, $x_1, x_2, \ldots, x_N$, into $K < N$ clusters. Clustering of data is one of the main applications of DPMs.

In practice, when using DPMs, we use advanced computational techniques such as **Gibbs sampling** to sample from the posterior. To implement this numerically, we must address the fact that our distribution, F, has an infinite number of point masses. As before, we can truncate the sum in *Eq. 22* to only include a large but finite number of point mass contributions. It turns out that the number of clusters needed to model N data points only grows logarithmically with N, so that the cap on the number of point masses we include in *Eq. 22* also only needs to grow logarithmically with N. This means that truncating the sum in *Eq. 22* at some large but tractable value is a valid heuristic approach for using DPMs.

That concludes this short section on DPs and how they are used in non-parametric Bayesian models, so, let's recap what we have learned in this section and this chapter overall.

What we learned

In this section, we have learned the following:

- How a DP is defined in terms of its base distribution, H, and its concentration parameter, α
- How a DP can be used as prior for a probability distribution

- How to sample a distribution from a DP

- How to sample data from the marginal distribution induced by a DP prior

- How the posterior of a distribution, F, conditional on the observed data, is also a DP if its prior was a DP

- How a DPM can be used to model a probability density function

Summary

This chapter was about non-parametric Bayesian methods. Non-parametric methods are extremely useful because we don't have to make parametric assumptions about the form of the relationship between our features and the target variable. That has required us to learn a new approach to probabilistic modeling and new concepts. Those new concepts include the following:

- Non-parametric Bayesian methods focus on specifying a modeling function as coming from a prior distribution over functions

- The priors in non-parametric Bayesian methods are stochastic processes such as GPs or DPs

- A GP is specified by a kernel (covariance) function and a mean function

- In GPR, our model is the mean function of the response (target) variable and we put a GP prior on that mean function

- In GPR, we can fit the kernel function parameters to the training data by maximizing the marginal likelihood

- A DP is defined in terms of its base distribution, H, and its concentration parameter, α

- A DP can be used as a prior for a probability distribution

- The posterior of a distribution, F, conditional on the observed data, is also a DP if its prior was a DP

- A DPM can be used to model a probability density function

Our next and final chapter is another advanced topic connected to probability distributions. It is about **random matrices**.

Exercises

The following is a series of exercises. Answers to all the exercises are given in the `Answers_to_Exercises_Chap14.ipynb` Jupyter Notebook in the GitHub repository:

1. The Matérn kernel function, $k(x, y)$, can be thought of as a generalization of the RBF kernel. It is of the following form:

$$k(x, y) = \frac{1}{\Gamma(v)\, 2^{v-1}} \left(\frac{\sqrt{2v}}{b} \left| x - y \right| \right)^v K_v \left(\frac{\sqrt{2v}}{b} \left| x - y \right| \right)$$

Eq. 29

$K_v(z)$ is the modified Bessel function of the second kind. The Matérn kernel is specified by the parameters, v and b. The lengthscale parameter, b, plays a similar role to the length-scale parameter, b, in the RBF kernel in *Eq. 10*. The parameter, v, controls how smooth the functions are when we use a GP prior with a Matérn covariance kernel.

Using the data from the code example in the main text and a Matérn kernel with the default value, $v = 1.5$, fit a GPR model to the data. Make predictions for a range of x values. Note that for the Matérn kernel, the parameter, v, is not optimized by the `scikit-learn` fitting process, so if you instantiate a Matérn kernel object with $v = 1.5$, the value of v will remain fixed at 1.5. The `scikit-learn` fitting will, however, optimize the length-scale parameter, b, of the Matérn kernel.

2. We have a discrete random variable, X, that takes integer values 1 to 9. Fifty observations of the random variable are given in the `benford_data.csv` file in the `Data` directory of the GitHub repository. Use the DP posterior given in *Eq. 26* and the Chinese restaurant process to sample six datasets, each of 100 data points, from the posterior DP and compare each of them by plotting each of them against the empirical distribution calculated from the data. You should use a concentration parameter, $\alpha = 10$, and the base distribution of the following form:

$$\text{Prob}(X = x) = \log_{10}\left(1 + \frac{1}{x}\right)$$

Eq. 30

15

Random Matrices

This is our final chapter. It is about random matrices. That name may sound esoteric or even abstract. It suggests that random matrices may not be a very useful thing to learn about. However, by now, you'll be very used to the idea that randomness is everywhere in data science. Dealing with matrices is a core part of data science as well, so maybe a random matrix is not so esoteric after all. This chapter will emphasize the idea that random matrices can be found everywhere in data science and are a useful way of representing large-scale interacting systems. That means that we must become familiar with the tools used to study random matrices. We will only very briefly introduce the main results and concepts connected to random matrices, so this will be a short chapter. In it, we will cover the following topics:

- *What is a random matrix*: In this section, we introduce the basic idea of what a random matrix is and why they are common in data science

- *Using random matrices to represent interactions in large-scale systems*: In this section, we introduce the idea that large random matrices are a natural way to represent and model large-scale interacting systems

- *Universal behavior of large random matrices*: In this section, we show how large random matrices start to behave in a similar way to each other

- *Random matrices and high-dimensional covariance matrices*: In this section, we show how universal behavior is also seen in covariance matrices, which are a core part of statistical and machine learning models, and where we also learn how random matrix theory is used to understand the behavior of large neural networks

Technical requirements

There are no code examples in this chapter, but Jupyter notebook-based answers to the exercises at the end of the chapter can be found at `https://github.com/PacktPublishing/15-Math-Concepts-Every-Data-Scientist-Should-Know/tree/main/Chapter15`. To run the Jupyter notebook, you will need a full Python installation, including the following packages:

- `numpy` (>= 1.24.3)
- `matplotlib` (>=3.7.2)

What is a random matrix?

A random matrix sounds like it is some esoteric mathematical object – the sort of thing that is studied by mathematicians for fun but is of no practical use. Why should you care about random matrices as a data scientist?

As a data scientist, you have already been working with matrices. You know that they are useful. You know that a matrix is made up of matrix elements and that those elements are often formed from data. By now, you're also most likely used to the idea that data has a random component, so a matrix formed from data must also have random matrix elements. This is what a random matrix is. A random matrix is just a matrix whose elements are drawn from a distribution. **Random Matrix Theory (RMT)** is the study of the properties of random matrices.

Usually, in RMT, the matrix elements are taken to be **independent and identically distributed random variables (iid)**, but recent research in the RMT field has extended this to looking at more structured randomness. Also, early work in the RMT field looked at square, that is, $N \times N$, matrices. In doing so, RMT discovered that as $N \to \infty$, these square random matrices develop some interesting properties.

Again, these RMT discoveries may sound esoteric. However, as data scientists, we often work with big data, so we often work with large matrices with a random component. This means that results from the RMT field can be particularly relevant to data science.

We have motivated this section by highlighting that as data scientists, we naturally work with large random matrices. However, the original motivation for RMT was to understand large-scale interacting systems. Since interacting systems are also often what we are tasked with analyzing as data scientists, it is instructive to understand this original motivation for RMT. This is what we will do in the next section, but for now, we'll recap what we have learned so far.

What we learned

In this section, we learned the following:

- A random matrix is a matrix whose matrix elements are drawn from a distribution.

- Since data contains a random element, a matrix formed from data can be thought of as a random matrix.

- RMT is the mathematical field that studies random matrices.

- When random matrices become large in terms of the number of matrix elements that they contain, they develop interesting properties.

Having thus learned what a random matrix is at a very basic level, in the next section, we'll see how they can be used to represent large-scale interacting systems.

Using random matrices to represent interactions in large-scale systems

In *Chapter 10*, we encountered the adjacency matrix method for representing a network. A network also represents a set of components, the nodes, and the network edges can be used to represent the pairwise interactions between the nodes. *Figure 15.1* gives an example of a network and its adjacency matrix representation.

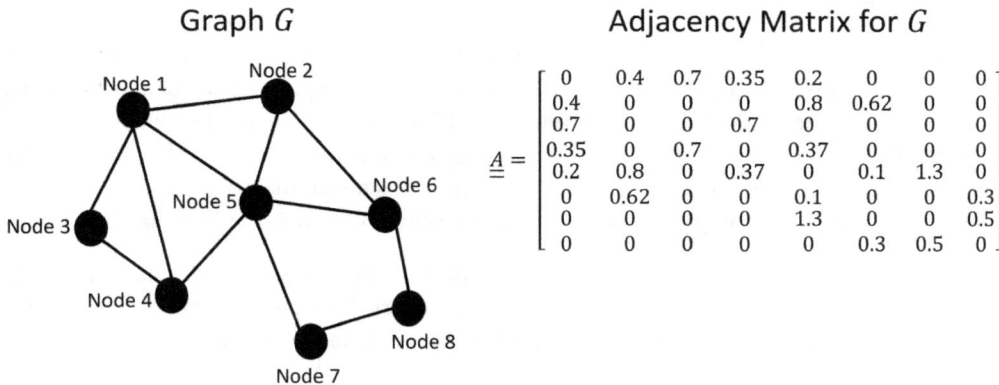

Figure 15.1: Network interactions represented as a matrix

It is natural to use a matrix to represent pairwise interactions between elements of any interacting system and not just between nodes in a network. Some of these systems may be physical, such as particles interacting via some force. They may also be non-physical, where the components of the system do not directly exert forces on each other but may influence each other, such as correlations between the share prices of different companies listed on a stock market.

Many of the systems that we study can be large, particularly real physical systems. Therefore, the interactions can be represented by a large matrix. The matrix elements that we use encode the strengths of the pairwise interactions between the components of the system. They can differ in strength, but aside from this, we believe that the interactions are all of the same qualitative type. Consequently, we can consider the interaction strengths as being values sampled from the same distribution. Therefore, the interaction matrix is a large random matrix, and we are using the large random matrix as a way of representing our real system. To study these systems, we need tools that study large random matrices, that is, RMT. Unsurprisingly, RMT has been applied to many different interacting systems in many different fields, including the following:

- **Energy levels of large atomic nuclei**: This was the original application of RMT and stimulated the early development of RMT as an area of mathematical research. Most of the ideas that we will introduce in the next section were discovered because of this RMT work on heavy nuclei.

- **Neuroscience**: Large-scale neuronal networks are modeled as pairwise connections, and hence dynamic interactions, between neurons in the brain.

- **Finance**: Correlations between the daily stock market movements of different stocks are modeled using a large random matrix.

Even if a system consists of more than pairwise interactions, it is not uncommon to model it with just pairwise interactions, as a pairwise interacting system is the simplest model of an interacting system we can have. The model interaction strengths in these cases represent effective interactions – they encapsulate the combined effects of the real pairwise interaction, as well as the real 3rd-order interactions, the real 4th-order interactions, and so on. Again, this means that we can understand a lot about the behavior of these real systems by looking at the behavior of their random matrix models.

Using RMT to study large-scale interacting systems is particularly insightful, as we find that large random matrices follow universal laws. It is RMT that uncovers those laws. We will introduce this universal behavior in the next section, so for now, let's recap what we have learned.

What we learned

In this section, we learned the following:

- A system with pairwise interactions can be represented as a matrix.

- A large-scale interacting system can be modeled as a large random matrix.

- We can use the tools of RMT to study the behavior of large-scale interacting systems.

Having learned that many large-scale interacting systems can be modeled as large random matrices and studied using the tools of RMT, in the next section, we will learn about some of the universal laws that large random matrices follow.

Universal behavior of large random matrices

We have already mentioned that when random matrices become large, they begin to display some interesting behaviors. However, what do we mean by this and why is it useful to us as data scientists?

The interesting behavior that we see is that the statistical properties of their eigen-decompositions or singular-value decompositions become **universal**. By universal, we mean that the same behavior is seen across many different matrices. In the case that we're going to illustrate, it means that the statistical characteristics of the eigen-decomposition of any large square random matrix are the same.

It is worth recalling from *Chapter 3* that eigen-decompositions of square matrices are an important and flexible way of representing any square matrix. So, universality in parts of the eigen-decomposition of large random matrices also means that calculations and algorithms involving the matrices will have universal aspects to them. It won't matter which matrix we have, we will get the same result for certain aspects of the calculation.

By now, you're probably curious to see what this universal behavior in the eigen-decomposition of large random matrices is in more detail, so we'll illustrate it with a numerical example. We'll illustrate what is known as the **Wigner semicircle law**.

The Wigner semicircle law

Matrix $\underline{M}_1$ is a 2000 × 2000 symmetric random matrix that I have generated. It has been created from matrix $\underline{A}_1$ using the following equation:

$$\underline{M}_1 = \tfrac{1}{2}\left(\underline{A}_1 + \underline{A}_1^\mathsf{T}\right)$$

Eq.1

The $\underline{A}_1$ matrix is a 2000 × 2000 random matrix. All of its matrix elements are i.i.d. and drawn from the standard normal distribution, $N(0,1)$. The calculation in *Eq.1* ensures that the $\underline{M}_1$ matrix is symmetric, even though $\underline{A}_1$ isn't.

Matrix $\underline{M}_2$ is also a 2000 × 2000 symmetric random matrix. I have generated it in a similar way to $\underline{M}_1$ but using a matrix $\underline{A}_2$ in *Eq.1* instead of $\underline{A}_1$. Like matrix $\underline{A}_1$, the $\underline{A}_2$ matrix is a 2000 × 2000 random matrix and all its matrix elements are i.i.d. and drawn from $N(0,1)$.

Although the two matrices, $\underline{M}_1$ and $\underline{M}_2$, are generated in the same way, they are two different matrices. They have different matrix elements that are generated independently from each other. Each matrix has 2000 × 2000 = 4 × 10⁶ matrix elements. There are a lot of ways in which matrix $\underline{M}_1$ can be different from matrix $\underline{M}_2$. *Figure 15.2* shows histograms of the scaled eigenvalues of each matrix – the left-hand panel shows the histogram for matrix $\underline{M}_1$, while the right-hand panel shows the histogram for matrix $\underline{M}_2$.

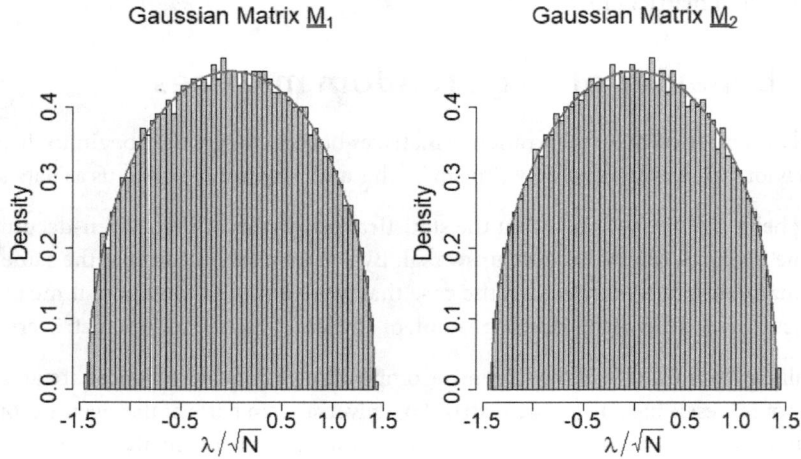

Figure 15.2: Histograms of scaled eigenvalues for two large symmetric random matrices with Gaussian matrix elements

The x axis in both plots shows the $\lambda/\sqrt{N}$ scaled eigenvalue, with N = 2000 in this case. So, for each matrix, we have taken its eigenvalues $\lambda_1, \lambda_2, ..., \lambda_N$ and divided them by $\sqrt{N}$. The histograms are an estimate of the probability density of the scaled eigenvalues.

What is remarkable is how similar the two histograms are, despite the matrices having a huge number of differences between them in terms of their individual matrix elements. The presence of the red line in *Figure 15.2*, which follows the shape of the two histograms, suggests that I expected the histograms to have this shape.

The red line is known as the semicircle law. Sometimes, you will see it referred to as the **Wigner semicircle distribution**, after Eugene Wigner, the famous theoretical physicist who used random matrices to study the properties of atomic nuclei. The semicircle law says that in the $N \to \infty$ limit, the probability density of the scaled eigenvalues tends to the density given by *Eq.2*.

$$p(x) = \frac{1}{\pi}\sqrt{2 - x^2} \quad -\sqrt{2} \le x \le \sqrt{2}$$

Eq.2

The red line in *Figure 15.2* is the equation in *Eq.2* with x = $\lambda/\sqrt{N}$. The $\sqrt{2 - x^2}$ mathematical form in *Eq.2* tells us that this function will have a shape related to a semicircle, hence the origin of the name *semicircle law* (or semicircle distribution).

There are a few things we should point out about the result in *Eq.2*:

- It is an asymptotic result. We have said that it is the probability density of $\lambda/\sqrt{N}$ that we get in the $N \to \infty$ limit. This may not seem like a very useful result, as no real-world matrix is of infinite size. However, it means that for large but finite values of N, the density in *Eq.2* will still be a very good approximation for the probability density of $\lambda/\sqrt{N}$. At large but finite values of N, we will see some deviations from the semicircle law. However, they will be small, and they will get smaller as N increases. As we can see from *Figure 15.2*, at $N = 2000$, the semicircle law is a very good approximation for the density of scaled eigenvalues. I have even seen the semicircle law being used as a reasonable approximation for 50×50 matrices.

- You'll also notice the finite support of the semicircular law in *Eq.2*. We can only get values between $-\sqrt{2}$ and $\sqrt{2}$ for the $\lambda/\sqrt{N}$ scaled eigenvalue. So, there is a minimum and maximum possible value for the scaled eigenvalue. Again, this is an asymptotic result. At finite N, we will see some departure from this rule, but at large values of N, the departures will be small, that is, $\pm\sqrt{2N}$ will be a good approximation to the range of eigenvalues that we would see from matrices generated in the way that matrix $\underline{M}_1$ and matrix $\underline{M}_2$ were.

The semicircle law of *Eq.2* is an example of what we mean by universal behavior or universality. Irrespective of the individual matrix, we will get the same statistical behavior. In this case, we get the same statistical behavior of the eigenvalue distribution of the matrix.

The semicircle law is the most well-known result from RMT. Deriving the shape of the eigenvalue distribution as $N \to \infty$ is the sort of task that RMT focuses on. We won't go into the methods used to derive the semicircle law or other RMT results, as they can be complex, but we will discuss more of these universal RMT results in the next section.

What does RMT study?

When we study ordinary random variables using the rules of probability, the mathematical results that we obtain are about expectation values, that is, statements about the average behavior of the random variable. It is the same when we use RMT to study random matrices. We can only derive formulae and laws about the expected behavior of random matrices. The semicircle law in *Eq.2* is actually a statement about what happens to the eigenvalue distribution on average as we go to the $N \to \infty$ limit. This means that if we were to generate lots of different matrices $\underline{M}$ using the process described in *Eq.1*, and we were to calculate their scaled eigenvalues and take the average of their empirical distributions, we would get something close to the semicircle law. As we made N larger and larger, this average scaled eigenvalue distribution would get closer and closer to the semicircle law.

However, this raises an interesting question. If the semicircle law is the average scaled eigenvalue distribution we'd get after averaging across all matrices $\underline{M}$ generated using *Eq.1* (and taking the $N \to \infty$ limit), how is the semicircle law such a good approximation to the eigenvalue distributions for the single $\underline{M}_1$ and $\underline{M}_2$ instances? It's like we had drawn two values from a random variable X and both just happened to be almost exactly the same as the expectation value $\mathbb{E}(X)$. Why did we not see any sampling variation (deviations) around the semicircle law when we generated $\underline{M}_1$ and $\underline{M}_2$? Were we just extraordinarily lucky? The answer is no, and the reason why has to do with the fact that N is large.

Self-averaging

For a single matrix $\underline{M}$ generated using the process in *Eq.1*, we'll use $\tilde{\lambda}$ to denote a scaled eigenvalue, so $\tilde{\lambda} = \lambda/\sqrt{N}$, and we'll use $\hat{p}\left(\tilde{\lambda}|\underline{M}\right)$ to denote the empirical density of scaled eigenvalues from $\underline{M}$. RMT says that the average of $\hat{p}\left(\tilde{\lambda}|\underline{M}\right)$ over all matrices $\underline{M}$ tends to the semicircle law in *Eq.2* as $N \to \infty$. In mathematical language, we would say this as follows:

$$\lim_{N\to\infty}\mathbb{E}_{\underline{M}}\left(\hat{p}\left(\tilde{\lambda}|\underline{M}\right)\right) = \tfrac{1}{\pi}\sqrt{2 - \tilde{\lambda}^2}$$

Eq.3

However, what we will also find is that the variance of $\hat{p}\left(\tilde{\lambda}|\underline{M}\right)$ around its expectation decreases to zero as $N \to \infty$. As $N \to \infty$, the empirical distribution of any single matrix $\underline{M}$ becomes the same as the average across all matrices of the same type as $\underline{M}$. In physics, we refer to this as **self-averaging**. The eigenvalue distribution of a matrix $\underline{M}$ generated from *Eq.1* is a self-averaging quantity. What we get from a single random instance is very close to the average across multiple instances and vice versa. It is this self-averaging behavior that makes the semicircle law a useful approximation even for single instances of a random matrix.

Ultimately, it is also this self-averaging behavior that makes it possible for RMT to derive formulae such as the semicircle law. Therefore, it is ultimately by considering **large** random matrices (that is, taking N to be large) that we can make progress in RMT. However, that is not to say that RMT doesn't also consider random matrices of small or intermediate finite size. It is just that deriving simple closed-form laws for such matrices can be considerably more challenging.

Universal is universal

You may be thinking that although matrices $\underline{M}_1$ and $\underline{M}_2$ in *Figure 15.2* are different matrices, they were generated according to the same process. How often are we going to be dealing with real-world matrices whose matrix elements are Gaussian distributed? Is the behavior illustrated in *Figure 15.2* really that universal? Here's the thing: universal behavior typically arises for reasons that are not related to the microscopic details of how a system is specified. This means that we can often change the microscopic details, such as which precise distribution the matrix elements are drawn from, and still get the same universal behavior. Let's try it. Let's take our matrix elements from another distribution, but still with mean zero and unit variance, and see what happens.

Figure 15.3 shows the scaled eigenvalue histograms for two 2000×2000 symmetric random matrices, $\underline{M}_3$ and $\underline{M}_4$, generated using *Eq.1* but using matrices $\underline{A}_3$ and $\underline{A}_4$ instead, respectively. The $\underline{A}_3$ and $\underline{A}_4$ matrices are 2000×2000 random matrices whose matrix elements have been drawn from a Laplace distribution that has zero mean and unit variance.

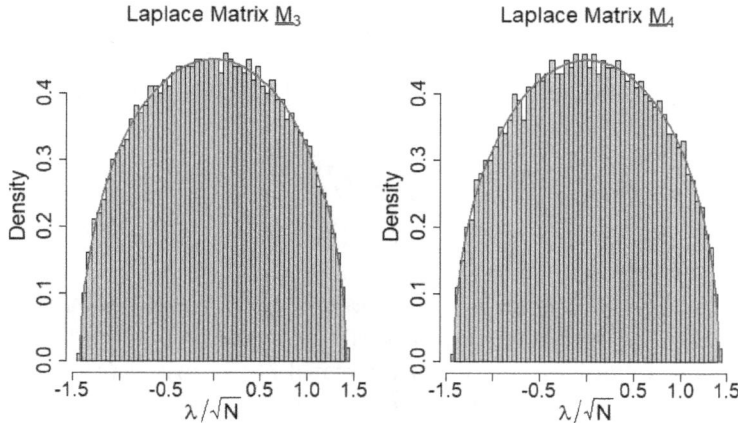

Figure 15.3: Histograms of scaled eigenvalues for two large symmetric random matrices with matrix elements drawn from a Laplace distribution

The zero mean unit variance Laplace distribution has a probability density of $p(x)$, which is given by the following equation:

$$p(x) = \frac{1}{\sqrt{2}} \exp\left(-\sqrt{2}\,|x|\right)$$

Eq.4

The density function of the Laplace distribution is different from the Gaussian distribution. It decays more slowly than the Gaussian distribution. Despite these big differences, the histograms in *Figure 15.3* are almost identical to those in *Figure 15.2*. Again, the red lines in *Figure 15.3* are the semicircle law from *Eq.2*. You can see that the semicircle law accurately describes the density of scaled eigenvalues for this Laplace distribution case. It turns out that the conditions on the distribution from which we can draw our matrix elements are very broad. Typically, for a real symmetric $N \times N$ matrix, as long as we have a zero mean unit variance distribution whose low-order moments are finite, then the resulting distribution of scaled eigenvalues will follow the semicircle law in the $N \to \infty$ limit. The universality of the semicircle law is indeed universal. You can even mix the distributions from which we draw our matrix elements and still get the semicircle law – this is one of the exercise questions at the end of this chapter.

The classical Gaussian matrix ensembles

The Gaussian matrices, which we defined via *Eq.1* and whose eigenvalue distributions we plotted in *Figure 15.2*, are part of a wider family of random matrices. It is time to meet that family.

You'll recall that we restricted our matrices in *Figure 15.2* to being real, square, symmetric, and generated according to *Eq.1*, with matrix $\underline{A}_1$ having all of its matrix elements drawn from the standard normal distribution. This group, or **ensemble**, of matrices is called the **Gaussian Orthogonal Ensemble (GOE)**. So, overall, the GOE is defined by the following:

$$\underline{M} \in \text{GOE if } \underline{M} = \tfrac{1}{2}(\underline{A} + \underline{A}^\top) \quad A_{ij} \sim N(0,1)$$

Eq.5

It is clear from *Eq.5* that $\underline{M}$ is symmetric and real. Consequently, its eigenvalues will be real. However, there are other ways in which we can construct a matrix $\underline{M}$, similar to how it is defined in *Eq.5* and such that it still has real eigenvalues. For example, if we make the matrix elements of $\underline{A}$ be complex numbers and draw the real and imaginary parts of the matrix elements from $N(0,1)$, then we can define $\underline{M}$ via the following equation:

$$\underline{M} = \tfrac{1}{2}(\underline{A} + \underline{A}^\text{H})$$

Eq.6

$\underline{A}^\text{H}$ means the Hermitian conjugate of $\underline{A}$. We encountered this operation in *Chapter 3*, where we used the symbol † for the Hermitian conjugate. It means taking the conjugate transpose of $\underline{A}$ and ensures that the matrix $\underline{M}$ is Hermitian and so has real eigenvalues. Matrices defined via *Eq.6* form the **Gaussian Unitary Ensemble (GUE)**. So, overall, the GUE is defined by the following:

$$\underline{M} \in \text{GUE if } \underline{M} = \tfrac{1}{2}(\underline{A} + \underline{A}^\text{H}) \quad A_{ij} = a + ib \quad a,b \sim N(0,1)$$

Eq.7

The matrix elements of a GUE matrix have a different number of components compared to the matrix elements of a GOE matrix. The matrix elements of GOE matrices are real numbers and so are defined by a single real number. Matrix elements of GUE matrices are complex numbers and so are defined by two real numbers – their real and imaginary parts. We can extend this pattern further to define numbers with four real components. These numbers are called **quaternions**. We won't go into quaternions any further here, but just like you can loosely think of complex numbers as being two-dimensional numbers because they live in a two-dimensional space called the **complex plane**, you can also think of quaternions as being four-dimensional numbers that live in a four-dimensional space. If we allow the matrix elements of $\underline{A}$ to be quaternions, then we can define a new ensemble of random matrices called the **Gaussian Symplectic Ensemble (GSE)**. So, overall, the GSE is defined by the following:

$$\underline{M} \in \text{GSE if } \underline{M} = \tfrac{1}{2}(\underline{A} + \underline{A}^\text{D}) \quad A_{ij} = \text{Quaternion}(a,b,c,d) \quad a,b,c,d \sim N(0,1)$$

Eq.8

In *Eq.8*, the $\underline{A}^D$ notation means taking the dual transpose of $\underline{A}$ and ensures that the matrix $\underline{M}$ is self-dual and so has real eigenvalues. Since a quaternion itself can be represented as a 2×2 matrix of complex numbers, an $N \times N$ GSE matrix can be represented as a $2N \times 2N$ matrix of complex numbers. This way of representing GSE matrices gives us a more practical way of generating them.

Generating matrices from the Gaussian Symplectic Ensemble

1. Generate two random complex $N \times N$ matrices, $\underline{X}$ and $\underline{Y}$, whose matrix elements have their real and imaginary parts independently drawn from $N(0,1)$.

2. Construct the $2N \times 2N$ matrix $\underline{A}$ defined in block form as follows:

$$\underline{A} = \begin{bmatrix} \underline{X} & \underline{Y} \\ -\underline{Y}^* & \underline{X}^* \end{bmatrix}$$

3. Calculate the GSE matrix $\underline{M}$ as $\underline{M} = \frac{1}{2}(\underline{A} + \underline{A}^H)$.

The GOE, GUE, and GSE differ in terms of the number of real numbers used to specify each matrix element. Traditionally, we use the β symbol to denote this number of real numbers. The GOE corresponds to $\beta = 1$, the GUE corresponds to $\beta = 2$, and the GSE corresponds to $\beta = 4$. Overall, these three ensembles form the **classical** ensembles of RMT.

Matrices from the GOE, GUE, or GSE all have real eigenvalues λ_i. For all three ensembles, if we calculate scaled eigenvalues as $\lambda_i/\sqrt{\beta N}$, then the distribution of scaled eigenvalues tends to the semicircle law of *Eq.2* in the $N \to \infty$ limit. The semicircle law is universal across all three of these classical random matrix ensembles.

It is not uncommon to see $\tilde{\lambda}_i = \lambda_i/\sqrt{N}$ used to define the scaled eigenvalues from any of the classical RMT ensembles. In this case, the limiting scaled eigenvalue distribution becomes the following:

$$p(\tilde{\lambda}) = \frac{1}{\beta\pi}\sqrt{2\beta - \tilde{\lambda}^2} \quad \text{for} \quad \tilde{\lambda} \in \left[-\sqrt{2\beta}, \sqrt{2\beta}\right]$$

Eq.9

Whichever classical ensemble, and hence value of β, we are working with, all the different variants of *Eq.9* tend to get referred to as the semicircle law, which can be confusing when you first see the different variants. This is why I prefer to absorb the $\sqrt{\beta}$ factor into the definition of the scaled eigenvalue rather than its probability density. That way, you get only one semicircle law for all three classical RMT ensembles: the one in *Eq.2*.

However, some authors also absorb the factor of $1/\sqrt{N}$ into the definition of the ensemble, that is, into the definition of how the matrix is constructed. You may also see differences in pre-factors of $\sqrt{2}$ between different authors when defining the classical ensembles. Again, these differences only lead to trivial variants of *Eq.9* for the form of the scaled eigenvalue density. However, it does require you to keep your eyes open and be aware of these differences when looking at different textbooks or research papers.

For much of the early history of RMT, the three classical ensembles were the focus of study, mainly because of their relevance to physics and quantum mechanics. As RMT progressed, researchers realized that the ideas and concepts of RMT could be applied to other research areas and other ensembles of matrices, including large data matrices of the kind that we encounter when building statistical and machine learning models. The matrices may be different (for example, a data matrix is not usually square), but they are still large and random. Unsurprisingly, we also find universal behavior in the statistical properties of these new matrix ensembles.

In the next section, we will look at ensembles of large random matrices that we encounter as data scientists, namely large covariance matrices and large weight matrices from neural networks. In both cases, we will highlight how RMT is used to study the properties of those classes of matrices. For now though, we'll wrap up this section by summarizing what we have covered.

What we learned

In this section, we have learned about the following:

- How the distribution of scaled eigenvalues of large symmetric random matrices can follow a universal pattern
- The Wigner semicircle law
- The GOE, GUE, and GSE families of matrices
- How the semicircle law is the distribution of scaled eigenvalues for the GOE, GUE, and GSE in the $N \to \infty$ limit
- How the semicircle law can apply to large square random matrices that are not drawn from the classical Gaussian random matrix ensembles

Having learned about the semicircle law, in the next section, we will extend it by looking at large covariance matrices that arise in statistics and machine learning.

Random matrices and high-dimensional covariance matrices

The examples of large random matrices in the previous section were all square matrices. However, in real-world data science, not all matrices are square. Take the $\underline{X}$ data matrix that we encountered in *Chapter 3* when doing **Principal Component Analysis (PCA)**. It is an $N \times d$ matrix, where N is the number of data points and d is the number of features. We will assume, for this section, that the data has already been mean-centered, so that the sum of each column of $\underline{X}$ is 0.

The $\underline{X}$ matrix is what we use to do PCA. It is also the design matrix that we use when building statistical models. So, the $\underline{X}$ matrix is non-square (unless $N = d$). However, in practice, we usually derive a square matrix from $\underline{X}$. For example, when doing PCA, we would calculate the sample covariance matrix $\widehat{\underline{C}}$, which is defined as follows:

$$\widehat{\underline{C}} = \frac{1}{N-1}\underline{X}^\mathsf{T}\underline{X}$$

Eq.10

The $\widehat{\underline{C}}$ matrix in *Eq.10* is $d \times d$ and symmetric. If we had many features, it would be a large matrix. Since $\widehat{\underline{C}}$ is derived from our data, which contains a random component, then $\widehat{C}$ is a large random matrix.

The eigenvalues of $\widehat{C}$ tell us about the principal components in our dataset. So, what would its eigenvalues look like? Given that $\widehat{\underline{C}}$ is a large random matrix, should we expect some universal behavior in the distribution of eigenvalues from $\widehat{\underline{C}}$? Let's do a numerical experiment to find out.

For illustration purposes, we'll use the simplest possible data matrix $\underline{X}$, where we first draw the matrix elements X_{ij} from $N(0,1)$ and then mean-center the columns. The histogram in *Figure 15.4* shows the approximate density of eigenvalues of a single sample covariance matrix generated in the way we have just described, with $N = 2000$ and $d = 1000$.

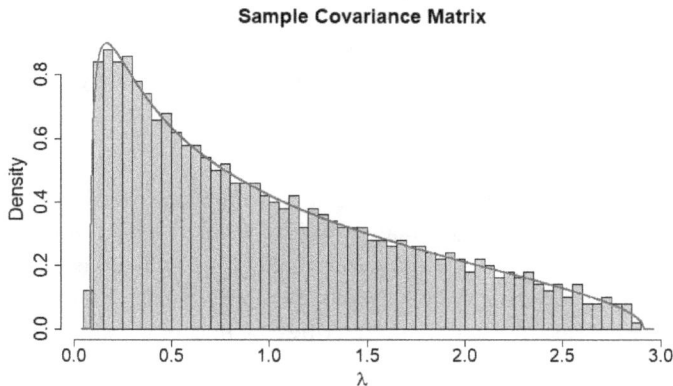

Figure 15.4: The distribution of sample covariance matrix eigenvalues

The red line in *Figure 15.4* is the **Marčenko-Pastur** distribution. It is the equivalent of the semicircle law but for covariance matrices formed from large random data matrices like $\underline{X}$. The Marčenko-Pastur formula for the density of eigenvalues, $p(\lambda)$, is given by *Eq.11*.

$$p(\lambda) = \begin{cases} \frac{\alpha}{2\pi\lambda}\sqrt{(\lambda_+ - \lambda)(\lambda - \lambda_-)} & \text{for } \lambda \in [\lambda_-, \lambda_+] \\ 0 & \text{otherwise} \end{cases}$$

Eq.11

The α value is the ratio of data to features. That is, $\alpha = N/d$, so α measures how non-square our matrix $\underline{X}$ is.

Like the semicircle law, the Marčenko-Pastur distribution has finite support. The λ_- and λ_+ values are the lower and upper ends of that support and are given by the following:

$$\lambda_- = \left(1 - \frac{1}{\sqrt{\alpha}}\right)^2 \quad \lambda_+ = \left(1 + \frac{1}{\sqrt{\alpha}}\right)^2$$

Eq.12

Strictly speaking, the Marčenko-Pastur formula in *Eq.11* is an asymptotic result. It is the distribution of eigenvalues that we would get in the $N \to \infty$ limit, while also ensuring that $N/d \to \alpha$. However, just like the semicircle law, it means we can also use the density in *Eq.11* as a very good approximation at finite N and where we set $\alpha = N/d$. Also, just like the semicircle law, the Marčenko-Pastur formula in *Eq.11* is the average eigenvalue distribution that we would get in the $N \to \infty$ limit. The reason why it is such a good approximation to the single instance of a sample covariance matrix shown in *Figure 15.4* is that the eigenvalue distribution of a large sample covariance matrix of the type generated in *Figure 15.4* is self-averaging.

The Marčenko-Pastur distribution is a bulk distribution

Okay, so the Marčenko-Pastur distribution is a universal law for large sample covariance matrices, but how universal is it exactly? The $\underline{X}$ data matrix generated in our numerical example contained only values drawn from $N(0,1)$. These would seem like just noise values, and so the result is not a very realistic data matrix.

Actually, it is realistic. In many real-world datasets and problems, we have lots of feature values. Typically, many of these features are uninformative or have limited signal in them. This means that in real-world datasets, the $\underline{X}$ data matrix can be overwhelmingly made up of noise. Consequently, most of the eigenvalues of the resulting sample covariance matrix $\widehat{\underline{C}}$ still follow the Marčenko-Pastur distribution. The Marčenko-Pastur distribution describes the *bulk* of the eigenvalues of $\widehat{\underline{C}}$, even for real-world data matrices $\underline{X}$.

Universality in the singular values of $\underline{X}$

We have already encountered the definition of $\widehat{\underline{C}}$ in *Eq.10*. We also came across it in *Chapter 3*, where we learned about the link between singular values from the SVD of the data matrix $\underline{X}$ and eigenvalues of the sample covariance matrix $\widehat{\underline{C}}$ calculated from $\underline{X}$. From this link, we know that if λ_i is an eigenvalue of $\widehat{\underline{C}}$, then $\sqrt{(N-1)\lambda_i}$ is a singular value of $\underline{X}$. It immediately follows that if all or most of the eigenvalues of $\widehat{\underline{C}}$ follow a universal distribution when the number of features d is large, then the singular values of $\underline{X}$ must also follow a universal distribution. See whether you can derive what this universal distribution should be using the relationship between the eigenvalues of $\widehat{\underline{C}}$ and the singular values of $\underline{X}$ and the rules for transforming probability densities that we learned about in *Chapter 2*.

Overall, this means that we can also see universal behavior in the singular values of some large non-square matrices, not just square ones. This has become particularly useful in the field of deep learning neural networks, as we shall learn about next.

The Marčenko-Pastur distribution and neural networks

One of the most interesting areas of RMT research that has emerged over the last decade or so, and that is of relevance to the field of data science, is the role that the Marčenko-Pastur distribution plays in characterizing the properties of weight matrices in large neural networks.

You're probably familiar with the idea that feed-forward neural networks consist of layers of nodes that are connected via weights. The weights, along with the output values from the preceding layer in the network, feed into a transfer function to compute the output values of the current layer. The schematic in *Figure 15.5* shows such a neural network structure.

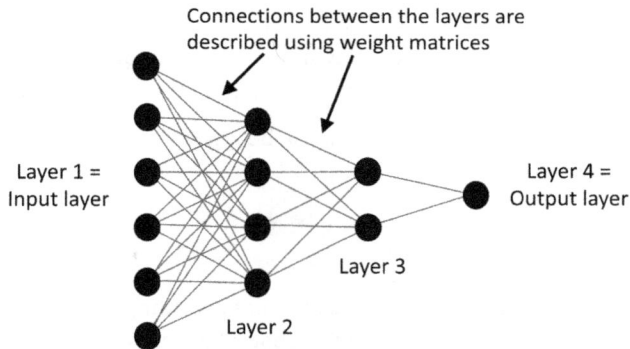

Figure 15.5: The schematic of a feed-forward neural network

The number of nodes in the initial input layer can be very large due to the large number of features that are input into the network. This is particularly true for modern deep learning neural networks, where large data volumes mean it is realistic to try and learn the relationship between high-dimensional feature vectors and the target variable. Therefore, the number of nodes, N_1, in the input layer will typically be large. As a consequence, for subsequent layers $l > 1$, the number of nodes N_l will also be large.

The number of nodes in a layer is usually less than the number of nodes in the preceding layer, as each layer is forced to learn an efficient representation of the information coming from the preceding layer. This means that we will also have $N_{l+1} < N_l$. Consequently, the weight matrix, $\underline{W}^{(l,l+1)}$, connecting layer l to layer $l + 1$ will be a large non-square matrix.

Since the weight matrix $\underline{W}^{(l,l+1)}$ represents the interactions between nodes in layer l and nodes in layer $l + 1$, you will not be surprised to find that the distribution of its singular values can be studied using the tools of RMT. In fact, it is usual to find that the distribution of the singular values of $\underline{W}^{(l,l+1)}$ can be described using the Marčenko-Pastur distribution. Again, the details are too extensive to go into in this short introduction – see, for example, the third point in the *Notes and further reading* section at the end of this chapter. However, it does illustrate once more that RMT is not an esoteric branch of mathematics, but a math concept of genuine relevance to data science and data science algorithms.

This has been a whirlwind tour of the Marčenko-Pastur distribution and its applications, so it's time to recap what we have learned in this section and wrap up the chapter as a whole.

What we learned

In this section, we have learned about the following:

- The distribution of all or most of the eigenvalues of the sample covariance matrix calculated from a large data matrix displays a universal behavior

- The Marčenko-Pastur distribution and how it is the equivalent of the semicircle law for sample covariance matrices

- How large non-square matrices can display universal behavior in the distribution of their singular values and how this distribution can be derived from the Marčenko-Pastur distribution

- The tools of RMT are applied to understand the behavior of weight matrices in large deep-learning neural networks

Summary

This chapter was about random matrices. What started out sounding like an esoteric plaything of mathematicians turned out to be a commonly occurring concept in data science with many applications. The tools to study random matrices come from RMT. These tools can be very mathematically advanced, so we have only given an overview of the main results of RMT and what the implications of those results are. We did not attempt to go into the derivations of those results. However, we had to learn about several new concepts. Those new concepts include the following:

- A random matrix is a matrix whose matrix elements are drawn from a distribution

- Random matrices are studied using RMT

- Large random matrices can display universal behavior

- A large-scale interacting system can be modeled as a large random matrix

- The Wigner semicircle law

- The GOE, GUE, and GSE families of matrices

- The Marčenko-Pastur distribution

- The bulk of the eigenvalues of a sample covariance matrix follow the Marčenko-Pastur distribution

- The tools of RMT are applied to understand the behavior of weight matrices in large deep-learning neural networks.

This has been our last chapter. Throughout this book, we covered a lot of material. However, there are a vast number of mathematical topics that we haven't covered. That is okay. By taking you on a journey through just 15 major math concepts that occur in data science and unpacking their nuances, we have equipped you with the skills to tackle any data science algorithm or idea on your own. Thank you for allowing me to be your guide for that math journey. I hope that you have enjoyed the journey as much as I have.

Exercises

The following is a series of exercises. Answers to all the exercises are given in the `Answers_to_Exercises_Chap15.ipynb` Jupyter notebook in the GitHub repository.

1. Create a 2000×2000 symmetric matrix $\underline{M}$ using the following relationship:

    $$\underline{M} = \tfrac{1}{2}(\underline{A} + \underline{A}^\top)$$

    Eq.13

    The $\underline{A}$ matrix should have its matrix elements drawn from the standard normal distribution with a probability of 0.5, and from the mean-zero unit-variance Laplace distribution in *Eq.4*, with a probability of 0.5. Calculate the eigenvalues, λ, of $\underline{M}$ and compute the empirical density of scaled eigenvalues $\lambda/\sqrt{N}$. Compare this empirical density to the semicircle law in *Eq.2*.

    > **Tip**
    > You can draw a value x from the mean-zero unit-variance Laplace distribution by first drawing a value u from the uniform distribution, uniform$(-0.5, 0.5)$, then calculating x as follows:
    > $$x = -\tfrac{1}{\sqrt{2}} \operatorname{sign}(u)\ln(1 - 2|u|)$$
    > Eq.14

    Alternatively, you can use the `numpy.random.laplace` NumPy function to sample the values directly.

2. From the definition of the GUE in *Eq.7*, generate a 2000×2000 GUE matrix and compute its eigenvalues λ. Compute the empirical probability density of scaled eigenvalues $\lambda/\sqrt{2N}$ and compare it to the semicircle law in *Eq.2*.

3. Assume that all the eigenvalues λ of a sample covariance matrix $\widehat{C}$ are distributed according to the Marčenko-Pastur distribution given in *Eq.11*. The sample covariance matrix has been calculated from a mean-centered data matrix $\underline{X}$. The singular values w of $\underline{X}$ are related to the eigenvalues λ via $\lambda = w^2/(N-1)$. Use this relationship to derive the probability density of the singular values w.

Notes and further reading

1. If you want to learn more about the mathematical details behind RMT, the book *Introduction to Random Matrices: Theory and Practice* by G. Livan, M. Novaes, and P. Vivo (ISBN: 978-3319708836, Springer, 2018), is a good start. You can find a copy of the material on the arXiv archive at `https://arxiv.org/pdf/1712.07903.pdf`. Be aware that the material can get mathematically advanced.

2. For a readable and short introduction to quaternions, I recommend *Introducing the Quaternions* by J. Huerta, which you can find at `https://math.ucr.edu/~huerta/introquaternions.pdf`.

3. A good recent paper on the application of RMT to studying neural network weight matrices is C.H. Martin and M.W. Mahoney's *Implicit Self-Regularization in Deep Neural Networks: Evidence from Random Matrix Theory and Implications for Learning* from the Journal of Machine Learning Research, 22:1-73, 2021.

4. The recent book, *Random Matrix Methods for Machine Learning* by R. Couillet and Z. Liao (ISBN: 978-1009123235, Cambridge University Press, 2022), also has a chapter on RMT and large neural networks, as well as on other areas of machine learning including some of the topics we have covered in earlier chapters in this book, such as kernel methods and community detection. You can find a copy of the material at one of the authors' GitHub sites at `https://zhenyu-liao.github.io/pdf/RMT4ML.pdf`.

Index

‹packt›

packtpub.com

Subscribe to our online digital library for full access to over 7,000 books and videos, as well as industry leading tools to help you plan your personal development and advance your career. For more information, please visit our website.

Why subscribe?

- Spend less time learning and more time coding with practical eBooks and Videos from over 4,000 industry professionals

- Improve your learning with Skill Plans built especially for you

- Get a free eBook or video every month

- Fully searchable for easy access to vital information

- Copy and paste, print, and bookmark content

Did you know that Packt offers eBook versions of every book published, with PDF and ePub files available? You can upgrade to the eBook version at packtpub.com and as a print book customer, you are entitled to a discount on the eBook copy. Get in touch with us at customercare@packtpub.com for more details.

At www.packtpub.com, you can also read a collection of free technical articles, sign up for a range of free newsletters, and receive exclusive discounts and offers on Packt books and eBooks.

Other Books You May Enjoy

If you enjoyed this book, you may be interested in these other books by Packt:

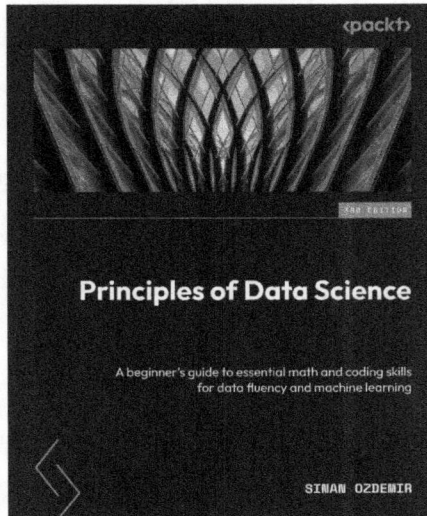

Principles of Data Science

Sinan Ozdemir

ISBN: 978-1-83763-630-3

- Master the fundamentals steps of data science through practical examples
- Bridge the gap between math and programming using advanced statistics and ML
- Harness probability, calculus, and models for effective data control
- Explore transformative modern ML with large language models
- Evaluate ML success with impactful metrics and MLOps
- Create compelling visuals that convey actionable insights
- Quantify and mitigate biases in data and ML models

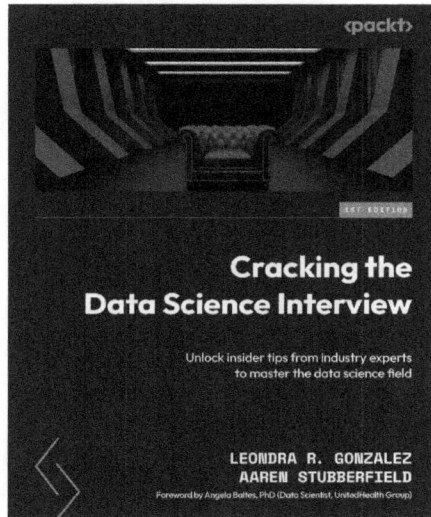

Cracking the Data Science Interview

Leondra R. Gonzalez, Aaren Stubberfield

ISBN: 978-1-80512-050-6

- Explore data science trends, job demands, and potential career paths
- Secure interviews with industry-standard resume and portfolio tips
- Practice data manipulation with Python and SQL
- Learn about supervised and unsupervised machine learning models
- Master deep learning components such as backpropagation and activation functions
- Enhance your productivity by implementing code versioning through Git
- Streamline workflows using shell scripting for increased efficiency

Packt is searching for authors like you

If you're interested in becoming an author for Packt, please visit `authors.packtpub.com` and apply today. We have worked with thousands of developers and tech professionals, just like you, to help them share their insight with the global tech community. You can make a general application, apply for a specific hot topic that we are recruiting an author for, or submit your own idea.

Share your thoughts

Now you've finished *15 Math Concepts Every Data Scientist Should Know*, we'd love to hear your thoughts! Scan the QR code below to go straight to the Amazon review page for this book and share your feedback or leave a review on the site that you purchased it from.

`https://packt.link/r/1-837-63418-1`

Your review is important to us and the tech community and will help us make sure we're delivering excellent quality content.

Download a free PDF copy of this book

Thanks for purchasing this book!

Do you like to read on the go but are unable to carry your print books everywhere?

Is your eBook purchase not compatible with the device of your choice?

Don't worry, now with every Packt book you get a DRM-free PDF version of that book at no cost.

Read anywhere, any place, on any device. Search, copy, and paste code from your favorite technical books directly into your application.

The perks don't stop there, you can get exclusive access to discounts, newsletters, and great free content in your inbox daily

Follow these simple steps to get the benefits:

1. Scan the QR code or visit the link below

https://packt.link/free-ebook/9781837634187

2. Submit your proof of purchase
3. That's it! We'll send your free PDF and other benefits to your email directly